Applications of Mathematics

Stochastic Modelling
and Applied Probability

49

Edited by B. Rozovskii
M. Yor

Advisory Board D. Dawson
D. Geman
G. Grimmett
I. Karatzas
F. Kelly
Y. Le Jan
B. Øksendal
E. Pardoux
G. Papanicolaou

Springer

Berlin
Heidelberg
New York
Hong Kong
London
Milan
Paris
Tokyo

Applications of Mathematics

Yuri Kabanov
Sergei Pergamenshchikov

Two-Scale Stochastic Systems

Asymptotic Analysis and Control

 Springer

Authors

Yuri Kabanov
Université de Franche-Comté
Département de Mathématiques
16 route de Gray
25030 Besançon Cedex, France
e-mail: kabanov@descartes.univ-fcomte.fr

Sergei Pergamenshchikov
Université de Rouen
LIFAR, UFR Sciences et Techniques
76821 Mont Saint Aignan Cedex, France
e-mail: serge.pergamenchtchikov@univ-rouen.fr

Managing Editors

B. Rozovskii
University of Southern California
Center for Applied Mathematical
Sciences
1042 West 36th Place,
Denney Research Building 308
Los Angeles, CA 90089, USA

M. Yor
Université de Paris VI
Laboratoire de Probabilités
et Modèles Aléatoires
175, rue du Chevaleret
75013 Paris, France

Mathematics Subject Classification (2000): 93-02, 60-02, 49-02

Cover pattern by courtesy of Rick Durrett (Cornell University, Ithaca)
Cover illustration by Margarita Kabanova

Library of Congress Cataloging-in-Publication Data applied for
Bibliographic information published by Die Deutsche Bibliothek
Die Deutsche Bibliothek lists this publication in the Deutsche Nationalbibliografie; detailed
bibliographic data is available in the Internet at <http://dnb.ddb.de>.

ISSN 0172-4568

ISBN 978-3-642-08467-6

Springer-Verlag Berlin Heidelberg New York
a member of BertelsmannSpringer Science + Business Media GmbH

http://www.springer.de

© Springer-Verlag Berlin Heidelberg 2010
Printed in Germany

Cover design: Erich Kirchner, Heidelberg

Printed on acid-free paper

Table of Contents

Introduction

In many complex systems one can distinguish "fast" and "slow" processes with radically different velocities. In mathematical models based on differential equations, such two-scale systems can be described by introducing explicitly a small parameter ε on the left-hand side of state equations for the "fast" variables, and these equations are referred to as *singularly perturbed*. Surprisingly, this kind of equation attracted attention relatively recently (the idea of distinguishing "fast" and "slow" movements is, apparently, much older). Robert O'Malley, in comments to his book, attributes the origin of the whole history of singular perturbations to the celebrated paper of Prandtl [79]. This was an extremely short note, the text of his talk at the Third International Mathematical Congress in 1904: the young author believed that it had to be literally identical with his ten-minute long oral presentation. In spite of its length, it had a tremendous impact on the subsequent development. Many famous mathematicians contributed to the discipline, having numerous and important applications. We mention here only the name of A.N. Tikhonov, who developed at the end of the 1940s in his doctoral thesis a beautiful theory for non-linear systems where the fast variables can almost reach their equilibrium states while the slow variables still remain near their initial values: the aerodynamics of a winged object like a plane or the "Katiusha" rocket may serve an example of such a system.

It is generally accepted that the probabilistic modeling of real-world processes is more adequate than the deterministic modeling. Needs of applications resulted in an increasing interest in the theory of two-scale stochastic systems where an essential progress has been achieved during the last 25 years. However, in comparison with the classical theory, many vast areas have never been explored and only a few research monographs have been available.

The main subject of this book is a stochastic version of the Tikhonov theory including some optimal control problems for systems with singular perturbations. Of course, we do not pretend to cover all aspects of singular perturbations: the absolute majority of the results presented here are our own. The principal model we deal with is given by the stochastic differential equations

$$dx_t^\varepsilon = f(t, x_t^\varepsilon, y_t^\varepsilon)dt + g(t, x_t^\varepsilon, y_t^\varepsilon)dw_t^x, \quad x_0^\varepsilon = x^o,$$

$$\varepsilon dy_t^\varepsilon = F(t, x_t^\varepsilon, y_t^\varepsilon)dt + \sigma(\varepsilon)G(t, x_t^\varepsilon, y_t^\varepsilon)dw_t^y, \quad y_0^\varepsilon = y^o,$$

where w^x and w^y are independent Wiener processes. We avoid in this brief introduction detailed discussions of needed assumptions on coefficients and mention only that for the linear case, where $F(t, x, y) = A_t x + B_t y$, we require the matrix-valued function A_t to be continuous and "exponentially stable". Sometimes, in problems where we feel that the full generality will make our study unreasonably complicated, we constrain ourselves by considering the model with only fast variables, given by a singularly perturbed SDE, which is, in certain cases, important and interesting in itself.

The assumption on the behavior of the coefficient when $\varepsilon \downarrow 0$ merits a discussion. In the literature, one may observe a dominance of studies where $\sigma(\varepsilon) = \varepsilon^{1/2}$ though there is a noticeable number of papers where models with $\sigma = \varepsilon^\delta$, $\delta > 1/2$, are also treated. No doubt, the case $\delta = 1/2$, where the typical random perturbation has an amplitude which behaves like a square root of the velocity, is worthy of attention. In general, models of this type fit the Bogoliubov averaging principle remarkably well. This principle prescribes to do the following. To get a description of the limiting behavior of the slow variable one should "freeze" it in the equation for the fast variable, i.e. consider the latter separately with a constant parameter x, representing a fixed point of the state space and replacing varying x_t^ε, and calculate the invariant measure (for time-dependent coefficients, "instantaneously invariant" measure) of the resulting dynamics. The coefficients of the limiting slow dynamics are obtained by averaging (in the y variable) the corresponding coefficients of the original prelimit equation with respect to these invariant measures (depending on x). Obviously, the ergodicity of the fast motion for the "frozen" slow variable can be postulated explicitly or via some sufficient conditions. Under various sets of hypotheses one may expect various types of convergence. The most developed theory concerns with the weak convergence; it is summarized in the deep treatise by Harold Kushner [57]. We avoid intersections with this excellent book by concentrating our efforts on models with $\delta > 1/2$. They arise in a natural way and we provide several typical examples widely discussed in the literature. At first glance, such models seem to be much simpler than those with $\delta = 1/2$ since the diffusion in the fast variable vanishes rapidly as ε tends to zero. However, the mathematics needed for their analysis is nontrivial (recall that the deterministic setting is just a particular case). If $\delta = 1/2$, the fast variable usually exhibits only convergence in distribution and, typically, a question posed to the model concerns either the convergence of the slow variable (with a marginal interest to the fast motion) or the convergence in distributions of the whole dynamics. We notice here that the slow variable may converge almost surely also if $\delta = 0$, though the fast one does not converge in any sense (recall earlier attempts of physicists to construct a theory of Brownian motion with velocity). In contrast to this, for models with $\delta > 1/2$ one can prove a uniform convergence in both variables; to get it for the fast motion on the whole time interval one needs "boundary layer" correction terms. So, the point $\delta = 1/2$ is critical since it corresponds to the situation of the Bogoliubov averaging principle, while above it there is a domain where the Tikhonov theory can be extended (of course, the philosophy of the averaging principle for the slow variable remains intact although the invariant measures are degenerated).

What can one say about other rates of $\sigma(\varepsilon)$? Our answer to this important and intriguing question (formulated by an anonymous referee of one of our papers) is that the Tikhonov theory still holds if $\sigma(\varepsilon) = o(\varepsilon^{1/2}/\sqrt{|\ln \varepsilon|})$. We believe that this fact, overlooked in previous studies, is of a deep nature, being related to the exact rate of growth of maximal function of the Ornstein–Uhlenbeck process.

The structure of the book is as follows.

Chapter 0 is a "warm-up": we present here several models that explicitly or not involve singular perturbed stochastic differential equations. The first example is a model where the drift coefficient of the Ito-type process X^ε with a constant diffusion coefficient is a function of a finite state Markov process with a large parameter ε^{-1} multiplying its transition intensity matrix which is assumed to be non-decomposable. A remarkable feature of this two-scale model is that the distribution of X^ε in the space of continuous type converges to the distribution of the Wiener process with the averaged drift (invariant measure of the Markov process) in the total variation norm and not only weakly. We prove the strong limit theorems for this model using inequalities based on the Hellinger processes suggested in [40]. Our analysis involves a filtering equation which happens to be a singularly perturbed one and exactly of the type we are interested in, being considered in the subspace orthogonal to the vector with unit components. We return to this model in Chapter 6 to show that $\varepsilon^{1/2}$ is the exact rate of convergence.

In Section 0.2 we discuss the Liénard oscillator driven by a random force. Using the Lapeyre inequality we prove a simple result on a rate of convergence for the Smoluchowski–Kramers approximation. This oscillator provides an example where $\sigma(\varepsilon) = const$ the convergence of slow variable takes place. Remarkably, the model can be treated within our approach: in the so-called Liénard coordinates the diffusion coefficients in the fast component vanish. We exploit such a representation later, in Section 4.3, to derive asymptotic expansions.

Section 0.3 deals with another example, namely, with an approximate filter arising in the scheme with a high signal-to-noise ratio. This filter is described by a singular perturbed stochastic differential equation (with $\delta = 1/2$) which can be treated in the context of the presented theory. We establish some its properties in Chapter 6.

Probably, the most serious applications are given in Section 0.4. It is well-known that models with infinite horizon and models with singular perturbations are intimately related and in some cases, where the coefficients are time-invariant, they can be transformed one into another. So, it is not a great surprise that continuous-time stochastic approximation procedures (designed to find a root of an unknown function by observing a controlled process in a long run) can be analyzed via a singular perturbation approach by an appropriate scaling. In Section 0.4 we consider a parametric family of stochastic approximation procedures of Robbins–Monro type and explain a reduction to the framework of our theory. A close look at the specific structure of asymptotic expansions for these procedures allows us to suggest a procedure with an asymptotically smaller bias. Proofs are given in Section 2.4.

Chapter 1 is our toolbox. We start with various moment inequalities for the uniform norm of solutions of linear stochastic differential equations assuming, basically, that the corresponding fundamental matrix admits an exponential bound. Section 1.1 is devoted to "easy results" under a more restrictive hypothesis formulated in terms of eigenvalues of the symmetrized drift coefficient. The central result is Theorem 1.1.7 on a Lapeyre-type bound for the growth rate of the L^p-norm of the maximal function of the solution. These results are extended in Section 1.2 to the case which is more appropriate to models with singular perturbation. In Section 1.3 we give a proof of the Lapeyre inequality following his original method. It provides, in particular, the exact rate of growth of the L^p-norm of supremum of the Ornstein–Uhlenbeck process.

Section 2.1 contains a stochastic version of the Tikhonov theorem asserting the uniform convergence of the slow variables on the whole interval $[0, T]$ and of the fast variables outside the boundary layer. The model considered is a direct generalization of the classical one. This theorem is proved under the assumption that the diffusion coefficient at the fast variable tends to zero faster than $\sqrt{\varepsilon / |\ln \varepsilon|}$. We emphasize once more that in the case where this coefficient is $\sqrt{\varepsilon}$ there is no convergence for the fast variable and the asymptotic behavior of the slow variable is the subject of another theory originating from the Bogoliubov averaging principle.

In Sections 2.2 and 2.3 we study successively the first- and the higher-order asymptotics for a fast homogeneous system. By the time stretching the problem is easily reduced to the famous "regularly" perturbed model, considered, e.g., by Freidlin and Wentzell, but in our case, of the stable stochastic equations, we have asymptotic expansions not only on fixed but on growing time intervals.

Chapter 3 deals with large deviations of the fast homogeneous system. We find exact logarithmic rates of large deviation probabilities in the uniform and L^2-metrics. This subject is purely probabilistic and has no analog in the deterministic theory.

The central problem of the Tikhonov theory is that of uniform expansions of both variables on the whole interval $[0, T]$, expansions involving boundary layer functions. It is solved in Chapter 4. In Section 4.1 we construct uniform expansions for a system with fast variables described by ordinary differential equations and prove an analog of the classical Vasil'eva theorem. It is worth noticing a new feature with respect to its deterministic prototype: only the first boundary layer function is exponentially decreasing at infinity. The general case, discussed in Section 4.2, is more delicate since there is a second small parameter at the diffusion coefficient of the fast component. As an example, we consider in Section 4.3 higher-order approximation in Narita's model of the Liénard oscillator driven by a random force.

Chapter 5 deals with controlled two-scale systems. It is inspired by the Donchev–Veliov theorem on convergence in the Hausdorff metrics of the attainability sets for linear deterministic model, but an extension of their result is far from trivial. having in mind the stochastic Mayer optimization problem, we define the attainability set as the set of distributions of all terminal random variables. The delicate point is that in stochastic optimal control there are two possibilities: control of strong solutions or control of weak solutions, defined, e.g., by the Girsanov theorem. As was shown by Beneš [9], under the Roxin condition the attainability set in the model with weak solutions is convex and closed in the weak topology but one cannot expect such nice properties in the model with strong solutions. But the latter has an important advantage: the deterministic techniques based on the Cauchy formula can be easily modified for stochastic differential equations. In our approach we consider both methods simultaneously. To our knowledge, the idea to combine these two different concepts of stochastic control theory is new. We use a result on a dense imbedding of the attainability set for the model with strong solutions into the attainability set for the model with weak solutions. It is worth noting that our basic model is similar to that considered previously in the books by Bensoussan [11] and [12] and Kushner [57], but the results and methods are quite different.

As is well-known, asymptotic results for stochastic differential equations can be translated, via probabilistic representations, into results for boundary value problems for second-order PDEs. In Section 6.1 we give theorems for the Cauchy and Dirichlet problems when the operators depend on small parameters in a particular way corresponding to singularly perturbed stochastic equations. A particular feature here is that the limit problems may have infinitely many solutions and conventional methods of PDEs are difficult to apply.

Surely, two-scale stochastic systems can be described by infinitesimal characteristics depending on a small parameter and singularly perturbed stochastic equations may appear when one filters a fast process observing a slow one (it seems that the interest for singular perturbations in SDEs was originated essentially by filtering problems of these type). In Section 6.2 we continue to study the model of a conditionally Gaussian process with the drift modulated by a Markov process with frequent switchings. To get a result on the exact rate of the asymptotics we study more attentively the convergence of the Radon–Nikodym densities using ideas and bounds developed in the theory presented.

In Sections 6.3 and 6.4 we study some properties of the approximate filter introduced in Section 0.3. In particular, we apply the large deviations result in the L^2-metric to evaluate its asymptotic performance.

In Section 6.5 we revise the LQG-problem on an infinite time interval using the so-called sensitive probabilistic criteria of optimality. We show that the positive part of the discrepancy between the time average of the running quadratic costs corresponding to the classical feedback control and to an arbitrary competing control tends to zero, even when being multiplied by any amplifying factor of order $o(T/\ln T)$ as $T \to \infty$. A simple example shows that this optimality property may fail with the amplifying factor $o(T/\ln T)$.

The book is intended for readers with a standard background in stochastic differential equations (based on the books by Liptser and Shiryaev [66] or Karatzas and Shreve [49]). In contrast to the majority of works in this field emphasizing aspects related to applications, we also address questions which may have interest as purely mathematical problems. To make the exposition self-contained we collect in the Appendix some useful but less well-known facts about stochastic equations, exponential bounds, measurable selection, Hellinger processes, Hausdorff metrics, compact sets in the space of probability measures (this part contains some new results), etc.

0 Warm-up

0.1 Processes with Fast Markov Modulations

0.1.1 Model Formulation

We consider here a two-scale system with the "slow" dynamics given by a one-dimensional conditionally Gaussian process X^ε with the drift modulated by a "fast" finite-state Markov process θ^ε. When θ^ε is in the state i the process X^ε behaves like the Wiener process with drift λ^i. If θ^ε is stationary, it is natural to expect that the process X^ε approximates in distribution the Wiener process with drift obtained by averaging of λ^i with weights proportional to the time spent by θ^ε in corresponding states.

Combining techniques based on the bounds for the total variation distance in terms of the Hellinger processes with methods of singular perturbations we prove a strong limit theorem for the slow variable even in the case of nonhomogeneous Markov modulations and establish a bound for the rate of convergence in the total variation norm. Notice that the model specification does not involve singular perturbed stochastic equations but they appear immediately when we look for an intrinsic description of the slow variable dynamics.

Let $(\Omega, \mathcal{G}, \mathbf{G} = (\mathcal{G}_t), P)$ be a stochastic basis with a one-dimensional Wiener process w and a nonhomogeneous Markov process $\theta^\varepsilon = (\theta^\varepsilon_t)_{t \leq T}$ taking values in the finite set $\{1, 2, \ldots, K\}$. The small parameter ε takes values in $]0, 1]$.

We shall consider the process X^ε given by

$$dX^\varepsilon_t = \lambda^\star J^\varepsilon_t dt + dw_t, \quad X^\varepsilon_0 = 0, \tag{0.1.1}$$

where $\lambda := (\lambda^1, \ldots, \lambda^K)^\star$ is a fixed (column) vector and $J^\varepsilon = (J^{1,\varepsilon}, \ldots, J^{K,\varepsilon})^\star$ is a vector with components $J^{i,\varepsilon} := I_{\{\theta^\varepsilon = i\}}$. In other words, (0.1.1) is just a convenient abbreviation for

$$X^\varepsilon_t = \int_0^t \sum_{i=1}^K \lambda^i I_{\{\theta^\varepsilon_s = i\}} \, ds + w_t. \tag{0.1.2}$$

Let $p^\varepsilon := (p^{1,\varepsilon}, \ldots, p^{K,\varepsilon})^\star := E J^\varepsilon_0$ be the initial distribution of θ^ε. Notice that in the theory of Markov processes it is convenient to represent distri-

butions as row vectors; to make notations of our model consistent with the
further development we deviate here from this tradition.

We assume that the transition intensity matrices of θ^ε have the form
$Q_t^\varepsilon := \varepsilon^{-1} Q_t$ where $Q = (Q_t)$ is a continuous matrix function with the
following properties:

(1) for any $t \in [0, T]$ there is a unique probability distribution

$$\pi_t = (\pi_t^1, \ldots, \pi_t^K)^\star$$

satisfying the equation

$$Q_t^\star \pi_t = 0, \tag{0.1.3}$$

i.e. zero is a simple eigenvalue and $\pi_t^\star$ is the corresponding left eigenvector of
the matrix Q_t;

(2) $\pi = (\pi_t)$ is a continuous function;

(3) there exists $\kappa > 0$ such that for any $t \in [0, T]$

$$\operatorname{Re} \lambda(Q_t) < -2\kappa \tag{0.1.4}$$

where $\lambda(Q_t)$ runs the set of nonzero eigenvalues of Q_t.

The above hypotheses need some comments. We recall that a transition
intensity matrix is a matrix with nonnegative elements except those in the
diagonal and the sum of the elements in each row is equal to zero (hence,
zero is always an eigenvalue). It is a well-known fact (see, e.g., [17]) that
all other eigenvalues of such a matrix have strictly negative real parts and
there are left eigenvectors which are probability distributions spanning the
eigenspace corresponding to the zero eigenvalue. Thus, the assumption (1) is,
actually, the requirement that zero is of multiplicity one while the properties
(2) and (3) follow from (1) and continuity of Q_t. In a probabilistic language
the property (1) means that for any fixed t the matrix Q_t can be viewed as
the transition intensity matrix of an irreducible homogeneous Markov process
and π_t is its invariant distribution. In particular, if Q does not depend on t,
the process θ^ε is ergodic.

0.1.2 Asymptotic Behavior of Distributions

Let P_T^ε be the distribution of X^ε in the space $C[0, T]$ and R_T be the distribution of the process $X = (X_t)_{t \leq T}$ given by

$$dX_t = \lambda^\star \pi_t dt + dw_t, \quad X_0 = 0, \tag{0.1.5}$$

i.e. of the Wiener process with drift $\lambda^\star \pi_t$.

Theorem 0.1.1 *(a)* $\lim_{\varepsilon \to 0} \operatorname{Var}(P_T^\varepsilon - R_T) = 0.$

(b) If $Q = (Q_t)$ is a continuously differentiable function then

$$\text{Var}\,(P_T^\varepsilon - R_T) \leq C(1 + \delta_\lambda)\delta_\lambda\varepsilon^{1/2} \tag{0.1.6}$$

where $\delta_\lambda := \max\lambda^i - \min\lambda^i$ and C is a constant depending only on Q and T.

(c) If Q does not depend on t and $\pi = p^\varepsilon$ there is a simpler bound

$$\text{Var}\,(P_T^\varepsilon - R_T) \leq C\delta_\lambda^2\varepsilon^{1/2}. \tag{0.1.7}$$

Proof. Let $\mathbf{F}^\varepsilon$ be the filtration generated by X^ε and null sets and let $\widehat{J}^\varepsilon$ be the $\mathbf{F}^\varepsilon$-optional projection of J^ε, i.e. $\mathbf{F}^\varepsilon$-optional process such that

$$\widehat{J}_\tau^\varepsilon = E(J_\tau^\varepsilon|\mathcal{F}_\tau^\varepsilon)$$

for any $\mathbf{F}^\varepsilon$-stopping time τ.

Put

$$\widetilde{w}_t := X_t^\varepsilon - \int_0^t \lambda^\star \widehat{J}_s^\varepsilon ds = w_t + \int_0^t \lambda^\star J_s^\varepsilon ds - \int_0^t \lambda^\star \widehat{J}_s^\varepsilon ds. \tag{0.1.8}$$

Then $\widetilde{w}$ is an $\mathbf{F}^\varepsilon$-adapted Wiener process (this simple observation is known as the innovation theorem) and X^ε can be represented as a diffusion-type process with

$$dX_t^\varepsilon = \lambda^\star \widehat{J}_t^\varepsilon dt + d\widetilde{w}_t, \quad X_0^\varepsilon = 0, \tag{0.1.9}$$

(see, e.g., [66], Th. 7.12). According to [66], Th. 9.1, $\widehat{J}^\varepsilon$ satisfies the filtering equation

$$d\widehat{J}_t^\varepsilon = \varepsilon^{-1}Q_t^\star\widehat{J}_t^\varepsilon dt + \phi(\widehat{J}_t^\varepsilon)d\widetilde{w}_t, \quad \widehat{J}_0 = p^\varepsilon, \tag{0.1.10}$$

where

$$\phi(\widehat{J}_t^\varepsilon) := \text{diag}\,\lambda\,\widehat{J}_t^\varepsilon - \widehat{J}_t^\varepsilon(\widehat{J}_t^\varepsilon\lambda) \tag{0.1.11}$$

and $\text{diag}\,\lambda$ is the diagonal matrix with $\lambda_{ii} := \lambda_i$.

Let $|.|_1$ be the absolute norm of a matrix (or a vector), that is, the sum of the absolute values of its components. It is easily seen that

$$|\phi(\widehat{J}_t^\varepsilon)|_1 = \sum_{i=1}^K J_t^{i,\varepsilon}|\lambda^i - J_t^\varepsilon\lambda| \leq \sum_{i=1}^K J_t^{i,\varepsilon}\delta_\lambda = \delta_\lambda$$

and hence

$$|\phi(\widehat{J}_t^\varepsilon)|^2 \leq |\phi(\widehat{J}_t^\varepsilon)|_1^2 \leq \delta_\lambda^2. \tag{0.1.12}$$

Applying for the pair of measures P^ε and R the upper bound in (A.3.3) we get that

$$\text{Var}\,(P_T^\varepsilon - R_T) \leq 4\sqrt{Eh_T^\varepsilon} \tag{0.1.13}$$

where the Hellinger process h^ε is given by

$$h_t^\varepsilon := \frac{1}{8}\int_0^t (\lambda^\star(\widehat{J}_s^\varepsilon - \pi_s))^2 ds. \tag{0.1.14}$$

For any $a = (a^1, \ldots, a^K)$ with $\sum a^i = 0$ we have

$$a\lambda = \sum a^i I_{\{a^i \geq 0\}} \lambda^i + \sum a^i I_{\{a^i < 0\}} \lambda^i \leq |a|/2(\max \lambda^i - \min \lambda^i) = (1/2)|a|\delta_\lambda.$$

Thus,

$$|\lambda^\star(\widehat{J}_s^\varepsilon - \pi_s)| \leq (1/2)\delta_\lambda |\widehat{J}_s^\varepsilon - \pi_s|_1 \leq (1/2)\delta_\lambda |\widehat{J}_s^\varepsilon - \pi_s|$$

and we get by virtue of (0.1.14) that

$$Eh_t^\varepsilon \leq \frac{1}{32}\delta_\lambda^2 E \int_0^t |\widehat{J}_s^\varepsilon - \pi_s|^2 \, ds. \tag{0.1.15}$$

Put $z^\varepsilon := \widehat{J}^\varepsilon - \pi$. It follows from (0.1.3) and (0.1.10) that

$$z_t^\varepsilon = z_0^\varepsilon + \varepsilon^{-1} \int_0^t Q_s^\star z_s^\varepsilon \, ds + \int_0^t \phi(\widehat{J}_s^\varepsilon) \, d\widetilde{w}_s - (\pi_t - \pi_0). \tag{0.1.16}$$

Let us consider the subspace $\mathcal{L} := \{x \in \mathbf{R}^K : x^\star \mathbf{1} = 0\}$ where

$$\mathbf{1} := (1, \ldots, 1)^\star.$$

Clearly, $\mathcal{L}$ is an invariant subspace for every operator Q_t and the restriction A_t of Q_t to $\mathcal{L}$ has the same eigenvalues as Q_t except zero. Thus, we can view (0.1.16) as the operator equation in $\mathbf{R}^{K-1}$. If the function Q_t is continuous differentiable, (0.1.16) can be written as

$$dz_t^\varepsilon = \varepsilon^{-1} A_t z_t^\varepsilon dt + \phi(\widehat{J}_t^\varepsilon) d\widetilde{w}_t + \dot{\pi}_t dt, \quad z_0^\varepsilon = p^\varepsilon - \pi_0, \tag{0.1.17}$$

and this we consider as a matrix equation in $\mathbf{R}^{K-1}$ (by choosing an orthonormal basis in $\mathcal{L}$). By the Cauchy formula we have

$$z_t^\varepsilon = \Phi^\varepsilon(t, 0) z_0^\varepsilon + \int_0^t \Phi^\varepsilon(t, s) \phi(\widehat{J}_s^\varepsilon) \, d\widetilde{w}_s + \int_0^t \Phi^\varepsilon(t, s) \dot{\pi}_s \, ds \tag{0.1.18}$$

where $\Phi^\varepsilon(t, s)$ is the fundamental (or transition) matrix corresponding to $\varepsilon^{-1}A$, i.e. the solution of the equation

$$\frac{\partial \Phi^\varepsilon(t, s)}{\partial t} = \varepsilon^{-1} A_t \Phi^\varepsilon(t, s) dt, \quad \Phi^\varepsilon(s, s) = I.$$

Using the exponential inequality

$$|\Phi^\varepsilon(t, s)| \leq ce^{-\kappa(t-s)/\varepsilon} \tag{0.1.19}$$

(see Proposition A.2.3) and taking into account (0.1.12) we easily obtain the following bounds:

$$|\Phi^\varepsilon(t,0)z_0^\varepsilon|^2 \le c^2|p^\varepsilon - \pi_0|e^{-2\kappa t/\varepsilon}, \tag{0.1.20}$$

$$E\left|\int_0^t \Phi^\varepsilon(t,s)\phi(\widehat{J}_s^\varepsilon)\,d\tilde{w}_s\right|^2 = E\int_0^t |\Phi^\varepsilon(t,s)|^2|\phi(\widehat{J}_s^\varepsilon)|^2\,ds \le \delta_\lambda^2 \frac{c^2\varepsilon}{2\kappa}, \tag{0.1.21}$$

$$\left|\int_0^t \dot{\pi}_s \Phi^\varepsilon(t,s)\,ds\right|^2 \le \|\dot{\pi}\|_T^2 \left(\int_0^t |\Phi^\varepsilon(t,s)|\,ds\right)^2 \le \|\dot{\pi}\|_T^2 \frac{c^2\varepsilon^2}{\kappa^2} \tag{0.1.22}$$

where $\|\dot{\pi}\|_T := \sup_{t \le T} |\dot{\pi}_t|$.

From (0.1.18) and (0.1.20)–(0.1.22) we get that for some constant C

$$\int_0^T E|z_s^\varepsilon|^2\,ds \le C^2(1 + \delta_\lambda^2)\varepsilon \tag{0.1.23}$$

and, in the homogeneous case with $p^\varepsilon = \pi$,

$$\int_0^T E|z_s^\varepsilon|^2\,ds \le C^2\delta_\lambda^2\varepsilon. \tag{0.1.24}$$

Now the assertion (b) is evident in view of (0.1.13) and (0.1.15).

In the case of the assertion (a) where the function Q (and hence π) is supposed to be only continuous the equation (0.1.16) cannot be written as (0.1.17) and the usual Cauchy formula is not applicable. Nevertheless, we can represent z as follows:

$$z_t^\varepsilon = \Phi^\varepsilon(t,0)z_0^\varepsilon + \int_0^t \Phi^\varepsilon(t,s)\phi(\widehat{J}_s^\varepsilon)\,d\tilde{w}_s + r_t^\varepsilon \tag{0.1.25}$$

where

$$r_t^\varepsilon := \Phi^\varepsilon(t,0)(\pi_t - \pi_0) + \int_0^t \frac{\partial \Phi^\varepsilon(t,s)}{\partial s}(\pi_t - \pi_s)\,ds. \tag{0.1.26}$$

Arguing as above we infer from (0.1.24) the bound

$$\int_0^T E|z_s^\varepsilon|^2\,ds \le C(\varepsilon + \|r^\varepsilon\|_T)$$

and it remains to show that $\|r^\varepsilon\|_T \to 0$ as $\varepsilon \to 0$. But taking into account the relation

$$\frac{\partial \Phi^\varepsilon(t,s)}{\partial s} = -\varepsilon^{-1} A_s \Phi^\varepsilon(t,s), \quad \Phi^\varepsilon(t,t) = I,$$

(which follows from the semigroup property) and using the exponential inequality we obtain that

$$|r_t^\varepsilon| \le ce^{-\kappa t/\varepsilon}|\pi_t - \pi_0| + c\varepsilon^{-1}\|A\|_T \int_0^t e^{-\kappa(t-s)/\varepsilon}|\pi_t - \pi_s|\,ds.$$

The uniform convergence of r^ε to zero follows now from Lemma A.2.4. $\square$

Conclusion. The bounds of Theorem 0.1.1 hold by virtue of properties of the equation (0.1.16) which is a **singularly perturbed stochastic equation** because it can be written as

$$\varepsilon dz_t^\varepsilon = A_t z_t^\varepsilon dt + \sigma(\varepsilon)G(z^\varepsilon)d\tilde{w}_t + \sigma(\varepsilon)b_t dt \qquad (0.1.27)$$

with $\sigma(\varepsilon) = o(1)$ as $\varepsilon \to 0$. It is important to note that the matrix function A_t admits the exponential bound (0.1.19). Another essential feature is that the parameter $\sigma(\varepsilon) := \varepsilon$ at the diffusion coefficient tends to zero faster than $\sqrt{\varepsilon}$ providing the convergence of trajectories to zero. Singularly perturbed stochastic equations of this type and, especially, more general systems involving also "slow" variables are the objects of principal interest of this book. Some techniques developed in the sequel will be used in Section 6.2 where we present a more deep analysis of our model and show that the bounds (0.1.6) and (0.1.7) give the correct order of convergence in ε by calculating the limit of $\varepsilon^{-1/2}\mathrm{Var}\,(P_T^\varepsilon - R_T)$.

0.2 The Liénard Oscillator Under Random Force

In this section we discuss briefly an important example of a two-scale stochastic system, namely, the Liénard oscillator driven by random force. This classic model arises in a mathematical description of the motion of a small particle in a viscous media. On an intuitive level, it can be described by the second-order equation

$$\varepsilon\ddot{x} + \dot{x} - h(x) = \dot{w} \qquad (0.2.1)$$

where $\dot{w}$ is a white noise and ε is a small positive parameter. The standard reduction transforms it to the system of two equations of the first order

$$\dot{x} = v, \qquad (0.2.2)$$
$$\varepsilon\dot{v} = -v + h(x) + \dot{w}. \qquad (0.2.3)$$

The rigorous formulation can be given by the following system of stochastic equations in the usual Ito sense where we exhibit explicitly the dependence on ε:

$$dx_t^\varepsilon = v_t^\varepsilon dt, \quad x_0^\varepsilon = x^o, \qquad (0.2.4)$$
$$\varepsilon dv_t^\varepsilon = -v_t^\varepsilon dt + h(t, x_t^\varepsilon)dt + dw_t, \quad v_0^\varepsilon = v^o. \qquad (0.2.5)$$

Here w is a Wiener process and the initial condition can be random.

It is worth noting that this system is quite specific: the equation for the position x^ε does not contain a diffusion term, while the equation for the velocity v^ε does not involve a small parameter at diffusion. For $h = 0$ the process v^ε is simply the Ornstein–Uhlenbeck process. Therefore, in general, we cannot expect the convergence of v^ε. Nevertheless, it is easy to prove that under mild assumptions on h the position process x^ε converges uniformly on any compact interval in probability to the process x (in the literature referred to as the Smoluchowski–Kramers approximation) satisfying the stochastic equation

$$dx_t = h(t, x_t)dt + dw_t, \quad x_0 = x^o. \tag{0.2.6}$$

We give here a bit more precise result of this kind assuming that the processes x^ε, v^ε, and w are n-dimensional.

Proposition 0.2.1 *Let $T \in \mathbf{R}_+$, $p \in [1, \infty[$. Assume that $x^o, v^o \in L^p(\Omega)$ and h satisfies on $[0, T] \times \mathbf{R}^n$ the global Lipschitz condition and the linear growth condition. Then there exists a constant C (depending on p and T) such that*

$$\lim_{\varepsilon \to 0} (E||x^\varepsilon - x||_T^p)^{1/p} \leq C\sqrt{\varepsilon|\ln \varepsilon|} \tag{0.2.7}$$

where $||.||_T$ is the norm in $C[0, T]$.

Proof. Put $\Delta_t^\varepsilon := x_t^\varepsilon - x_t$. Using the Cauchy formula we "resolve" (0.2.5) and obtain the representation

$$v_s = e^{-s/\varepsilon}v^o + \frac{1}{\varepsilon}\int_0^s e^{-(s-u)/\varepsilon}h(u, x_u^\varepsilon)du + \frac{1}{\varepsilon}\int_0^s e^{-(s-u)/\varepsilon}dw_u. \tag{0.2.8}$$

Substituting it into (0.2.4), we get from (0.2.4) and (0.2.6) that

$$\Delta_t^\varepsilon = \varepsilon(1 - e^{-t/\varepsilon})v^o + \zeta_t^\varepsilon + \xi_t^\varepsilon + \eta_t^\varepsilon \tag{0.2.9}$$

where

$$\zeta_t^\varepsilon := \int_0^t \left(\frac{1}{\varepsilon}\int_0^s e^{-(s-u)/\varepsilon}(h(u, x_u^\varepsilon) - h(u, x_u))du\right)ds,$$

$$\xi_t^\varepsilon := \int_0^t \left(\frac{1}{\varepsilon}\int_0^s e^{-(s-u)/\varepsilon}h(u, x_u)du\right)ds - \int_0^t h(s, x_s)ds,$$

$$\eta_t^\varepsilon := \frac{1}{\varepsilon}\int_0^t \left(\int_0^s e^{-(s-u)/\varepsilon}dw_u\right)ds - w_t.$$

By virtue of the Fubini theorems

$$\zeta_t^\varepsilon = \int_0^t (1 - e^{-(t-u)/\varepsilon})(h(u, x_u^\varepsilon) - h(u, x_u))du, \tag{0.2.10}$$

$$\xi_t^\varepsilon = \int_0^t e^{-(t-u)/\varepsilon}h(u, x_u)du, \tag{0.2.11}$$

$$\eta_t^\varepsilon = -\int_0^t e^{-(t-u)/\varepsilon}dw_u. \tag{0.2.12}$$

By assumption, there is a constant L such that

$$|h(t, y_1) - h(t, y_2)| \leq L|y_1 - y_2| \quad \forall t \in [0, T], \ y_1, y_2 \in \mathbf{R}^n,$$

and

$$|h(t, y)| \leq L(1 + |y|) \quad \forall t \in [0, T], \ y \in \mathbf{R}^n.$$

Using this we deduce from (0.2.9)–(0.2.11) that for every $t \leq T$

$$||\Delta^\varepsilon||_t \leq \varepsilon|v^o| + L \int_0^t ||\Delta^\varepsilon||_u du + L\varepsilon(1 + ||x||_T) + ||\eta^\varepsilon||_T$$

and, hence, by the Gronwall–Bellman lemma

$$||\Delta^\varepsilon||_T \leq \left(\varepsilon|v^o| + \varepsilon L(1 + ||x||_T) + ||\eta^\varepsilon||_T\right)e^{LT}. \tag{0.2.13}$$

This bound implies the result. Indeed, we assume that the initial conditions are random variables belonging to $L^p(\Omega)$. Due to the linear growth and global Lipschitz conditions $||x||_T \in L^p(\Omega)$ for all finite p. It remains to notice that for some constant C_p we have

$$(E||\eta^\varepsilon||_T^p)^{1/p} \sim C_p\sqrt{\varepsilon|\ln \varepsilon|}, \quad \varepsilon \to 0, \tag{0.2.14}$$

see Chapter 1. $\square$

Remark. Notice that the equation (0.2.5) for the fast variable does not contain the small parameter at diffusion and, thus, the model looks different from the basic one considered in this book. However, if we choose for analysis of (0.2.1) the so-called Liénard coordinates by putting $u = \varepsilon\dot{x} + x$, the resulting system will be

$$du_t = h(x_t)dt + dw_t, \tag{0.2.15}$$

$$\varepsilon dx_t = (-x_t + u)dt, \tag{0.2.16}$$

with the diffusion coefficient of the fast variable equal to zero. The general theory of Chapter 4 includes this case if h is Lipschitz.

Comment. It would be more consistent with the modern methodology to start with the models (0.2.4), (0.2.5) each defined on its own probability space and indexed (together with its Wiener process) by the parameter ε. The physically meaningful question is the convergence of the distribution of x^ε in the space $C[0, T]$ to that of x. To do this, there is a powerful method to construct a realization of processes on a single probability space and prove the convergence of x^ε to x (as random variables with values in $C[0, T]$) in probability. We start, actually, from the point where this transfer has been done. Does $||x^\varepsilon - x||_T$ converge to zero almost surely? The positive answer, in view of the bound (0.2.13), seems obvious. But take care: (0.2.13) holds only a.s. and the exceptional set may depend on ε. In fact, for the process η^ε we cannot ensure even the convergence $||\eta^\varepsilon||_T$ to zero a.s. because stochastic integrals are defined up to P-null sets. To make this question mathematically correct (and admitting the positive answer) we should construct a good realization of the whole family to ensure the continuity of paths $||\Delta^\varepsilon||_T$ and $||\eta^\varepsilon||_T$ for $\varepsilon > 0$. For this simple model it is not difficult. For instance, integrating by parts we get that for every $\varepsilon > 0$

$$\eta_t^\varepsilon = w_t - \varepsilon^{-1} \int_0^t e^{-(t-s)/\varepsilon} w_s ds$$

almost surely for all t. The right-hand side of this formula can be used to define the appropriate version of η^ε.

0.3 Filtering of Nearly Observed Processes

The problem of nonlinear filtering consists in estimating a stochastic process (a signal) that is not directly observed. A lot of studies are devoted to the practically important case where the process is *nearly* observed. This is an asymptotic setting in which computable asymptotic filters can be easily studied. The aim of this section is to provide a simple illustrative example where the singular perturbed stochastic equations appear in a natural way.

Let us consider the model described by two processes x (unobservable signal) and y^ε (observations), both, for simplicity, n-dimensional, given by

$$dx_t = f_t dt + \sigma_t dw^x, \quad x_0 = x^0, \tag{0.3.1}$$

$$dy_t^\varepsilon = x_t dt + \varepsilon dw^y, \quad y_0^\varepsilon = y^0, \tag{0.3.2}$$

where w^x and w^y are independent Wiener processes in $\mathbf{R}^n$, and f and σ are continuous processes of corresponding dimensions adapted to the filtration generated by w^x. The parameter $\varepsilon \in \,]0,1]$ is small; it formalizes the fact that noises in the signal and observations are of different scales and the signal-to-noise ratio is large.

A filter is any process adapted with respect to the filtration of y^ε. Engineers are looking often for filters which approximate x in some sense. Such filters may not perform so well as but are easier to implement.

Let us consider the filter $\widehat{x}^\varepsilon$ admitting the following representation:

$$d\widehat{x}_t^\varepsilon = \widehat{f}_t^\varepsilon dt - \varepsilon^{-1} A_t (dy_t^\varepsilon - \widehat{x}_t^\varepsilon dt), \tag{0.3.3}$$

where the continuous vector-valued process $\widehat{f}^\varepsilon$ (assumed to be a function of y^ε), the continuous function A with values in the set of $n \times n$ matrices, and the initial condition can be viewed as filter parameters. We assume that there is a constant $\kappa > 0$ such that for Re $\lambda(A_t) < -2\kappa$ for all t. For the error process $\Delta^\varepsilon := \widehat{x}^\varepsilon - x$ we get from (0.3.1)–(0.3.3) the equation

$$d\Delta_t^\varepsilon = \varepsilon^{-1} A_t \Delta_t^\varepsilon dt + (\widehat{f}_t^\varepsilon - f_t)dt + G_t d\tilde{w}_t, \quad \Delta_0^\varepsilon = \widehat{x}_0^\varepsilon - x^0, \tag{0.3.4}$$

where $G_t := (A_t A_t^\star + \sigma_t \sigma_t^\star)^{1/2}$ and

$$\tilde{w}_t := -\int_0^t G_s^{-1} A_s dw_s^y - \int_0^t G_s^{-1} \sigma_s dw_s^x$$

is a Wiener process in $\mathbf{R}^n$.

Let $\Phi^\varepsilon(t, s)$ be the fundamental matrix defined by the linear equation

$$\varepsilon \frac{\partial \Phi^\varepsilon(t, s)}{\partial t} = A_t \Phi^\varepsilon(t, s), \quad \Phi^\varepsilon(s, s) = I. \tag{0.3.5}$$

Using the Cauchy formula we can write the solution of (0.3.4) as

$$\Delta_t^\varepsilon = \Phi^\varepsilon(t, 0)(\widehat{x}_0^\varepsilon - x^0) + \int_0^t \Phi^\varepsilon(t, s)(\widehat{f}_s^\varepsilon - f_s) ds + \xi_t^\varepsilon \tag{0.3.6}$$

where

$$\xi_t^\varepsilon := \int_0^t \Phi^\varepsilon(t, s) G_s d\tilde{w}_s. \tag{0.3.7}$$

The process ξ_t^ε is the solution of

$$\varepsilon d\xi_t^\varepsilon := A_t \xi_t^\varepsilon dt + \varepsilon G_t d\tilde{w}_t, \quad \xi_0^\varepsilon = 0. \tag{0.3.8}$$

For us it is important to note that the asymptotic behavior of the approximate filter is determined by properties of solutions of a singularly perturbed stochastic equation (with a small parameter at the diffusion term of order ε).

Using the exponential bound for $|\Phi^\varepsilon(t, s)|$ (see Lemma A.2.2) we obtain that

$$|\Delta_t^\varepsilon| \leq C e^{-\kappa t/\varepsilon} |\widehat{x}_0^\varepsilon - x^0| + C \int_0^t e^{-\kappa(t-s)/\varepsilon} |\widehat{f}_s^\varepsilon - f_s| ds + |\xi_t^\varepsilon|. \tag{0.3.9}$$

This implies the following less precise but simpler inequality which gives a clear idea of the filter behavior:

$$|\Delta_t^\varepsilon| \leq C e^{-\kappa t/\varepsilon} |\widehat{x}_0^\varepsilon - x^0| + C \varepsilon \kappa^{-1} \|\widehat{f}^\varepsilon - f\|_t + |\xi_t^\varepsilon| \tag{0.3.10}$$

where $\|.\|_t$ denotes the uniform norm on $[0, t]$.

In the particular case of constant A and σ we have $\Phi^\varepsilon(t, s) = e^{(t-s)A/\varepsilon}$, the process ξ^ε is Gaussian, and $\varepsilon^{-1/2} \xi_t^\varepsilon$ converges in distribution as $\varepsilon \to 0$ to the centered Gaussian vector with covariance matrix

$$S_A = \int_0^\infty e^{rA}(AA^\star + \sigma\sigma^\star)e^{rA^\star} dr. \tag{0.3.11}$$

Assuming, e.g., that $|\widehat{x}_0^\varepsilon|$ is bounded and $\varepsilon^{1/2} \|\widehat{f}^\varepsilon - f\|_t$ converges to zero in probability as $\varepsilon \to 0$ we infer from (0.3.6) that $\varepsilon^{-1/2}(\widehat{x}_t^\varepsilon - x_t)$ is asymptotically Gaussian with zero mean and covariance S_A.

In a more specific situation of scalar processes ($n = 1$) with the filter parameter $A = -\gamma > 0$ we have $S_A := (\gamma^2 + \sigma^2)/(2\gamma)$. If $\sigma > 0$ is known one can attain the smallest value of asymptotic variance $S_A = \sigma$ by choosing $\gamma = \sigma$.

In the vector model with a known nondegenerated matrix σ it is reasonable to choose $A = -(\sigma\sigma^\star)^{1/2}$ and get the limit covariance $S_A = (\sigma\sigma^\star)^{1/2}$. To justify such a choice we notice that for any symmetric negative definite matrix B commuting with $\sigma\sigma^\star$ the difference $S_B - (\sigma\sigma^\star)^{1/2}$ is positive definite because, in this case,

$$S_B = -\frac{1}{2}B^{-1}(B^2 + (\sigma\sigma^\star)) = (\sigma\sigma^\star)^{1/2} - \frac{1}{2}B^{-1}(B + (\sigma\sigma^\star)^{1/2})^2 \quad (0.3.12)$$

and the last term is negative definite.

One can expect that the filter will exhibit a better performance also in a non-asymptotic sense (i.e. for realistic values of the signal-to-noise ratio) if $\widehat{f}_t^\varepsilon$ tracks f_t. We return to this model in Sections 6.3 and 6.4.

0.4 Stochastic Approximation

The stochastic approximation theory, initially developed for discrete-time models but now treated more and more often in a very general semimartingale setting, deals with the problems of estimating a root of an unknown function F on the basis of observations of a controlled random process $\theta = \theta^\gamma$. We consider here a rather particular continuous-time white-noise model which, nevertheless, covers several approximation procedures studied in the literature. Our aim is to show that, being rescaled, it comes into the framework of the theory of singularly perturbed stochastic differential equations which allows us to analyze stochastic approximation procedures in a systematic and transparent way and get asymptotic expansions of estimators.

Let $\theta = \theta^\gamma$ be given on $[t_0, \infty[$ by the SDE

$$d\theta_t = \gamma_t F(\theta_t)dt + \gamma_t dw_t, \quad \theta_{t_0} = \theta_0, \quad (0.4.1)$$

where w is a Wiener process in $\mathbf{R}^n$, the function $F : \mathbf{R}^n \to \mathbf{R}^n$ is continuously differentiable, the "control" $\gamma = (\gamma_t)_{t>0}$ is a nonnegative continuous deterministic function, and the initial condition is posed at some point $t_0 > 0$.

We assume that F satisfies the following hypotheses:

H.0.4.1 There is a unique root θ_* of the equation

$$F(\theta) = 0$$

and

$$(\theta - \theta_*)^\star F(\theta) < 0 \quad \forall \theta \in \mathbf{R}^n \setminus \{\theta_*\}. \quad (0.4.2)$$

H.0.4.2 The real parts of all eigenvalues of the matrix

$$A := F'(\theta_*)$$

are strictly negative: $\operatorname{Re} \lambda(A) < -2\kappa < 0$.

Resembling the standard problems of optimal control, the model has specific features: it is not completely specified since F is unknown (but some extra information on F may be available) and the class of controls is quite restrictive. For these reasons, the traditional paradigm of stochastic approximation does not formulate the optimal control problem by stipulating in a precise way an objective function but uses instead the ideology and concepts of mathematical statistics. There is a vast literature devoted to analysis, for particular stochastic procedures γ, of the asymptotic behavior as $T \to \infty$ of θ_T or, more recently, of the average

$$\hat{\theta}_T := \frac{1}{T - t_1} \int_{t_1}^{T} \theta_s ds, \tag{0.4.3}$$

as statistical estimators of θ_*. For instance, the continuous-time version of the classic Robbins–Monro procedure claims that θ_T is a strongly consistent estimator of θ_*. Its precise formulation is as follows.

Proposition 0.4.1 *Assume that* **H.0.4.1** *holds and*

$$\int_{t_0}^{\infty} \gamma_u\, du = \infty, \qquad \int_{t_0}^{\infty} \gamma_u^2\, du < \infty. \tag{0.4.4}$$

Then

$$\lim_{t \to \infty} \theta_t = \theta_* \quad a.s.\ and\ in\ L^4. \tag{0.4.5}$$

Proof. Put $U_t := \theta_t - \theta_*$. Then

$$dU_t = F(\theta_t)\gamma_t dt + \gamma_t dw_t, \quad \theta_{t_0} = \theta_0,$$

and, by the Ito formula,

$$|U_t|^2 = |U_{t_0}|^2 + 2 \int_{t_0}^{t} U_s^* F(\theta_s)\gamma_s ds + 2M_t + \int_{t_0}^{t} \gamma_s^2 ds \tag{0.4.6}$$

where

$$M_t := \int_{t_0}^{t} U_s^* \gamma_s dw_s.$$

Notice that $U_s^* F(\theta_s) \le 0$ by (0.4.2) and hence the first integral in the right-hand side of (0.4.6) defines a decreasing process. Localizing the stochastic integral M and taking the expectation, we get, with help of the Fatou lemma, that

$$E|U_t|^2 \leq E|U_{t_0}|^2 + \int_{t_0}^{\infty} \gamma_s^2 ds.$$

It follows that

$$E\langle M\rangle_\infty = \int_{t_0}^{\infty} E|U_s|^2 \gamma_s^2 ds \leq \left(E|U_{t_0}|^2 + \int_{t_0}^{\infty} \gamma_s^2 ds\right) \int_{t_0}^{\infty} \gamma_s^2 ds < \infty$$

by the second relation in (0.4.4). The square integrable martingale M bounded in L^2 converges a.s. to a finite limit. Thus, the processes on the right-hand side of (0.4.6) converge at infinity to finite limits (a.s.). The continuity of F and the relation **H.0.4.2** imply that for every $r \in \,]0,1[$ there is a constant $c_r > 0$ such that

$$(\theta - \theta^*)F(\theta) \leq -c_r$$

when $r \leq |\theta - \theta^*| \leq 1/r$. The divergence of the integral of γ implies that on the set $\{\lim |U_t| > 0\}$ the first integral in (0.4.6) diverges to $-\infty$. Hence, U_t converges to zero a.s. At last, $||M||_{t_0,\infty} \in L^2$ and the process $|U|^2$, being bounded by a square integrable random variable, converges to zero in L^2. $\square$

We consider here two stochastic approximation procedures and study asymptotic expansions of the estimator (0.4.3). The first procedure, depending on a parameter $\rho \in \,]1/2,1[$, corresponds to the choice

$$\gamma_t := t^{-\rho} \tag{0.4.7}$$

and $t_1 = t_1(T) = Tr_1(T)$ with

$$r_1(T) := \frac{1}{\ln(\gamma_T T)} = \frac{1}{(1-\rho)\ln T}. \tag{0.4.8}$$

The second one, with the characteristics marked by the superscript ^{o}, is given by

$$\gamma_t^o := e\frac{\ln t}{t \ln_3 t} \tag{0.4.9}$$

with

$$r_1^o(T) := \frac{1}{\ln_2 T} \tag{0.4.10}$$

where $\ln_n$ denotes the n-times-iterated logarithm.

Theorem 0.4.2 *Suppose that $F \in C^3$ and* **H.0.4.1**, **H.0.4.2** *are fulfilled. Then for the procedure given by (0.4.7), (0.4.8) we have*

$$\widehat{\theta}_T = \theta_* + \xi_T \frac{1}{T^{1/2}} + h\frac{1}{1-\rho}\frac{1}{T^\rho} + R_T\frac{1}{T^\rho} \tag{0.4.11}$$

where $h \in \mathbf{R}^n$, ξ_T is a centered Gaussian random vector with covariance matrix converging to $(A^\star A)^{-1}$, and $R_T \to 0$ in probability as $T \to \infty$.

Theorem 0.4.3 *Suppose that $F \in C^3$ and* **H.0.4.1**, **H.0.4.2** *are fulfilled. Then for the procedure given by (0.4.9), (0.4.10) we have*

$$\widehat{\theta}_T^o = \theta_* + \xi_T^o \frac{1}{T^{1/2}} + he\frac{\ln T}{T} + R_T^o \frac{\ln T}{T} \qquad (0.4.12)$$

where $h \in \mathbf{R}^n$, ξ_T^o is a centered Gaussian random vector with covariance matrix converging to $(A^\star A)^{-1}$, and $R_T^o \to 0$ in probability as $T \to \infty$.

Remark 1. The vector h in the above theorems depends only on F. In the scalar case we have $h = (1/4)A^{-2}F''(\theta_*)$. The explicit expression in the general case can be found in Section 2.4.

Remark 2. An inspection of (0.4.11) makes plausible the idea that the third term on its right-hand side is responsible for the bias of the estimator. Obviously, for sufficiently large T

$$\max_{\rho \in]1/2,1[} (1 - \rho)T^\rho = \frac{1}{e}\frac{T}{\ln T}.$$

Thus, the minimum over $\rho \in]1/2, 1[$ of the third term on the right-hand side of (0.4.11) coincides with the corresponding terms in (0.4.12). This observation explains our interest in the second procedure. Indeed, under a certain auxiliary condition $E|R_T|$ and $E|R_T^o|$ converge to zero, see Theorem 0.4.6 below.

We prove the above results in Chapter 2 providing here only the reduction to the framework of singular perturbations.

First, let us consider the procedure with the function γ_t defined by (0.4.7). A simple rescaling leads to a problem on the interval with a fixed right extremity. Indeed, put $\tilde{\theta}_r := \theta_{rT}$. Then, by virtue of (0.4.1), on the interval $[t_0/T, 1]$

$$d\tilde{\theta}_r = \gamma_T T F(\tilde{\theta}_r)\gamma_r dr + \gamma_T T^{1/2}\gamma_r d\tilde{w}_r, \quad \tilde{\theta}_{t_0/T} = \theta_0, \qquad (0.4.13)$$

where $\tilde{w}_r := T^{-1/2}w_{rT}$ is a Wiener process. Obviously,

$$\widehat{\theta}_T = \frac{1}{1 - r_1} \int_{r_1}^1 \tilde{\theta}_r dr.$$

Now we reparameterize the problem by introducing instead of the large parameter T the small parameter

$$\varepsilon := \frac{1}{\gamma_T T} = \frac{1}{T^{1-\rho}}. \qquad (0.4.14)$$

Then

$$T = T(\varepsilon) = \frac{1}{\varepsilon^{1/(1-\rho)}}. \qquad (0.4.15)$$

Setting $y_r^\varepsilon := \tilde{\theta}_{rT(\varepsilon)}$, we rewrite (0.4.13) as the singularly perturbed stochastic equation

$$\varepsilon dy_r^\varepsilon = F(y_r^\varepsilon)\gamma_r dr + \beta\varepsilon^{1/2}\gamma_r d\tilde{w}_r, \quad y_{t_0/T(\varepsilon)} = \theta_0, \qquad (0.4.16)$$

where

$$\beta := \sqrt{\gamma_{T(\varepsilon)}} = \varepsilon^{(1/2)\rho/(1-\rho)}. \qquad (0.4.17)$$

With this new parameterization $\widehat{\theta}_T$ becomes equal to

$$\widehat{y}_1^\varepsilon := \frac{1}{1-r_1^\varepsilon}\int_{r_1^\varepsilon}^1 y_r^\varepsilon dr \qquad (0.4.18)$$

where

$$r_1^\varepsilon = -\frac{1}{\ln\varepsilon}. \qquad (0.4.19)$$

Theorem 0.4.3 has the following equivalent form:

Theorem 0.4.4 *Suppose that $F \in C^3$ and* **H.0.4.1, H.0.4.2** *are fulfilled. Then for the model (0.4.14)–(0.4.19)*

$$\widehat{y}_1^\varepsilon = \theta_* + \xi^\varepsilon\varepsilon^{1/2}\beta + h\frac{1}{1-\rho}\beta^2 + R^\varepsilon\beta^2 \qquad (0.4.20)$$

where $h \in \mathbf{R}^n$, ξ^ε is a centered Gaussian random vector with covariance matrix converging to $(A^\star A)^{-1}$, and $R^\varepsilon \to 0$ in probability as $\varepsilon \to 0$.

One can notice that the small parameters are involved in (0.4.16) in a very simple, multiplicative, way. The only particular feature is that the starting time depends on ε and the function γ has a singularity at zero which is integrable. The coefficient $1/(1-\rho)$ in (0.4.20) is equal to the integral of γ over $[r_1^\varepsilon, 1]$ up to $o(1)$.

Similarly, the rescaling of the model with γ^o defined in (0.4.9) results in the stochastic equation on $[t_0/T, 1]$

$$d\tilde{\theta}_r = \gamma_{rT}^o TF(\tilde{\theta}_r)dr + \gamma_{rT}^o T^{1/2}d\tilde{w}_r, \quad \tilde{\theta}_{t_0/T} = \theta_0. \qquad (0.4.21)$$

For sufficiently large T we define the function $\varepsilon = \varepsilon(T)$ by putting

$$\varepsilon := \frac{1}{\gamma_T^o T} = \frac{1}{e}\frac{\ln_3 T}{\ln T}. \qquad (0.4.22)$$

Let $T(\varepsilon)$ be the inverse of the above function. We rewrite (0.4.21) as

$$\varepsilon dy_r^\varepsilon = F(y_r^\varepsilon)\gamma_r^\varepsilon dr + \beta\varepsilon^{1/2}\gamma_r^\varepsilon d\tilde{w}_r, \quad y_{t_0/T(\varepsilon)} = \theta_0, \qquad (0.4.23)$$

where

$$\gamma_r^\varepsilon := \frac{1}{r} \frac{\ln(rT(\varepsilon))}{\ln T(\varepsilon)} \frac{\ln_3 T(\varepsilon)}{\ln_3(rT(\varepsilon))} \tag{0.4.24}$$

and

$$\beta := \sqrt{\gamma_{T(\varepsilon)}^o} = \frac{1}{\sqrt{\varepsilon T(\varepsilon)}}. \tag{0.4.25}$$

Again $\widehat{\theta}_T$ is equal to

$$\widehat{y}_1^\varepsilon := \frac{1}{1 - r_1^\varepsilon} \int_{r_1^\varepsilon}^1 y_r^\varepsilon dr \tag{0.4.26}$$

but now

$$r_1^\varepsilon = \frac{1}{\ln_2 T(\varepsilon)}. \tag{0.4.27}$$

The corresponding equivalent version of Theorem 0.4.3 is

Theorem 0.4.5 *Suppose that $F \in C^3$ and* **H.0.4.1**, **H.0.4.2** *are fulfilled. Then for the model (0.4.22)–(0.4.27)*

$$\widehat{y}_1^\varepsilon = \theta_* + \xi^{o,\varepsilon}\varepsilon^{1/2}\beta + h\beta^2 \ln_3 T(\varepsilon) + R^{o,\varepsilon}\beta^2 \ln_3 T(\varepsilon) \tag{0.4.28}$$

where $h \in \mathbf{R}^n$, $\xi^{o,\varepsilon}$ is a centered Gaussian random vector with covariance matrix converging to $(A^\star A)^{-1}$, and $R^{o,\varepsilon} \to 0$ in probability as $\varepsilon \to 0$.

Of course, more systematic notations require the superscript ^{o} at $T(\varepsilon)$, β, etc, but we skip it for obvious reasons.

Remark. Clearly, the equations (0.4.16) and (0.4.23) are of the same structure. However, in the latter case the function γ_r^ε has a singularity at zero like $1/r$ which is not integrable and which yields in the term $\ln_3 T(\varepsilon)$ after integrating over the interval $[r_1^\varepsilon, 1]$.

To get a convergence of residual terms we add to our assumption the following hypothesis on a "global" behavior of F:

H.0.4.3 There exists a bounded matrix-valued function $A(y_1, y_2)$ such that for all y_1, y_2

$$F(y_1) - F(y_2) = A(y_1, y_2)(y_1 - y_2) \tag{0.4.29}$$

and

$$z^* A(y_1, y_2)z \leq -\kappa|z|^2 \qquad \forall z \in \mathbf{R}^n \tag{0.4.30}$$

for some constant $\kappa > 0$.

Clearly, **H.0.4.3** implies the Lipschitz and linear growth condition. In the one-dimensional case this hypothesis holds if $F \in C^1$ and $F' \leq -\kappa < 0$.

Theorem 0.4.6 *Suppose that $F \in C^3$, the second derivative F'' is bounded and satisfies the Lipschitz condition, and the conditions* **H.0.4.1**–**H.0.4.3** *are fulfilled. Then $E|R^\varepsilon| = o(1)$ and $E|R^{o,\varepsilon}| = o(1)$ as $\varepsilon \to 0$.*

As a corollary we obtain, under the assumptions of Theorem 0.4.6, that

$$E\widehat{\theta}_T = \theta_* + h\frac{1}{1-\rho}\frac{1}{T^\rho} + \frac{1}{T^\rho}o(1), \qquad (0.4.31)$$

$$E\widehat{\theta}_T^o = \theta_* + he\frac{\ln T}{T} + \frac{\ln T}{T}o(1) \qquad (0.4.32)$$

as $T \to \infty$.

1 Toolbox: Moment Bounds for Solutions of Stable SDEs

In this chapter we present results on the growth of solutions of SDEs

$$dy_t = f(t, y_t)dt + G_t dW_t, \quad y_0 = 0, \tag{1.0.1}$$

driven by the multidimensional Wiener process $W = (W_t)$ under the assumption of exponential asymptotic stability at zero of solutions of the ordinary differential equation

$$dx_t = f(t, x_t)dt. \tag{1.0.2}$$

We are interested in bounds for moments of the uniform norm $||y||_T = y_T^*)$ of solutions of (1.0.1) on the time interval $[0, T]$. For us especially important is the "linear" case where $f(t, y) = A_t y$ (the quotation marks here mean that A may depend on y). We shall use mainly a direct method which is based on the Ito formula only. Alternative techniques, involving the law of iterated logarithms as well as some ideas from the theory of Gaussian processes also will be discussed (Section 1.3).

When the matrix A is constant, the requirement of asymptotic stability means that A is stable (Hurvitz), i.e. the real parts of all eigenvalues of A are negative: $\mathrm{Re}\,\lambda(A) \leq -\gamma < 0$, implying that the solutions of (1.0.2) have an exponential decay with a rate γ. The rate of exponential decay is an essential parameter and plays an important role in our applications of the moment bounds. It is well-known that for time-dependent A, even smooth, it is not possible to guarantee the asymptotic stability only in spectral terms. There is a simple two-dimensional example where all A_t have only one eigenvalue (of multiplicity 2) -1 but (1.0.2) has a solution of the exponential growth. It is rather easy to obtain good moment bounds when the eigenvalues of the symmetrized matrices $A_t + A_t^*$ are bounded from above by a strictly negative constant. These moment bounds ensure that the growth rate of L^p-norms of y_T^* is $\sqrt{\ln T}$ (see Theorem 1.1.7). To relax the spectral assumption, more delicate work is done in Section 1.2; results of the latter are helpful in our stochastic version of the Tikhonov theory. In Section 1.3 we show also that $\sqrt{\ln T}$ is the exact growth rate for the Ornstein–Uhlenbeck process.

1.1 Moment Bounds for Nonlinear Equations

1.1.1 Key Lemma

Let $(\Omega, \mathcal{F}, \mathbf{F} = (\mathcal{F}_t), P)$ be a stochastic basis with an l-dimensional Wiener process W adapted to the filtration $\mathbf{F}$.

Let $y = (y_t)$ be a process satisfying (1.0.1) where $f : \Omega \times \mathbf{R}_+ \times \mathbf{R}^q \to \mathbf{R}^q$ is a $\mathcal{P} \times \mathcal{B}(\mathbf{R}^q)$-measurable function, $\mathcal{P}$ is the predictable σ-algebra in $\Omega \times \mathbf{R}_+$, and G is a bounded predictable process with values in the set of $q \times l$ matrices.

Of course, (1.0.1) suggests that for finite T

$$\int_0^T |f(t, y_t)| dt < \infty.$$

We assume that the function f satisfies the following hypothesis:

H.1.1.1 There exists $\gamma > 0$ such that

$$y^* f(t, y) \le -\gamma |y|^2 \quad \forall\, t \in \mathbf{R}_+,\ y \in \mathbf{R}^q.$$

Notice that the relation (1.0.1) is, actually, a representation of the given process y rather than an equation to determine y. So, we have no need to discuss here the problems of existence and uniqueness for this equation.

In the important linear case where $f(t, y) = A_t y$ and

$$dy_t = A_t y_t dt + G_t dW_t, \quad y_0 = 0, \tag{1.1.1}$$

with a bounded predictable process $A = (A_t)$ taking values in the set of $q \times q$ matrices the hypothesis **H.1.1.1** is read as

$$y^* A_t y \le -\gamma |y|^2 \quad \forall\, t \in \mathbf{R}_+,\ y \in \mathbf{R}^q.$$

Obviously, it is equivalent to

H.1.1.2 There exists $\gamma > 0$ such that

$$\lambda(A_t + A_t^*)/2 \le -\gamma \quad \forall t,$$

i.e. all eigenvalues of the symmetric matrices $(A_t + A_t^*)/2$ are negative and bounded from zero by a constant.

Of course, **H.1.1.2** holds if all eigenvalues of A_t are real and

$$\lambda(A_t) \le -\gamma\ \forall t.$$

Let $\mathcal{T}_T$ be the set of stopping times taking values in the interval $[0, T]$.

Lemma 1.1.1 *Let y be the solution of the equation (1.0.1) with f satisfying* **H.1.1.1** *and $\|G\|_T \leq M$ where $T \in \mathbf{R}_+$ and $M = M_G$ is a constant.*

Then for every integer $m \geq 1$ the following inequalities hold:

$$E|y_t|^{2m} \leq k_m(t) \quad \forall\, t \in \mathbf{R}_+, \tag{1.1.2}$$
$$E|y_\tau|^{2m} \leq c_m(T) \quad \forall\, \tau \in T_T, \tag{1.1.3}$$

where

$$k_m(t) := (2m-1)!! M^{2m} \left(\frac{1 - e^{-2\gamma t}}{2\gamma} \right)^m, \tag{1.1.4}$$

$$c_m(T) := m(2m-1)M^2 \int_0^T k_{m-1}(u)du. \tag{1.1.5}$$

Proof. By the Ito formula

$$|y_t|^{2m} = 2m \int_0^t |y_s|^{2(m-1)} y_s^\star f(s, y_s)ds + 2m \int_0^t |y_s|^{2(m-1)} y_s^\star G_s dW_s$$
$$+ 2m(m-1) \int_0^t |y_s|^{2(m-2)} y_s^\star G_s G_s^\star y_s ds$$
$$+ m \int_0^t |y_s|^{2(m-1)} |G_s|^2 ds. \tag{1.1.6}$$

Put $S_t(m) := E|y_t|^{2m}$. As a first step, we prove the result assuming that $S_t(m)$ is bounded and the stochastic integral above is a martingale. Taking the expectation of the both sides of (1.1.6) we conclude that $S_t(m)$ has a continuous derivative. Using the hypothesis **H.1.1.1** we infer the differential inequality

$$\frac{dS_t(m)}{dt} \leq -2m\gamma S_t(m) + m(2m-1)M^2 S_t(m-1), \quad S_0(m) = 0.$$

We have from here by the Gronwall–Bellman lemma the following recurrent sequence of inequalities:

$$S_t(m) \leq m(2m-1)M^2 \int_0^t S_u(m-1)e^{-2m\gamma(t-u)}du. \tag{1.1.7}$$

In particular,

$$S_t(1) \leq 2M^2 \left(\frac{1 - e^{-2\gamma t}}{2\gamma} \right) = k_1(t),$$

i.e., (1.1.2) holds for $m = 1$. The general case follows from (1.1.7) by induction.

In the representation (1.1.6) the first integral is negative while the second one is a martingale. Thus, the relations (1.1.6), (1.1.2) imply that for any stopping time τ with values in the interval $[0, T]$

$$E|y_\tau|^{2m} \le m(2m-1)M^2 \int_0^T k_{m-1}(u)du. \qquad (1.1.8)$$

To remove the auxiliary assumptions we notice that they are fulfilled for the process $y^{(N)} := y I_{[0,\tau_N]} + Y^{(N)} I_{]\tau_N, T]}$ where $\tau_N := \inf\{t : |y_t| \ge N\}$ and

$$dY_t^{(N)} = -\gamma Y_t^{(N)} dt + G_t dW_t, \qquad y_{\tau_N}^{(N)} = y_{\tau_N}.$$

Letting $N \to \infty$ in the moment bounds for $y^{(N)}$ we get the result by the Fatou lemma. $\square$

1.1.2 Bounds Efficient on Small Intervals

Lemma 1.1.1 leads easily to moment bounds of simpler structure which are more convenient to use on small time intervals. We summarize them in

Proposition 1.1.2 *Let y be the solution of (1.0.1). Under the assumptions of Lemma 1.1.1*

$$E|y_t|^{2m} \le m!(2M^2 t)^m, \qquad (1.1.9)$$
$$E|y_\tau|^{2m} \le m!(2M^2 T)^m \qquad \forall \tau \in \mathcal{T}_T, \qquad (1.1.10)$$
$$E\|y\|_T^{2m} \le (3 + m \ln m)m!(2M^2 T)^m. \qquad (1.1.11)$$

Proof. The first two claims follow directly from (1.1.2) and (1.1.3): since $1 - e^{-x} \le x$ we have the bound $k_m(t) \le (2m-1)!!M^{2m}t^m$ which implies via (1.1.5) the similar one for $c_m(T)$; clearly, $(2m-1)!! \le m!2^m$.

To get (1.1.11) we introduce the stopping time

$$\tau_a := \inf\{t : |y_t|^{2m} \ge a\} \wedge T.$$

Obviously,

$$E\|y\|_T^{2m} = \int_0^\infty P(\|y\|_T^{2m} \ge a)da = \int_0^\infty P(|y_{\tau_a}|^{2m} \ge a)da. \qquad (1.1.12)$$

Splitting the interval of integration in three parts by the points

$$u := m!(2M^2 T)^m, \qquad v := m^m u$$

and applying twice the Chebyshev inequality and the bound (1.1.10), we have:

$$\int_0^\infty P(|y_{\tau_a}|^{2m} \geq a)da \leq u + \int_u^v \frac{E|y_{\tau_a}|^{2m}}{a}da + \int_v^\infty \frac{E|y_{\tau_a}|^{4m}}{a^2}da$$

$$\leq u + u\ln\frac{v}{u} + \frac{(2m)!}{(m!)^2}u^2\frac{1}{v}$$

$$\leq (1 + m\ln m)u + \frac{(2m!)}{(m!)^2 m^m}u$$

$$\leq (3 + m\ln m)u. \qquad (1.1.13)$$

The inequality (1.1.11) follows from (1.1.12) and (1.1.13). $\square$

Corollary 1.1.3 *Let y be the solution of (1.0.1). Under the assumptions of Lemma 1.1.1 for all $\lambda \in [0, 1/(6M^2T)]$*

$$Ee^{\lambda\|y\|_T^2} \leq (1 - 6\lambda M^2 T)^{-1}. \qquad (1.1.14)$$

Proof. Substituting the moment bound (1.1.11) into the Taylor expansion for the exponential and noticing that $3 + m\ln m \leq 3^m$ we have

$$Ee^{\|y\|_T^2} \leq 1 + \sum_{m=1}^\infty (3 + m\ln m)(2\lambda M^2 T)^m \leq 1 + \sum_{m=1}^\infty (6\lambda M^2 T)^m$$

and the result follows. $\square$

Remark. The above inequalities hold with obvious modifications with arbitrary $t_0 \geq 0$ as the initial point. Moreover, in this case one can change the expectation by the conditional expectation with respect to the σ-algebra $\mathcal{F}_{t_0}$.

We apply immediately this observation to conclude the subsection with a bound which will be needed further in the study of large deviations.

Let $N \in \mathbf{N}$, $\Delta := T/N$, $t_k := k\Delta$. We consider on $[0, T]$ the process ξ_t, which coincides on any interval $]t_{k-1}, t_k]$, $k \leq N$, with the solution of the stochastic differential equation

$$d\xi_t = A_t\xi_t dt + G_t dW_t, \quad \xi_{t_{k-1}+} = 0, \qquad (1.1.15)$$

i.e.,

$$\xi_t = \sum_{k=1}^N \left(\int_{t_{k-1}}^t \Phi(t, s)G_s dW_s \right) I_{]t_{k-1}, t_k]}(t)$$

where $\Phi(t, s)$ is the fundamental matrix corresponding to A.

Put $u_k := \sup_{t \in]t_{k-1}, t_k]} |\xi_t|$.

Lemma 1.1.4 *Assume that $A = (A_t)$ satisfies **H.1.1.2** and $\|G\|_T \leq M$. Then for all $\lambda \in [0, 1/(6M^2\Delta)]$*

$$E\exp\left\{ \lambda \sum_{k=1}^N u_k^2 \right\} \leq (1 - 6\lambda M^2\Delta)^{-N}. \qquad (1.1.16)$$

Proof. Notice that if ζ and η are positive random variables such that ζ is $\mathcal{G}$-measurable and $E(\eta \mid \mathcal{G}) \le c = \text{const}$ then $E\zeta\eta \le cE\zeta$. With this remark, using the "conditional" versions of (1.1.14) and (1.1.16) with $\mathcal{F}_{t_{N-1}}$, $\mathcal{F}_{t_{N-2}}$, ..., we get the claim. $\square$

1.1.3 Bounds Efficient on Large Intervals

Of course, the bounds of the previous subsection are not helpful for large T. To deduce from Lemma 1.1.1 moment bounds with explicit linear dependence on T it is sufficient to simplify the expression for $k_m(t)$ using the inequality $1 - e^{-\gamma t} \le 1$.

Proposition 1.1.5 *Let y be the solution of (1.0.1). Under the assumptions of Lemma 1.1.1*

$$E|y_t|^{2m} \le m!(M^2/\gamma)^m, \tag{1.1.17}$$

$$E|y_\tau|^{2m} \le 2mm!(M^2/\gamma)^m\gamma T \qquad \forall \tau \in \mathcal{T}_T, \tag{1.1.18}$$

$$E\|y\|_T^{2m} \le (3 + m\ln m)2mm!(M^2/\gamma)^m(1 \vee (\gamma T)). \tag{1.1.19}$$

Proof. Clearly,

$$k_m(t) \le (2m - 1)!!M^{2m}(2\gamma)^{-m}$$

implying by virtue of (1.1.5) that

$$c_m(T) \le m(2m - 1)!!M^{2m}(2\gamma)^{-m+1}T.$$

So, the first two bounds are obvious corollaries of (1.1.2) and (1.1.3).

To establish (1.1.19) we use again, as in the proof of Proposition 1.1.2, the stopping time $\tau_a := \inf\{t : |y_t|^{2m} \ge a\} \wedge T$ together with representation (1.1.12) but splitting now the interval of integration by points

$$u := 2mm!(M^2/\gamma)^m, \qquad v := m^m u.$$

In this case the bound (1.1.18) works and we get that

$$\int_0^\infty P(|y_{\tau_a}|^{2m} \ge a)da \le u + \int_u^v \frac{E|y_{\tau_a}|^{2m}}{a}da + \int_v^\infty \frac{E|y_{\tau_a}|^{4m}}{a^2}da$$

$$\le u + \gamma Tu \ln \frac{v}{u} + \frac{(2m)!}{m(m!)^2}\gamma Tu^2\frac{1}{v}$$

$$\le u + \gamma Tum \ln m + \frac{(2m)!}{(m!)^2m^{m+1}}u$$

$$\le (3 + m\ln m)(1 \vee (\gamma T))u$$

completing the proof. $\square$

In the same way as in the previous subsection we infer

Corollary 1.1.6 *Let y be the solution of (1.0.1). Under the assumptions of Lemma 1.1.1 for all $\lambda \in [0, 6\gamma/M^2]$*

$$Ee^{\lambda \|y\|_T^2} \leq (1 - 6\lambda M^2/\gamma)^{-1}(1 \vee (\gamma T)). \qquad (1.1.20)$$

Individually, the bound (1.1.19) implies that the M^{2m}-norm of $\|y\|_T$ increases as $T \to \infty$ not faster than $T^{1/2m}$. The information accumulated in the bound (1.1.20) for Laplace transform provides the exact rate.

We shall denote by $\|.\|_p$ the $L^p(\Omega)$-norm. To avoid awkward expressions like $\|\|y\|_T\|_p$ we shall use the alternative notation for $\|y\|_T$, namely, $y_T^* := \sup_{s \leq T} |y_s|$, which is a standard one in the literature on stochastic calculus.

Theorem 1.1.7 *Let y be the solution of (1.0.1). Under the assumptions of Lemma 1.1.1 for every $p \geq 1$ and $\gamma T \geq 1$*

$$\|y_T^*\|_p \leq C_p \sqrt{M^2/\gamma}\Big(1 + \sqrt{\ln(\gamma T)}\Big) \qquad (1.1.21)$$

where the constant C_p depends only on p.

Proof. Take in (1.1.20) the parameter $\lambda = 1/\mu^2$, where $\mu^2 := 12M^2/\gamma$. Then

$$Ee^{\|y\|_T^2/\mu^2} \leq 2\gamma T$$

and the claim follows from the lemma below. $\square$

Lemma 1.1.8 *Let ξ be a scalar random variable such that*

$$Ee^{\xi^2} \leq K. \qquad (1.1.22)$$

Then for every $p \geq 1$

$$\|\xi\|_p \leq C_p\Big(1 + \sqrt{\ln K}\Big). \qquad (1.1.23)$$

Proof. Without loss of generality we may assume that p is even integer. Using on the interval $[u, \infty[$ the exponential Chebyshev inequality together with the assumed bound (1.1.22) we infer that

$$E|\xi|^p = p \int_0^\infty a^{p-1} P(|\xi| \geq a)da \leq u^p + Kp \int_u^\infty a^{p-1} e^{-a^2} da.$$

It remains to put $u := \sqrt{\ln K}$ and observe that the last integral is the product of e^{-u^2} by a polynomial of order $p - 2$. $\square$

In Section 1.3 we show that the growth rate $\sqrt{\ln T}$ cannot be improved.

1.2 Bounds for Linear Equations

1.2.1 Assumption on the Fundamental Matrix

In this section we concentrate our efforts to the study of moment bounds for the processes y satisfying the "linear equation" (1.1.1) under the hypothesis

H.1.2.1 There exist positive constants L and κ such that for all s, $t \in \mathbf{R}_+$, $s \le t$, we have

$$|\Phi(t, s)| \le L e^{-\kappa(t-s)} \tag{1.2.1}$$

where $\Phi(t, s)$ is the fundamental matrix corresponding to $A = (A_t)$, i.e., the solution of the linear matrix equation

$$\frac{\partial \Phi(t, s)}{\partial t} = A_t \Phi(t, s), \quad \Phi(s, s) = I,$$

where $I = I_q$ is the unit matrix.

Of course, **H.1.2.1** is weaker than **H.1.1.1**. Indeed, if the latter hypothesis is fulfilled then

$$\frac{\partial |\Phi(t, s)|^2}{\partial t} = \operatorname{tr} \frac{\partial}{\partial t} \left(\Phi(t, s) \Phi(t, s)^\star \right)$$
$$= \operatorname{tr} A_t \Phi(t, s) \Phi(t, s)^\star + \operatorname{tr} \Phi(t, s) \Phi(t, s)^\star A_t^\star \le -2\gamma |\Phi(t, s)|^2.$$

Since $|\Phi(s, s)|^2 = |I|^2 = q$, the Gronwall–Bellman lemma implies that

$$|\Phi(t, s)|^2 \le q e^{-2\gamma(t-s)} \tag{1.2.2}$$

and we have **H.1.2.1** with $L = q^{1/2}$ and $\kappa = \gamma$.

Notice that if A is a constant matrix then **H.1.2.1** holds if and only if A is stable, i.e., the real parts of all its eigenvalues are strictly negative (but the hypothesis **H.1.1.2** on the spectrum of the matrix $A + A^\star$ may fail).

Another sufficient condition for **H.1.2.1** is

H.1.2.2 The function $A = (A_t)_{t \in [0, \infty]}$ is deterministic, continuous, and the matrix A_∞ are stable.

The claim follows from the representation

$$\Phi(t, s) = e^{tA_\infty} + \int_s^t e^{(t-u)A_\infty} (A_u - A_\infty) \Phi(t, u) \, du.$$

Details are given in Appendix A.1 where we show that the claim holds also for a bounded measurable A converging at infinity to a stable matrix A_∞.

In general, it is not possible to give a condition expressed exclusively in spectral terms of A_t to guarantee **H.1.2.1**, assuming only the continuity of A on $[0, \infty[$ and not on $[0, \infty]$. The classic example of the theory of Lyapunov's exponents is:

$$A_t := \begin{pmatrix} -(1 + 2\cos 4t) & 2(1 + \sin 4t) \\ 2(\sin 4t - 1) & -1 + 2\cos 4t \end{pmatrix}.$$

All matrices A_t have -1 as the eigenvalue of multiplicity 2 but the function $y = (y_t)$ with

$$y_t = e^t (\sin 2t, \, \cos 2t)^\star$$

satisfies the equation $dy_t = A_t y_t dt$, in obvious contradiction to **H.1.2.1**.

Remarkably, in singular perturbation theory, where equations are considered on a finite interval $[0, T]$, the fundamental matrix $\Phi^\varepsilon(t, s)$ corresponding to $A^\varepsilon = \varepsilon^{-1}A$, where $A = (A_t)$ is continuous, admits a suitable exponential bound (at least, for sufficiently small ε) if $\operatorname{Re} \lambda(A_t) < -2\gamma < 0$ (see Appendix A.1).

1.2.2 Differential Equations with Random Coefficients

The main idea of how to extend the above results to the case where we have only the hypothesis **H.1.2.1** consists in comparing the solution of (1.1.1) with the solution of the equation with the drift coefficient $-\kappa I$: the difference of these two processes satisfies an ordinary differential equation.

We prove two lemmas on moment bounds of solutions of linear differential equations with random coefficients.

Lemma 1.2.1 *Let y be the solution of the linear differential equation*

$$dy_t = (A_t y_t + \alpha_t)dt, \quad y_0 = 0, \tag{1.2.3}$$

*where A is a bounded measurable process satisfying **H.1.1.2** and α is a measurable process. Then for any $p > 1$*

$$E\|y\|_T^p \le p\gamma^{1-p} \sup_{t \le T} E|\alpha_t|^p \, T. \tag{1.2.4}$$

Proof. Put $S_t = S_t(p) := E|y_t|^p$, $\bar{S}_T := \sup_{t \le T} S_t$. Without loss of generality we assume that $V := \sup_{t \le T} E|\alpha_t|^p < \infty$ and hence by virtue of the Cauchy formula $\bar{S}_T < \infty$. Taking the expectation of both sides of the representation

$$|y_t|^p = p \int_0^t |y_s|^{p-2} y_s^\star A_s y_s ds + p \int_0^t |y_s|^{p-2} y_s^\star \alpha_s ds \tag{1.2.5}$$

we get the identity

$$\frac{dS_t}{dt} = pE|y_t|^{p-2} y_t^\star A_t y_t + pE|y_t|^{p-2} y_t^\star \alpha_t.$$

Using the bounds $y^\star A y \le -\gamma|y|^2$, $y^\star \alpha \le |y||\alpha|$, and estimating the last term by the Hölder inequality we obtain that

$$\frac{dS_t}{dt} \le -p\gamma S_t + pV^{1/p} S_t^{1/p'}$$

where $p' := p/(1-p)$. The Gronwall–Bellman lemma yields the bound

$$S_t \le pV^{1/p} \int_0^t S_u^{1/p'} e^{-p\gamma(t-u)} du.$$

Hence, $\bar{S}_T \le \gamma^{-1} V^{1/p} \bar{S}_T^{1/p'}$ implying $\bar{S}_T \le V\gamma^{-p}$. We infer from (1.2.5), by neglecting the negative first term on its right-hand side and using again the Hölder inequality to estimate the second, that

$$E\|y\|_T^p \le pV^{1/p} \bar{S}_T^{1/p'} T \le p\gamma^{-p/p'} VT = p\gamma^{1-p} \sup_{t \le T} E|\alpha_t|^p T.$$

The desired inequality is proved. $\square$

Lemma 1.2.2 *Let z be the solution of the linear differential equation*

$$dz_t = (A_t z_t + \alpha_t)dt, \quad z_0 = 0, \tag{1.2.6}$$

where A is a measurable process satisfying **H.1.2.1** *with $\|A\|_T \le M_A$ and α is a measurable process. Then for every $p > 1$*

$$E\|z\|_T^p \le 2^p pK\kappa^{1-p} \sup_{t \le T} E|\alpha_t|^p T \tag{1.2.7}$$

where

$$K = K(L, p, M_A/\kappa) := 1 + L^p(1 + M_A/\kappa)^p. \tag{1.2.8}$$

Proof. As in the previous proof, assume V finite. Using the Cauchy formula and the exponential bound (1.2.1) we get easily that

$$\|z_t\|_p \le \int_0^t \|\Phi(t,s)\alpha_s\|_p ds \le L \int_0^t e^{-\kappa(t-s)} \|\alpha_s\|_p ds \le (L/\kappa) V^{1/p}. \tag{1.2.9}$$

Let us consider the equation

$$dy_t = (-\kappa y_t + \alpha_t)dt, \quad y_0 = 0.$$

According to Lemma 1.2.1 its solution admits the bound

$$E\|y\|_T^p \le p\kappa^{1-p} VT. \tag{1.2.10}$$

Put $\Delta := y - z$. Then

$$d\Delta_t = (-\kappa\Delta_t - \nu_t)dt, \quad \Delta_0 = 0,$$

where $\nu_t := \kappa z_t + A_t z_t$ and by virtue of (1.2.9)

$$\sup_{t \leq T} E|\nu_t|^p \leq (\kappa + M_A)^p (L/\kappa)^p V.$$

Thus, by Lemma 1.2.1

$$E||\Delta||_T^p \leq pL^p(1 + M_A/\kappa)^p \kappa^{1-p} VT. \tag{1.2.11}$$

The needed assertion is a consequence of (1.2.10) and (1.2.11). □

Corollary 1.2.3 *Let z be the solution of (1.2.6) where A is a bounded measurable process satisfying* **H.1.2.1** *and α is a measurable process such that for some constant c_α we have*

$$\sup_{t \leq T} E|\alpha_t|^{2m} \leq m! c_\alpha^{2m} \quad \forall \ m \in \mathbf{N}.$$

Assume that $\kappa T \geq 1$. Then

$$||z_T^*||_p \leq C_p c_\alpha D\left(1 + \sqrt{\ln(\kappa T)}\right) \tag{1.2.12}$$

where

$$D := 4(1 + L)(1 + M_A/\kappa)/\kappa$$

and C_p is a constant depending only on p.

Proof. According to (1.2.7) for every $m \geq 1$

$$E||z||_T^{2m} \leq m! c_\alpha^{2m} D^{2m} \kappa T.$$

Considering the nontrivial case where $c_\alpha > 0$ we get from here

$$Ee^{||z||_T^2/(2c_\alpha^2 D^2)} \leq 1 + \kappa T \sum_{m=1}^{\infty} 2^{-m} \leq 2\kappa T$$

and the result holds by virtue of Lemma 1.1.8. □

1.2.3 The Continuity Theorem

Corollary 1.2.3 allows us to extend Theorem 1.1.7 to linear SDEs with the drift coefficient satisfying only **H.1.2.1**. Though the resulting bound in this case has a slightly more complicated structure, it catches the dependence on parameters in a way we need further for the asymptotic analysis.

Proposition 1.2.4 *Let y be a solution of linear SDE (1.1.1) where A is a bounded predictable process satisfying* **H.1.2.1** *with $||A||_T \leq M_A$ and G is a bounded predictable process with $||G||_T \leq M_G$. Assume that $\kappa T \geq 1$. Then for each $m \geq 1$ and $p \geq 1$*

$$E|z_t|^{2m} \leq m! L_1^{2m}, \tag{1.2.13}$$

$$\|y_T^*\|_p \le C_p K \sqrt{M_G^2/\kappa} \left(1 + \sqrt{\ln(\kappa T)}\right) \tag{1.2.14}$$

where

$$L_1 := 2(1+L)(1+M_A/\kappa)\sqrt{M_G^2/\kappa}, \qquad K := (1+M_A/\kappa)^2(1+L),$$

and C_p is a constant depending only on p.

Proof. Let us consider the solution z of the stochastic equation

$$dz_t = -\kappa z_t dt + G_t dW_t, \quad z_0 = 0,$$

admitting by Proposition 1.1.5 and Theorem 1.1.7 the bounds

$$E|z_t|^{2m} \le m! \left(\sqrt{M_G^2/\kappa}\right)^{2m}, \quad \forall \, m \in \mathbf{N}, \, t \in [0, T], \tag{1.2.15}$$

$$\|z_T^*\|_p \le C\sqrt{M_G^2/\kappa}\left(1 + \sqrt{\ln(\kappa T)}\right) \tag{1.2.16}$$

with C depending only on p.

Put $Z := y - z$. Then

$$dZ_t = (AZ_t + \alpha_t)dt, \quad Z_0 = 0,$$

where $\alpha_t := \kappa z_t + A_t z_t$. By virtue of (1.2.15)

$$\sup_{t \le T} E|\alpha_t|^{2m} \le m! c_\alpha^{2m}$$

with $c_\alpha := (\kappa + M_A)\sqrt{M_G^2/\kappa}$. The bound (1.2.13) holds by virtue of (1.2.9) and (1.2.15). Corollary 1.2.3 implies that

$$\|Z_T^*\|_p \le C(1+L)(1+M_A/\kappa)^2\sqrt{M_G^2/\kappa}\left(1 + \sqrt{\ln(\kappa T)}\right). \tag{1.2.17}$$

The claim follows from (1.2.16) and (1.2.17). $\square$

From the above bounds it is easy to obtain the following continuity result:

Theorem 1.2.5 Let Y^ε be solutions of the linear SDE

$$dY_t^\varepsilon = (A_t Y_t^\varepsilon + \phi_t^\varepsilon)dt + \psi_t^\varepsilon dW_t, \quad Y_0^\varepsilon = 0, \tag{1.2.18}$$

and let $\rho = \rho^\varepsilon$ be stopping times with values in the interval $[0, T/\varepsilon]$. Assume that the bounded predictable process A satisfies **H.1.2.1** and the predictable processes ϕ^ε and ψ^ε are such that

$$\phi_\rho^{\varepsilon*} \le o\left(1/\sqrt{|\ln \varepsilon|}\right), \tag{1.2.19}$$

$$\psi_\rho^{\varepsilon*} \le o\left(1/\sqrt{|\ln \varepsilon|}\right), \tag{1.2.20}$$

where $o\left(1/\sqrt{|\ln \varepsilon|}\right)$ is a (deterministic) function converging to zero faster than $1/\sqrt{|\ln \varepsilon|}$ as $\varepsilon \to 0$. Then for any $p \geq 1$

$$\lim_{\varepsilon \to 0} ||Y_\rho^{\varepsilon*}||_p = 0. \tag{1.2.21}$$

Proof. Obviously, on $[0, \rho]$ we have $Y^\varepsilon = y^\varepsilon + z^\varepsilon$ where

$$dy_t^\varepsilon = A_t y_t^\varepsilon dt + \psi_t^\varepsilon I_{[0,\rho]}(t) dW_t, \quad y_0^\varepsilon = 0,$$
$$dz_t^\varepsilon = (A_t z_t^\varepsilon + \phi_t^\varepsilon I_{[0,\rho]}(t)) dt, \quad z_0^\varepsilon = 0.$$

The result holds because by virtue of Corollary 1.2.3 and Proposition 1.2.4 $||z_{T/\varepsilon}^{\varepsilon*}||_p$ and $||y_{T/\varepsilon}^{\varepsilon*}||_p$ converges to zero as $\varepsilon \to 0$. $\square$

1.2.4 Linear SDEs with Unbounded Coefficients

In the study of asymptotic equations we need moment bounds for the solutions of linear stochastic equations with nonhomogeneous drift when bounded deterministic A but other coefficients are random and, eventually, unbounded. Our methods allow us easily to obtain necessary extensions.

Lemma 1.2.6 *Let y be the solution of the linear SDE*

$$dy_t = (A_t y_t + h_t) dt + G_t dW_t, \quad y_0 = \xi, \tag{1.2.22}$$

where A is a bounded function with values in the set of $n \times n$ matrices satisfying the condition **H.1.2.1**, *h and G are predictable processes with values in $\mathbf{R}^q$ and in the space of $q \times l$ matrices, respectively, and W is a Wiener process in $\mathbf{R}^l$.*

Then for every $p \geq 2$ there exists a constant $K = K_p$ such that for all T

$$\sup_{t \leq T} E|y_t|^p \leq K \left(E|\xi|^p + \sup_{t \leq T} E|h_t|^p + \sup_{t \leq T} E|G_t|^p \right). \tag{1.2.23}$$

In particular, if

$$E|\xi|^p + \sup_{t \leq T} E|h_t|^p + \sup_{t \leq T} E|G_t|^p < \infty, \tag{1.2.24}$$

then $\sup_{t \leq T} E|y_t|^p < \infty$.

Proof. Without loss of generality we assume (1.2.24) is finite.

Let us introduce the processes $y_t^0 := \Phi(t, 0)\xi$,

$$y_t^1 := \int_0^t \Phi(t, s) h_s ds, \quad y_t^2 := \int_0^t \Phi(t, s) G_s dW_s. \tag{1.2.25}$$

The fundamental matrix $\Phi(t, s)$ satisfies (1.2.1) and hence

$$\sup_t E|y_t^0|^p \le E||y^0||_T^p \le L^p E|\xi|^p. \tag{1.2.26}$$

As usual, put $1/p' := 1 - 1/p$.

Due to the Hölder inequality

$$|y_t^1|^p \le \left(\int_0^t |\Phi(t,s)|^{p'} e^{-\kappa p'(t-s)/2} ds \right)^{p/p'} \int_0^t e^{-\kappa p(t-s)/2} |h_s|^p ds.$$

The bound (1.2.1) implies that

$$\sup_{t \le T} E|y_t^1|^p \le C \sup_{t \le T} E|h_t|^p. \tag{1.2.27}$$

By the Burkholder–Gundy inequality

$$E|y_t^2|^p \le C_p E \left(\int_0^t |\Phi(t,s)|^2 |G_s|^2 ds \right)^{p/2}$$

with an absolute constant C_p. Applying the Hölder inequality and (1.2.1) we get that

$$\sup_{s \le T} E|y_s^2|^p \le C \sup_{s \le T} E|G_s|^p < \infty. \tag{1.2.28}$$

Since $y = y^0 + y^1 + y^2$, (1.2.26)–(1.2.28) imply the assertion. $\square$

Proposition 1.2.7 *Let y be the solution of the linear SDE*

$$dy_t = (A_t y_t + h_t)dt + G_t dW_t, \quad y_0 = \xi, \tag{1.2.29}$$

where A is a bounded (deterministic) function with values in the set of $n \times n$ matrices satisfying the condition **H.1.2.1**, *h and G are predictable processes with values in $\mathbf{R}^q$ and in the space of $q \times l$ matrices, respectively.*

Then there are constants C_0 (depending only on m and L) and C_1 (depending only on m, L, and κ) such that for all $T \ge 1$ we have

$$E||y||_T^{2m} \le L^{2m} E|\xi|^{2m} + C_1 (V_T(m,h) + V_T(2m,G)) T \tag{1.2.30}$$

where

$$V_T(m,h) := \sup_{s \le T} E|h_s|^{2m},$$

$$V_T(2m,G) := \sup_{s \le T} E|G_s|^{4m}.$$

Proof. Assume that all characteristics in the right-hand side of (1.2.30) are finite. As in the proof of Lemma 1.2.6, we use the decomposition

$$y = y^0 + y^1 + y^2$$

where $y_t^0 := \Phi(t,0)\xi$ and the processes y^1 and y^2 are defined by (1.2.25). It follows from the hypothesis **H.1.2.1** that

$$E||y^0||_T^{2m} \leq L^{2m} E|\xi|^{2m}. \tag{1.2.31}$$

Since y^1 is the solution of the equation

$$dy_t^1 = (A_t y_t^1 + h_t)dt, \quad y_0^1 = 0,$$

we have, by virtue of Lemma 1.2.2, the bound

$$E||y^1||_T^{2m} \leq C \sup_{s \leq T} E|h_s|^{2m} T. \tag{1.2.32}$$

Thus, it remains to consider only the process y^2, or, the same, the process y assuming that $\xi = 0$ and $h = 0$. In this case, according to (1.2.28)

$$S_t(m) := E|y_t|^{2m} \leq C_m \sup_{s \leq t} E|G_s|^{2m} \tag{1.2.33}$$

where the constant depends only on m, L, and κ.

Assume for a moment that for every t the matrix A_t is symmetric and all its eigenvalues are strictly negative. We may use now the arguments at the beginning of the proof of Lemma 1.1.1.

In the Ito formula (1.1.6), written with $f(t,y) = A_t y$, first integral in the right-hand side is negative, while the stochastic integral is a local square-integrable martingale. Thus, for any stopping time τ with values in $[0,T]$ we have the bound

$$E|y_\tau|^{2m} \leq (2m-1)m \int_0^T E|y_s|^{2(m-1)}|G_s|^2 ds.$$

By the Hölder inequality

$$E|y_s|^{2(m-1)}|G_s|^2 \leq (E|y_s|^{2m})^{1-1/m}(E|G_s|^{2m})^{1/m}$$

and we get, using (1.2.33), that

$$E|y_\tau|^{2m} \leq C_m \sup_{s \leq T} E|G_s|^{2m} T = C_m V_T(m,G)T. \tag{1.2.34}$$

Applying the formula (1.1.12) and splitting the integration interval at the point

$$u = \sqrt{C_{2m} V_T(2m,G)} T,$$

we have, using the Chebyshev inequality and the above bound, that

$$E||y||_T^{2m} \leq u + \int_u^\infty \frac{E|y_\tau|^{4m}}{a^2} da \leq 2\sqrt{C_{2m} V_T(2m,G))} T. \tag{1.2.35}$$

In the general case we introduce the auxiliary process z with

$$dz_t = -\kappa z_t dt + G_t dW_t, \quad z_0 = 0.$$

By above

$$E||z||_T^{2m} \leq C \sqrt{V_T(2m, G)} T. \tag{1.2.36}$$

Applying Lemma 1.2.2 and (1.2.33) it is easy to check that

$$E||z - y||_T^{2m} \leq C \sup_{s \leq t} E|G_s|^{2m} T \leq C \sqrt{V_T(2m, G)} T. \tag{1.2.37}$$

It follows from (1.2.36), (1.2.37) that

$$E||y||_T^{2m} \leq C \sqrt{V_T(2m, G)} T \tag{1.2.38}$$

and we get the result. $\square$

1.3 On the Growth Rate of the Maximal Function

Consider the Ito process y with

$$dy_t = f_t dt + G_t dW_t, \quad y_0 = y^\circ, \tag{1.3.1}$$

where f and G are predictable processes such that the corresponding integrals are well-defined and the following hypothesis holds:

H.1.3.1 There exists $\gamma > 0$ such that

$$y_t^\star f_t \leq -\gamma |y_t|^2 \qquad P(d\omega) \otimes dt\text{-a.s} \tag{1.3.2}$$

It is easily seen that the arguments in the proof of the key Lemma 1.1.1 use exactly the above property of $f_t = f(t, y_t)$. Therefore, the important Theorem 1.1.7 is proved, in fact, for the Ito processes satisfying **H.1.3.1**. We present here an alternative approach to the problem and obtain a bound for the L^p-norm of the maximal function $y_T^* = ||y||_T$, which is, basically, the same as that of Theorem 1.1.7 and asserts that the growth rate is not faster than $C\sqrt{\ln T}$. We show here that for the Ornstein–Uhlenbeck process this is the correct growth rate of any L^p-norm of the maximal function. We allow now the nonzero initial value, which is, of course, an easily treated generalization.

1.3.1 Lapeyre's Inequality

Theorem 1.3.1 *Assume* **H.1.3.1** *and* $||G||_T \leq L$. *Then for any* $p \geq 1$ *there is a constant* C *depending on* L *and* γ *such that*

$$||y_T^*||_p \leq K^{1/2} + 1 + Cp^{1/2}(\ln_2(KR^2 e^{4\gamma T} + e^e \vee e^{p-2}))^{1/2} \tag{1.3.3}$$

where $R^2 := L^2/\gamma$, $K := |y^\circ|^2 + R^2/2$, *and* C *is an absolute constant.*

Proof. First, notice that for any t

$$E|y_t|^2 \le |y^o|^2 + L^2 t. \tag{1.3.4}$$

Indeed, let $\tau_N := \inf\{t : |y_t| \ge N\}$. By the Ito formula

$$|y_{t \wedge \tau_N}|^2 = |y^o|^2 + 2 \int_0^{t \wedge \tau_N} y_s^\star f_s ds + 2 \int_0^{t \wedge \tau_N} y_s^\star G_s dW_s + \int_0^{t \wedge \tau_N} |G_s|^2 ds.$$

Since the first integral on the right-hand side is negative,

$$E|y_{t \wedge \tau_N}|^2 \le |y^o|^2 + L^2 t$$

and (1.3.4) follows by the Fatou lemma.

Again by the Ito formula

$$e^{2\gamma t}|y_t|^2 = |y^o|^2 + 2 \int_0^t e^{2\gamma s}(\gamma |y_s|^2 + y_s^\star f_s) ds + \int_0^t e^{2\gamma s}|G_s|^2 ds + M_t$$

with

$$M_t := 2 \int_0^t e^{2\gamma s} y_s^\star G_s dW_s.$$

Taking into account (1.3.2) and the bound $\|G\|_T \le L$ we come to the inequality

$$|y_t|^2 \le K + e^{-2\gamma t} M_t. \tag{1.3.5}$$

Due to (1.3.4) the process M is a square integrable martingale with

$$\langle M \rangle_t = 4 \int_0^t e^{4\gamma s}|y_s^\star G_s|^2 ds \le e^{4\gamma t} R^2 y_t^{\star 2}. \tag{1.3.6}$$

By virtue of (1.3.5) $E|y_t|^2 \le K$ and

$$E\langle M \rangle_t \le R^2 K e^{4\gamma t}. \tag{1.3.7}$$

Using the inequality $(1+y)^{1/2} \le 1 + y^{1/2}$ we get from (1.3.6) that

$$e^{-2\gamma t}(1 + \langle M \rangle_t)^{1/2} \le 1 + R y_t^\star. \tag{1.3.8}$$

Define the process

$$Y_t := M_t / (1 + \langle M \rangle_t)^{1/2}.$$

Taking the supremum up to T in (1.3.5) and making use of (1.3.8) we come to the quadratic inequality with respect to $y_T^\star$:

$$y_T^{\star 2} \le K + Y_T^\star + R Y_T^\star y_T^\star.$$

So, $y_T^\star$ is always between the two roots of the equation

$$x^2 - Y_T^* Rx - K - Y_T^* = 0$$

and hence does not exceed the larger one. It follows that

$$y_T^* \leq Y_T^* R + \sqrt{K} + \sqrt{Y_T^*} \leq 1 + \sqrt{K} + (1 + R)Y_T^*. \qquad (1.3.9)$$

An appropriate bound for $||Y_T^*||_p$ follows from

Lemma 1.3.2 *Let M be a scalar continuous square integrable martingale such that $M_0 = 0$ and $\langle M \rangle_\infty = \infty$. Let $Y_t := M_t/(1 + \langle M \rangle_t)^{1/2}$. Then there is an absolute constant C such that for any $p \geq 1$*

$$||Y_T^*||_p \leq Cp^{1/2}(\ln_2(E\langle M \rangle_T + e^e \vee e^{p-2}))^{1/2}. \qquad (1.3.10)$$

Proof. We use the representation of a continuous martingale as a time change of a Wiener process. The precise statement we need is the following: there is a Wiener process $w = (w_t)$ (defined, maybe, on a some extension of the probability space) such that $M = w_{\langle M \rangle}$.

Put $\phi(t) := (1 + t)^{1/2}$ for $t \leq e^e$ and $\phi(t) := ((1 + t) \ln_2 t)^{1/2}$ for $t > e^e$. Define also $\psi(t) := \phi(t)/(1 + t)^{1/2}$. Notice that for $t \geq e^e \vee e^{p-2}$

$$\frac{d}{dt^2}\psi^{2p}(t) = p\frac{(\ln_2 t)^{p-2}}{(t \ln t)^2}[p - 1 - (\ln t + 1)\ln_2 t] \leq 0$$

and the function $\psi^{2p}(t)$ is concave on $[e^e \vee e^{p-2}, \infty[$. Thus, by the Jensen inequality

$$E\psi^{2p}(\langle M \rangle) \leq E\psi^{2p}(\langle M \rangle \vee e^e \vee e^{p-2}) \leq \psi^{2p}(E(\langle M \rangle \vee e^e \vee e^{p-2}))$$
$$\leq \psi^{2p}(E\langle M \rangle + e^e \vee e^{p-2}). \ (1.3.11)$$

From the above definitions we have

$$Y := M/(1 + \langle M \rangle)^{1/2} = \frac{M}{\phi(\langle M \rangle)}\psi(\langle M \rangle) = \frac{w_{\langle M \rangle}}{\phi(\langle M \rangle)}\psi(\langle M \rangle)$$

and hence

$$Y_T^* \leq ||\xi||\psi(\langle M \rangle_T) \qquad (1.3.12)$$

where $||\xi|| := \sup_{t \geq 0}|\xi_t|$ for $\xi_t := w_t/\phi(t)$.

By the law of iterated logarithms for the Wiener process the random variable $||\xi|||$ is finite a.s. Notice that $||\xi|| = \sup_i|\xi_{t_i}|$ where t_i runs through a countable dense subset of $\mathbf{R}_+$. By Fernique's lemma given at the end of this section there is a constant $\alpha > 0$ such that $E\exp\{\alpha||\xi||^2\} < \infty$.

Since $x^n \leq n!e^x$ for $x > 0$ we deduce from here using the Stirling formula that

$$(E||\xi||^{2p})^{1/p} \leq Cp. \qquad (1.3.13)$$

We have from (1.3.11) by the Cauchy–Schwarz inequality that

$$EY_T^{*p} \leq \sqrt{E\|\xi\|^{2p}} \sqrt{E\psi^{2p}(\langle M \rangle_T)} \tag{1.3.14}$$

and the result of Lemma 1.3.2 is an immediate corollary of (1.3.14), (1.3.11), and (1.3.13). □

The assertion of the theorem follows from (1.3.9), (1.3.10), and (1.3.7). □

1.3.2 Ornstein–Uhlenbeck Process

By virtue of Theorem 1.1.7 (or Theorem 1.3.1) for any Ito process y satisfying **H.1.3.1** and with bounded diffusion coefficient we have that

$$\|y_T^*\|_p = O(\sqrt{\ln T}), \quad T \to \infty.$$

To show that the rate $\sqrt{\ln T}$ cannot be improved for the Ornstein–Uhlenbeck process we need the following result:

Lemma 1.3.3 *Let w be a Wiener process, $r > 1$. Then there is a constant $C(r)$ such that for all integers $k \geq 1$*

$$E \sup_{j \leq k} \frac{w_{r^j}}{\sqrt{r^j}} \geq C(r)\sqrt{\ln k}.$$

Proof. Put $\zeta_j := w_{r^j}$, $j \geq 1$, $\zeta_0 := 0$. Since $\sup a_n \geq \sup(a_n - b_n) - \sup(-b_n)$ we have

$$E \sup_{j \leq k} \frac{\zeta_j}{\sqrt{r^j}} \geq E \sup_{j \leq k} \frac{\zeta_j - \zeta_{j-1}}{\sqrt{r^j}} - r^{-1/2} E \sup_{j \leq k} \frac{-\zeta_{j-1}}{\sqrt{r^{j-1}}}.$$

But $-w$ is again a Wiener process and it follows that

$$E \sup_{j \leq k} \frac{\zeta_j}{\sqrt{r^j}} \geq \frac{1}{1 + r^{-1/2}} E \sup_{j \leq k} \frac{\zeta_j - \zeta_{j-1}}{\sqrt{r^j}} = CE \sup_{j \leq k} \eta_j$$

where $C := (1 + r^{-1/2})^{-1}(1 - 1/r)^{1/2}$ and η_j are independent standard normal random variables. The obtained bound and the first inequality of Lemma 1.3.4 given below yield the result. □

Lemma 1.3.4 *Let η_j be independent standard normal random variables. Then there are positive constants C_1 and C_2 such that for all integers $k \geq 2$*

$$C_1 \sqrt{\ln k} \leq E \sup_{j \leq k} \eta_j \leq C_2 \sqrt{\ln k}. \tag{1.3.15}$$

Proof. We have

$$E \sup_{j \leq k} \eta_j = \int_{-\infty}^{\infty} x \, d(\Phi^k(x)) = \int_{-\infty}^{0} x \Phi^{k-1}(x)\varphi(x)dx + \int_{0}^{\infty} (1 - \Phi^k(x))dx$$

where $\varphi(x) = \Phi'(x)$ is the standard normal density.

Since $\Phi(x) \leq 1/2$ for $x \leq 0$ the first term on the right-hand side of the above representation converges to zero and has no significance in our question. Put $a_k := \sqrt{2\ln k}$. Then

$$\int_0^\infty (1 - \Phi^k(x))dx \geq \int_2^{a_k} (1 - \Phi^k(x))dx = a_k - 2 - \int_2^{a_k} \Phi^k(x)dx$$

and it remains to show that the integral on the right-hand side is bounded. Indeed, $1 - \Phi(x) \geq (x^{-1} - x^{-3})\varphi(x)$ for $x > 0$ and hence $\Phi(x) \leq 1 - x^{-1}\varphi(x)/2$ for $x \geq 2$. Using this observation we have:

$$\int_2^{a_k} \Phi^k(x)dx \leq \int_2^{a_k} (1 - x^{-1}\varphi(x)/2)^k dx$$

$$= 2\int_2^{a_k} (1 - x^{-1}\varphi(x)/2)^k \frac{1}{(x^{-2}+1)\varphi(x)} d(1 - x^{-1}\varphi(x)/2)$$

$$\leq 2\frac{1}{\varphi(a_k)} \int_2^{a_k} (1 - x^{-1}\varphi(x)/2)^k d(1 - x^{-1}\varphi(x)/2)$$

$$= C\frac{k}{k+1}[(1 - (8\pi)^{-1/2}a_k^{-1}k^{-1})^{k+1} - (1 - \varphi(2)/4)^{k+1}] \to C.$$

To prove the second inequality in (1.3.15), notice that

$$\int_0^\infty (1 - \Phi^k(x))dx \leq a_k + \int_{a_k}^\infty (1 - \Phi^k(x))dx.$$

Since $(1 - z)^k \geq 1 - kz$ for $z \in [0,1]$ and $1 - \Phi(x) \leq x^{-1}\varphi(x)$ for $x > 0$,

$$\int_{a_k}^\infty (1 - \Phi^k(x))dx \leq k\int_{a_k}^\infty x^{-1}\varphi(x)dx \leq ka_k^{-2}\int_{a_k}^\infty d(-\varphi(x))$$

$$= ka_k^{-2}\varphi(a_k) = Ca_k^{-2} \to 0$$

and we obtain the result. $\square$

Let us consider the scalar Ornstein–Uhlenbeck process y with

$$dy_t = -\frac{1}{2}y_t dt + dW_t, \quad y_0 = 0.$$

Obviously,

$$y_t = e^{-t/2}\int_0^t e^{s/2}dW_s = e^{-t/2}w_{e^t - 1}$$

where w is the Wiener process. Let $T_k := \ln(r^k + 1)$ with a fixed $r > 1$. It follows from Lemma 1.3.3 that

$$Ey_{T_k}^* \geq E\sup_{j\leq k} \frac{|w_{r^j}|}{\sqrt{r^j + 1}} \geq [r/(r+1)]^{1/2}E\sup_{j\leq k} \frac{|w_{r^j}|}{\sqrt{r^j}} \geq C\sqrt{\ln k}$$

where C is a constant depending on r. Since $T_k \sim k\ln r$ as $k \to \infty$ we have the bound $E|y_{T_k}^*| \geq C\sqrt{\ln k}$ showing that the rate of convergence in T following from Theorem 1.1.7 cannot be improved. $\square$

1.3.3 Sample Path Growth

We complete this section by showing that the growth rate of trajectories of the multidimensional Ornstein–Uhlenbeck process is also $\sqrt{\ln T}$.

Proposition 1.3.5 *Let x be the solution of the linear stochastic equation with constant coefficients*

$$dy_t = By_t dt + GdW_t, \quad y_0 = 0, \tag{1.3.16}$$

where B is a stable matrix with $\operatorname{Re}\lambda(B) < -\gamma < 0$. Then

$$\sup_{T \geq 2} \frac{y_T^*}{\sqrt{\ln T}} < \infty \quad a.s. \tag{1.3.17}$$

Proof. First, we prove (1.3.17) for the process $\tilde{y}$ which corresponds to the matrix $B = -\kappa I$ where $\kappa > 0$ and I is the identity matrix of order $q \times q$. Clearly, it is sufficient to consider only the case $q = 1$ where

$$d\tilde{y}_t = -\kappa \tilde{y}_t dt + dW_t, \quad \tilde{y}_0 = 0.$$

Using the well-known time change, we can write that

$$\tilde{y}_t = e^{-\kappa t} \int_0^t e^{-\kappa s} dW_s = e^{-\kappa t} w_{S_t}$$

where w is a Wiener process and $S_t = (2\kappa)^{-1}(e^{2\kappa t} - 1)$. The law of iterated logarithms for the Wiener process implies the existence of a random variable $\xi < \infty$ a.s. such that for all $T \geq 2$ we have $|w_{S_T}|/h(S_T) \leq \xi$ where $h(t) = (2t \ln_2 t)^{1/2}$. The result for $\tilde{y}$ follows from this observation immediately.

For the general case we put $\Delta := y - \tilde{y}$. Then

$$d\Delta_t = B\Delta_t dt + (B + \kappa I)\tilde{y}_t dt, \quad \Delta_0 = 0.$$

Hence,

$$\Delta_t = \int_0^t e^{(t-s)B}(B + \kappa I)\tilde{y}_s ds$$

and the exponential inequality (1.2.1) yields the bound $\|\Delta\|_T \leq C\|\tilde{y}\|_T$ for some constant C implying the assertion. $\square$

Remark. Of course, for the Ornstein–Uhlenbeck process more delicate results are available. For instance, if B is symmetric and $G = I$ then, independently of the starting point, for the normalized process $z_t := y_t/\sqrt{\ln t}$ the set of limit points (as $t \to \infty$) coincides a.s. with the ellipsoid

$$\{x \in \mathbf{R}^q : -x^* Bx \leq 1\};$$

see [4].

1.3.4 Fernique's Lemma

Lemma 1.3.6 *Let $\eta = (\eta_1, \eta_2, \ldots)$ be a Gaussian sequence, $||\eta|| := \sup_i |\eta_i|$. Assume that $||\eta|| < \infty$ a.s. Then there is a constant $\alpha > 0$ such that*

$$Ee^{\alpha ||\eta||^2} < \infty. \tag{1.3.18}$$

Proof. We show that for any $t > s > 0$

$$P(||\eta|| \le s)P(||\eta|| > t) \le P^2(||\eta|| > (t-s)/\sqrt{2}). \tag{1.3.19}$$

Let η' and η'' be two independent copies of η, i.e., two independent vectors with the same distributions as η. Since η is Gaussian, $(\eta' - \eta'')/\sqrt{2}$ and $(\eta' + \eta'')/\sqrt{2}$ are also independent copies of η. It follows that

$$P(||\eta|| \le s)P(||\eta|| > t) = P(||\eta' - \eta''||/\sqrt{2} \le s,\ ||\eta' + \eta''||/\sqrt{2} > t). \tag{1.3.20}$$

From the identities

$$(x+y)/\sqrt{2} = (x-y)/\sqrt{2} + \sqrt{2}y, \quad (x+y)/\sqrt{2} = (y-x)/\sqrt{2} + \sqrt{2}x$$

we get that

$$||x+y||/\sqrt{2} \le ||x-y||/\sqrt{2} + \sqrt{2}||x|| \wedge ||y||.$$

Thus, if $||x-y||/\sqrt{2} \le s$ and $||x+y||/\sqrt{2} > t$ then $||x|| \wedge ||y|| > (t-s)/\sqrt{2}$. It follows that the right-hand side of (1.3.18) is less than or equal to

$$P(||\eta'|| > (t-s)/\sqrt{2},\ ||\eta''|| > (t-s)/\sqrt{2}) = P^2(||\eta|| > (t-s)/\sqrt{2})$$

and (1.3.19) holds.

Since $||\eta|| < \infty$ a.s. we can choose s such that $q := P(||\eta|| \le s) > 1/2$. Take $t_0 := s$ and define $t_n := s + \sqrt{2}t_{n-1}$ for $n \ge 1$; clearly,

$$t_n = (\sqrt{2} + 1)(2^{(n+1)/2} - 1)s.$$

By virtue of (1.3.19)

$$P(||\eta|| > t_n) \le q^{-1}P^2(||\eta|| > t_{n-1}).$$

Obviously, $P(||\eta|| > t_0) = 1 - q$ and it follows by induction that

$$P(||\eta|| > t_n) \le q\left(\frac{1-q}{q}\right)^{2^n}.$$

Using this bound and the explicit formula for t_n we get that

$$Ee^{\alpha\|\eta\|^2} = Ee^{\alpha\|\eta\|^2} I_{\{\|\eta\|\leq t_0\}} + \sum_{n\geq 0} Ee^{\alpha\|\eta\|^2} I_{\{t_n < \|\eta\| \leq t_{n+1}\}}$$

$$\leq qe^{\alpha s^2} + \sum_{n\geq 0} e^{\alpha t_{n+1}^2} P(\|\eta\| > t_n)$$

$$\leq qe^{\alpha s^2} + q \sum_{n\geq 0} \exp\left\{2^n \left[\ln(1/q - 1) + 4(\sqrt{2}+1)^2 s^2 \alpha\right]\right\}.$$

Since $\ln(1/q-1) < 0$, the expression in the square brackets is strictly negative for sufficiently small $\alpha > 0$ and the series converges. $\square$

2 The Tikhonov Theory for SDEs

Let us consider the following initial value problem for the system of ordinary differential equations

$$dx_t^\varepsilon = f(t, x_t^\varepsilon, y_t^\varepsilon)dt, \quad x_0^\varepsilon = x^o, \tag{2.0.1}$$
$$\varepsilon dy_t^\varepsilon = F(t, x_t^\varepsilon, y_t^\varepsilon)dt, \quad y_0^\varepsilon = y^o, \tag{2.0.2}$$

where the "slow" variable x takes values in $\mathbf{R}^k$ and the "fast" variable y takes values in $\mathbf{R}^n$, $\varepsilon \in]0,1]$ is a small parameter. The *reduced problem* corresponding to the formal substitution of the zero value of ε has the form

$$dx_t^0 = f(t, x_t^0, y_t^0)dt, \quad x_0^0 = x^o, \tag{2.0.3}$$
$$0 = F(t, x_t^0, y_t^0). \tag{2.0.4}$$

Let us suppose that the behavior of the system is such that at the very beginning of the time interval, when the slow variable does not deviate essentially from the initial point, the fast variable almost reaches its rest point $\tilde{y}_\infty$ depending on x^o. Mathematically speaking, this means that the solution of the so-called *associated* equation (or the "inner" problem)

$$d\tilde{y}_\tau = F(0, x^o, \tilde{y}_\tau)d\tau, \quad \tilde{y}_0 = y^o, \tag{2.0.5}$$

converges at infinity to the point $\tilde{y}_\infty$.

Assume, moreover, that there exists the solution $y = \varphi(t, x)$ of the algebraic equation

$$F(t, x, \varphi(t, x)) = 0 \tag{2.0.6}$$

such that $\tilde{y}_\infty = \varphi(0, x^o)$. It is natural to expect that if t is relatively small but y_t^ε is yet close to $\tilde{y}_\infty$ then the derivative $\dot{x}_t^\varepsilon$ is approximately equal to

$$f(0, x^o, \varphi(0, x^o)) \approx f(t, x_t^\varepsilon, \varphi(t, x_t^\varepsilon)).$$

Thus, on a small time interval the trajectory x^ε can be approximated by the solution

$$dx_t^0 = f(t, x_t^0, \varphi(t, x_t^0))dt, \quad x_0^0 = x^o, \tag{2.0.7}$$

and one may expect that such an approximation holds on a finite interval provided that some stability assumption is made, for example, if

$F_y(t, x_t^0, \varphi(t, x^0))$ is a strictly stable matrix for all $t \in [0, T]$. This is indeed the case and the famous Tikhonov theorem asserts that

$$||x^\varepsilon - x^0||_T \to 0, \tag{2.0.8}$$

$$||y^\varepsilon - y^0||_{S,T} \to 0, \tag{2.0.9}$$

where x^0 is the solution of (2.0.7), $y_t^0 := \varphi(t, x_t^0)$, S is an arbitrary point in $]0, T]$ and $||h||_{S,T} := \sup_{S \le t \le T} |h_t|$.

We present here a version of the Tikhonov theorem for the system of stochastic differential equations

$$dx_t^\varepsilon = f(t, x_t^\varepsilon, y_t^\varepsilon)dt + g(t, x_t^\varepsilon, y_t^\varepsilon)dw_t^x, \quad x_0^\varepsilon = x^o, \tag{2.0.10}$$

$$\varepsilon dy_t^\varepsilon = F(t, x_t^\varepsilon, y_t^\varepsilon)dt + G(\varepsilon, t, x_t^\varepsilon, y_t^\varepsilon)dw_t^y, \quad y_0^\varepsilon = y^o, \tag{2.0.11}$$

where w^x and w^y are independent Wiener processes. The new important feature of the stochastic setting is the presence of the small parameter at the diffusion coefficient of the equation for the fast variable. We shall assume that

$$G(\varepsilon, t, x, y) = \sigma(\varepsilon)G(t, x, y)$$

and show that (2.0.8), (2.0.9) holds with convergence in probability for the case where $\sigma(\varepsilon) = \sqrt{\varepsilon}\beta_\varepsilon$ with $\beta_\varepsilon = o(1/\sqrt{|\ln \varepsilon|})$ and this hypothesis is important if one wants to get a result on a convergence of the fast variable. Notice that our theorem is nontrivial even for $\sigma(\varepsilon) = 0$, i.e., without the diffusion term in the equation for the fast variable; it reduces to the classic Tikhonov theorem in the absence of both diffusion terms.

It is worth mentioning that there is another class of intensively studied models with $\sigma(\varepsilon) = \sqrt{\varepsilon}$ usually related to the Bogoliubov average principle. In such a case the fast variable may be "oscillatory" and it does not converge in probability: consider as an example the system with only the fast process y^ε with

$$\varepsilon dy_t^\varepsilon = -\gamma y_t^\varepsilon dt + \sqrt{\varepsilon}dw_t^y, \quad y_0^\varepsilon = y^o.$$

Nevertheless, the limit of the slow variable may exist under assumptions of ergodicity of the fast process with the "frozen" x-argument. The systematic study of the model with $\sigma(\varepsilon) = \sqrt{\varepsilon}$ requires a different technique and is beyond the scope of our book.

In this chapter we start the study of asymptotic expansions in the simplest case of only fast variables to cover the needs of stochastic approximation procedures. The special structure of coefficients allows us to avoid some technicalities by using time stretching. The more complicated general case is considered in Chapter 4.

2.1 The Stochastic Tikhonov Theorem

2.1.1 Setting

Consider in $\mathbf{R}^k \times \mathbf{R}^n$ the system of stochastic differential equations

$$dx_t^\varepsilon = f(t, x_t^\varepsilon, y_t^\varepsilon)dt + g(t, x_t^\varepsilon, y_t^\varepsilon)dw_t^x, \quad x_0^\varepsilon = x^o, \tag{2.1.1}$$

$$\varepsilon dy_t^\varepsilon = F(t, x_t^\varepsilon, y_t^\varepsilon)dt + \beta\varepsilon^{1/2}G(t, x_t^\varepsilon, y_t^\varepsilon)dw_t^y, \quad y_0^\varepsilon = y^o, \tag{2.1.2}$$

where w^x and w^y are independent Wiener processes with values in $\mathbf{R}^k$ and $\mathbf{R}^n$ respectively.

Let E_1 and E_2 be two Euclidean spaces. We say that a measurable function $h : [0, T] \times E_1 \to E_2$ satisfies the *linear growth and local Lipschitz conditions in* x if:

(1) there is a constant L such that

$$|h(t, x)| \leq L(1 + |x|) \quad \forall\, t \in [0, T],\ x \in E_1;$$

(2) for any $N > 0$ there is a constant L_N such that

$$|h(t, x_1) - h(t, x_2)| \leq L_N |x_1 - x_2| \quad \forall\, t \in [0, T],\ x_i \in E_1,\ |x_i| \leq N.$$

Remark. The local Lipschitz condition implies the continuity of the function $H : C([0, T], E_1) \to \mathbf{R}_+$ where $H(x) := \sup_{t \leq T} |h(t, x_t)|$.

To formulate the result we introduce the following set of hypotheses:

H.2.1.1 The functions f, F (with values in $\mathbf{R}^k$ and $\mathbf{R}^n$), and g, G (with values in the sets of $k \times k$ and $n \times n$ matrices) are continuous in all variables and satisfy the linear growth and local Lipschitz conditions in (x, y).

H.2.1.2 There is a function $\varphi : [0, T] \times \mathbf{R}^k \to \mathbf{R}^n$ satisfying the linear growth and local Lipschitz conditions in x such that

$$F(t, x, \varphi(t, x)) = 0 \quad \forall\, t \in [0, T],\ x \in \mathbf{R}^k.$$

H.2.1.3 The solution of the problem

$$d\tilde{y}_s = F(0, x^o, \tilde{y}_s)dt, \quad \tilde{y}_0 = y^o, \tag{2.1.3}$$

tends to $\varphi(0, x^o)$ as $s \to \infty$:

$$\lim_{s \to \infty} \tilde{y}_s = \varphi(0, x^o). \tag{2.1.4}$$

One can say that **H.2.1.3** means that the initial value y^o belongs to the domain of influence of the root $\varphi(0, x^o)$ of the equation $F(0, x^o, y) = 0$.

H.2.1.4 The derivative F_y exists, it is a continuous function on the set $[0, T] \times \mathbf{R}^k \times \mathbf{R}^n$ and for any $N > 0$ there is a constant $\kappa_N > 0$ such that for every $(t, x) \in [0, T] \times \mathbf{R}^k$ with $|x| \leq N$

$$z^{\star} F_y(t, x, \varphi(t, x)) z \leq -\kappa_N |z|^2 \quad \forall z \in \mathbf{R}^n.$$

H.2.1.5 The function $\beta = \beta_\varepsilon = o\left(1/\sqrt{|\ln \varepsilon|}\right)$ as $\varepsilon \to 0$.

Consider the stochastic differential equation

$$dx_t = f(t, x_t, \varphi(t, x_t)) dt + g(t, x_t, \varphi(t, x_t)) dw_t^x, \quad x_0 = x^o. \tag{2.1.5}$$

Let

$$y_t = \varphi(t, x_t). \tag{2.1.6}$$

Theorem 2.1.1 *Under the hypotheses* **H.2.1.1–H.2.1.5**

$$P\text{-}\lim_{\varepsilon \to 0} ||x^\varepsilon - x||_T = 0, \tag{2.1.7}$$

$$P\text{-}\lim_{\varepsilon \to 0} ||y^\varepsilon - y||_{S,T} = 0 \tag{2.1.8}$$

where S is an arbitrary point from $]0, T]$.

The proof is rather straightforward though a bit lengthy. It consists of two independent stages. In the first stage we study the "boundary layer" and show, applying the standard theorem on the continuous dependence on a parameter for a "regularly" perturbed SDE in the stretched time, that the "essential" part of trajectories falls quickly into a small neighborhood of the point $(x^o, \varphi(0, x^o))$. In the second stage we check that the stability assumption guarantees that the solution is close to the process (x_t, y_t) given by (2.1.5), (2.1.6).

2.1.2 Boundary Layer Behavior

We set

$$\delta_t^\varepsilon := x_t^\varepsilon - x_t,$$
$$\Delta_t^\varepsilon := y_t^\varepsilon - \varphi(t, x_t^\varepsilon).$$

Since

$$|y_t^\varepsilon - y_t| \leq |y_t^\varepsilon - \varphi(t, x_t^\varepsilon)| + |\varphi(t, x_t^\varepsilon) - \varphi(t, x_t)|$$

and φ satisfies the local Lipschitz condition it follows from our remark that it is sufficient to establish the relations

$$P\text{-}\lim_{\varepsilon \to 0} ||\delta^\varepsilon||_T = 0, \tag{2.1.9}$$

$$P\text{-}\lim_{\varepsilon \to 0} ||\Delta^\varepsilon||_{S,T} = 0. \tag{2.1.10}$$

We introduce the stopping times

$$\tau_N^\varepsilon := \inf\{t : |x_t^\varepsilon| + |y_t^\varepsilon| \geq N\} \wedge T,$$
$$\tau_N := \inf\{t : |x_t| + |y_t| \geq N\} \wedge T,$$
$$\sigma_N^\varepsilon := \tau_N^\varepsilon \wedge \tau_N.$$

In what follows, it will always be supposed that $N > |x^o| + |y^o|$.

We show that for a suitable choice of a point $t_0 = t_0(\varepsilon)$ the random variables $||\delta^\varepsilon||_{t_0}$ and $|\Delta_{t_0}^\varepsilon|$ become small as $\varepsilon \to 0$. The precise statement is

Lemma 2.1.2 *For any $\gamma > 0$ there exists $r_0 = r_0(\gamma)$ such that for $t_0 := r_0\varepsilon$ and arbitrary N*

$$\lim_{\varepsilon \to 0} EI_{\{\sigma_N^\varepsilon \geq t_0\}}||\delta^\varepsilon||_{t_0}^2 = 0, \tag{2.1.11}$$

$$\lim_{\varepsilon \to 0} P(|\Delta_{t_0}^\varepsilon| > \gamma, \ \sigma_N^\varepsilon \geq t_0) = 0. \tag{2.1.12}$$

Proof. In the stretched time $s := t/\varepsilon$ the system (2.1.1), (2.1.2) in terms of the variables $\tilde{x}_s^\varepsilon := x_{\varepsilon s}^\varepsilon$ and $\tilde{y}_s^\varepsilon := y_{\varepsilon s}^\varepsilon$ can be written as follows:

$$d\tilde{x}_s^\varepsilon = \varepsilon f(\varepsilon s, \tilde{x}_s^\varepsilon, \tilde{y}_s^\varepsilon)ds + \varepsilon^{1/2}g(\varepsilon s, \tilde{x}_s^\varepsilon, \tilde{y}_s^\varepsilon)d\tilde{w}_s^x, \quad \tilde{x}_0^\varepsilon = x^o,$$
$$d\tilde{y}_s^\varepsilon = F(\varepsilon s, \tilde{x}_s^\varepsilon, \tilde{y}_s^\varepsilon)ds + \beta_\varepsilon G(\varepsilon s, \tilde{x}_s^\varepsilon, \tilde{y}_s^\varepsilon)d\tilde{w}_s^y, \quad \tilde{y}_0^\varepsilon = y^o, \tag{2.1.13}$$

where $\tilde{w}_s^x := w_{\varepsilon s}^x/\varepsilon^{1/2}$ and $\tilde{w}_s^y = w_{\varepsilon s}^y/\varepsilon^{1/2}$ are again independent Wiener processes, $\beta_\varepsilon \to 0$ due to **H.2.1.5**. By the theorem on continuous dependence of the solution on a parameter we have for any fixed s that

$$P\text{-}\lim_{\varepsilon \to 0} ||\tilde{x}^\varepsilon - x^o||_s = 0, \tag{2.1.14}$$

$$P\text{-}\lim_{\varepsilon \to 0} ||\tilde{y}^\varepsilon - \tilde{y}||_s = 0 \tag{2.1.15}$$

where $\tilde{y}$ is the solution of (2.1.3). The dominated convergence theorem implies that

$$\lim_{\varepsilon \to 0} E||\tilde{x}^\varepsilon - x^o||_{s \wedge \tilde{\tau}_N^\varepsilon} = 0, \tag{2.1.16}$$

$$\lim_{\varepsilon \to 0} E||\tilde{y}^\varepsilon - \tilde{y}||_{s \wedge \tilde{\tau}_N^\varepsilon} = 0 \tag{2.1.17}$$

where

$$\tilde{\tau}_N^\varepsilon := \varepsilon^{-1}\tau_N^\varepsilon = \inf\{s : |\tilde{x}_s^\varepsilon| + |\tilde{y}_s^\varepsilon| \geq N\} \wedge (T/\varepsilon).$$

It follows from **H.2.1.3** that for any $\gamma > 0$ there exists $r_0 = r_0(\gamma) > 0$ such that

$$|\tilde{y}_{r_0} - \varphi(0, x^o)| \leq \gamma/4.$$

By virtue of continuity of φ for sufficiently small ε

$$|\varphi(0, x^o) - \varphi(r_0\varepsilon, x^o)| \leq \gamma/4.$$

Therefore,

$$\begin{aligned}
|\Delta_{t_0}^\varepsilon| = |y_{t_0}^\varepsilon - \varphi(t_0, x_{t_0}^\varepsilon)| &\le |\tilde{y}_{r_0}^\varepsilon - \tilde{y}_{r_0}| + |\tilde{y}_{r_0} - \varphi(0, x^o)| \\
&\quad + |\varphi(0, x^o) - \varphi(r_0\varepsilon, x^o)| + |\varphi(t_0, x^o) - \varphi(t_0, x_{t_0}^\varepsilon)| \\
&\le |\tilde{y}_{r_0}^\varepsilon - \tilde{y}_{r_0}| + |\varphi(t_0, x^o) - \varphi(t_0, x_{t_0}^\varepsilon)| + \gamma/2.
\end{aligned}$$

By the local Lipschitz condition in the set $\{\sigma_N^\varepsilon \ge t_0\}$ (where $|x_{t_0}^\varepsilon| \le N$) we have with $L = L_N$ that

$$|\varphi(t_0, x^o) - \varphi(t_0, x_{t_0}^\varepsilon)| \le L|x^o - x_{t_0}^\varepsilon| \le L|x^o - \tilde{x}_{r_0}^\varepsilon|.$$

Consequently,

$$\begin{aligned}
P(|\Delta_{t_0}^\varepsilon| \ge \gamma, \ \sigma_N^\varepsilon \ge t_0) &\le P(|\tilde{y}_{r_0}^\varepsilon - \tilde{y}_{r_0}| \ge \gamma/4, \ \sigma_N^\varepsilon \ge t_0) \\
&\quad + P(L|x^o - \tilde{x}_{r_0}^\varepsilon| \ge \gamma/4, \ \sigma_N^\varepsilon \ge t_0) \\
&\le P(\|\tilde{y}^\varepsilon - \tilde{y}\|_{r_0 \wedge \tilde{\tau}_N^\varepsilon} \ge \gamma/4) \\
&\quad + P(L\|\tilde{x}^\varepsilon - x^o\|_{r_0 \wedge \tilde{\tau}_N^\varepsilon} \ge \gamma/4).
\end{aligned}$$

According to (2.1.14) and (2.1.15) the probabilities on the right-hand side tend to zero as $\varepsilon \to 0$ and we get (2.1.12).

For the proof of (2.1.11) we note that

$$\begin{aligned}
EI_{\{\sigma_N^\varepsilon \ge t_0\}} \|\delta^\varepsilon\|_{t_0} &\le E\|x^\varepsilon - x^0\|_{t_0 \wedge \sigma_N^\varepsilon} + E\|x - x^0\|_{t_0 \wedge \sigma_N^\varepsilon} \\
&\le E\|\tilde{x}^\varepsilon - x^0\|_{t_0 \wedge \tilde{\tau}_N^\varepsilon} + E\|x - x^0\|_{t_0 \wedge \tau_N}.
\end{aligned}$$

Here the first term on the right-hand side tends to zero by (2.1.16); the convergence to zero of the second term follows from the continuity of x. $\square$

We use below the following auxiliary assertion:

Lemma 2.1.3 *Let $N > N_0 := 1 + |x^o| + \sup|\tilde{y}_s|$. Then*

$$P\text{-}\lim_{\varepsilon \to 0} \varepsilon^{-1}\tau_N^\varepsilon = \infty.$$

Proof. Let $M > 0$ be arbitrary. Put again $\tilde{\tau}_N^\varepsilon = \varepsilon^{-1}\tau_N^\varepsilon$ and define

$$\Gamma := \{\|\tilde{y}^\varepsilon - \tilde{y}\|_{M \wedge \tilde{\tau}_N^\varepsilon} \le 1/2, \ \|\tilde{x} - x^o\|_{M \wedge \tilde{\tau}_N^\varepsilon} \le 1/2\}.$$

On the set Γ for any $s \le M \wedge \tilde{\tau}_N^\varepsilon$ we have

$$|\tilde{x}_s^\varepsilon| + |\tilde{y}_s^\varepsilon| \le |x^o| + |\tilde{x}_s - x^o| + |\tilde{y}_s^\varepsilon - \tilde{y}_s| + |\tilde{y}_s| \le N_0$$

and hence $\{\tilde{\tau}_N^\varepsilon \le M\} \subseteq \bar{\Gamma}$. Consequently,

$$P(\tilde{\tau}_N^\varepsilon \le M) \le P(\|\tilde{y}^\varepsilon - \tilde{y}\|_{M \wedge \tilde{\tau}_N^\varepsilon} > 1/2) + P(\|\tilde{x} - x^o\|_{M \wedge \tilde{\tau}_N^\varepsilon} > 1/2).$$

But by (2.1.14) and (2.1.15) both probabilities on the right-hand side tend to zero as $\varepsilon \to 0$, implying the result. $\square$

For sufficiently large N the exit time τ_N is strictly positive and hence

$$\limsup_{\varepsilon \to 0} P(\tau_N < r_0 \varepsilon) = 0.$$

By Lemma 2.1.3 we have

$$\limsup_{\varepsilon \to 0} P(\tau_N^\varepsilon < r_0 \varepsilon) = 0.$$

Thus,

$$P(\sigma_N^\varepsilon < r_0 \varepsilon) \le P(\tau_N < r_0 \varepsilon) + P(\tau_N^\varepsilon < r_0 \varepsilon) \to 0, \quad \varepsilon \to 0.$$

The above arguments lead to the following conclusion which we formulate as

Proposition 2.1.4 *For every $\gamma > 0$ there exists a function $t_0 = t_0(., \gamma)$ on $]0,1]$ such that $t_0(\varepsilon, \gamma) \to 0$ as $\varepsilon \to 0$ and*

$$\lim_{\varepsilon \to 0} P(|\Delta_{t_0(\varepsilon,\gamma)}^\varepsilon| \ge \gamma) = 0,$$

$$\lim_{\varepsilon \to 0} E\|\delta^\varepsilon\|_{t_0(\varepsilon,\gamma)} = 0.$$

Remark. It is easy to understand that in results which do not concern the boundary layer (like the Tikhonov theorem) the only important property is a way in which the solutions "leave" the latter formalized in the above proposition and which can be taken as a starting hypothesis for some useful generalization of the model being considered. In particular, in the study of the stochastic approximation we meet a specification with singularly perturbed SDEs defined on the interval $[t_0, T]$ with this hypothesis fulfilled.

2.1.3 Large Scale Behavior

We continue the proof of the theorem by establishing some inequalities which control the deviation of solutions outside the boundary layer. To simplify formulae we shall skip usually the argument t in the notation for functions.

Lemma 2.1.5 *For any $t_0 \in]0,T]$ and a stopping time τ with values in the interval $[t_0, (\sigma_N^\varepsilon \wedge T) \vee t_0]$*

$$E(\|\delta^\varepsilon\|_{t_0,\tau}^2 | \mathcal{F}_{t_0}) \le C \left(|\delta_{t_0}^\varepsilon|^2 + E(\|\Delta^\varepsilon\|_{t_0,\tau}^2 | \mathcal{F}_{t_0}) \right) \tag{2.1.18}$$

where the constant C depends on T and N.

Proof. Since $\delta_t^\varepsilon := x_t^\varepsilon - x_t$, we get from the equations (2.1.1) and (2.1.5) that

$$\delta_t^\varepsilon = \delta_{t_0}^\varepsilon + \int_{t_0}^t \widehat{f}_s ds + \int_{t_0}^t \widehat{g}_t dw_t^x \tag{2.1.19}$$

where

$$\widehat{f_t^\varepsilon} := f(x_t^\varepsilon, y_t^\varepsilon) - f(x_t, \varphi(x_t)),$$
$$\widehat{g_t^\varepsilon} := g(x_t^\varepsilon, y_t^\varepsilon) - g(x_t, \varphi(x_t)).$$

It is easy to deduce from (2.1.19), using the Doob inequality to estimate the supremum of the stochastic integral, that

$$E(||\delta^\varepsilon||_{t_0, \tau \wedge t}^2 | \mathcal{F}_{t_0}) \leq C \left(|\delta_{t_0}^\varepsilon| + \int_{t_0}^{t \wedge \tau} E(|\widehat{f_s^\varepsilon}|^2 + |\widehat{g_s^\varepsilon}|^2 | \mathcal{F}_{t_0}) ds \right). \quad (2.1.20)$$

Notice that the local Lipschitz condition for f and φ implies that for $s \leq \sigma_N^\varepsilon$

$$|\widehat{f_s^\varepsilon}| \leq |f(x_s^\varepsilon, y_s^\varepsilon) - f(x_s^\varepsilon, \varphi(x_s^\varepsilon))| + |f(x_s^\varepsilon, \varphi(x_s^\varepsilon)) - f(x_s, \varphi(x_s))|$$
$$\leq C(|\delta_s^\varepsilon| + |\Delta_s^\varepsilon|).$$

The function $\widehat{g}^\varepsilon$ admits a similar bound. Thus, if $\sigma_N^\varepsilon \geq t_0$ and $s \in [t_0, \tau \wedge t]$ then

$$|\widehat{f_s^\varepsilon}|^2 + |\widehat{g_s^\varepsilon}|^2 \leq C(|\delta_s^\varepsilon|^2 + |\Delta_s^\varepsilon|^2) \leq C(||\Delta^\varepsilon||_{t_0, \tau}^2 + ||\delta^\varepsilon||_{t_0, \tau \wedge s}^2)$$

and we obtain from (2.1.20) that on the set $\{\sigma_N^\varepsilon \geq t_0\}$

$$E(||\delta^\varepsilon||_{t_0, \tau \wedge t}^2 | \mathcal{F}_{t_0}) \leq C \left(|\delta_{t_0}^\varepsilon| + E(||\Delta^\varepsilon||_{t_0, \tau}^2 | \mathcal{F}_{t_0}) + \int_{t_0}^{t} E(||\delta^\varepsilon||_{t_0, \tau \wedge s}^2 | \mathcal{F}_{t_0}) ds \right)$$

and by the Gronwall–Bellman lemma

$$E(||\delta^\varepsilon||_{t_0, \tau}^2 | \mathcal{F}_{t_0}) \leq C e^{CT} \left(|\delta_{t_0}^\varepsilon|^2 + E(||\Delta^\varepsilon||_{t_0, \tau}^2 | \mathcal{F}_{t_0}) \right).$$

The inequality (2.1.18) on the set $\{\sigma_N^\varepsilon < t_0\}$ is trivial. $\square$

The next lemma which is a crucial step in the proof of the theorem uses the stability assumption **H.2.1.4**.

For $h > 0$ let

$$\rho_N(h) := \sup \left| \frac{\partial F_j}{\partial y_i}(t, x, y_1) - \frac{\partial F_j}{\partial y_i}(t, x, y_2) \right|$$

the sup is taken over all $t \in [0, T]$ and x, y_1, y_2 such that $|x| + |y_1| + |y_2| \leq N$ and $|y_1 - y_2| \leq h$. Let

$$\phi_N(h) := \sup \left| \frac{\partial F_j}{\partial y_i}(t, x_1, \varphi(t, x_1)) - \frac{\partial F_j}{\partial y_i}(t, x_2, \varphi(t, x_2)) \right|,$$

the sup is taken over all $t \in [0, T]$, and x_1, x_2 such that $|x_1| \leq N$, $|x_2| \leq N$, and $|x_1 - x_2| \leq h$.

By virtue of continuity of F_y and φ

$$\rho_N(h) \to 0, \quad \phi_N(h) \to 0 \text{ as } h \to 0. \quad (2.1.21)$$

Lemma 2.1.6 *For any $t_0 \in]0, T]$, Δ, $\delta > 0$, $\eta \in]0, \Delta]$, $\gamma \in]0, \eta[$, $m > 1/\eta$, and any stopping time τ with values in the interval $[t_0, T]$ we have that*

$$P(\|\Delta^\varepsilon\|_{t_0,\tau} \geq \Delta) \leq P(\sigma_N^\varepsilon < \tau) + P(|\Delta_{t_0}^\varepsilon| > \gamma, \ \sigma_N^\varepsilon \geq t_0) + R_1 + R_2 \quad (2.1.22)$$

where

$$R_1 := C_N \delta^{-2}(E\|\delta^\varepsilon\|_{t_0}^2 + \eta^2), \tag{2.1.23}$$
$$R_2 := (\eta - 1/m)^{-1}[\gamma + C_N/m + (\rho_N(\eta) + \phi_N(\delta))\eta + C_{N,m}o(1)] \tag{2.1.24}$$

with $o(1) = (\varepsilon^{1/2} + \beta_\varepsilon)|\ln \varepsilon|^{1/2} \to 0$.

Proof. We introduce the stopping time

$$\theta = \theta_\eta^{\varepsilon,N} := \inf\{t \geq t_0 : |\Delta_t^\varepsilon| \geq \eta\} \wedge \tau \wedge \sigma_N^\varepsilon.$$

Put

$$\Gamma := \{\|\delta^\varepsilon\|_{t_0,\theta} \leq \delta, \ |\Delta_{t_0}^\varepsilon| \leq \gamma, \ \sigma_N^\varepsilon \geq t_0\}.$$

We have:

$$\begin{aligned}
P(\|\Delta^\varepsilon\|_{t_0,\tau} \geq \Delta) &\leq P(\|\Delta^\varepsilon\|_{t_0,\tau} \geq \eta) \\
&\leq P(\sigma_N^\varepsilon < \tau) + P(\|\Delta^\varepsilon\|_{t_0,\tau} \geq \eta, \ \sigma_N^\varepsilon \geq t_0) \\
&\leq P(\sigma_N^\varepsilon < \tau) + P(|\Delta_{t_0}^\varepsilon| > \gamma, \ \sigma_N^\varepsilon \geq t_0) \\
&\quad + P(\|\delta^\varepsilon\|_{t_0,\theta} > \delta, \ |\Delta_{t_0}^\varepsilon| \leq \gamma, \ \sigma_N^\varepsilon \geq t_0) + P(I_\Gamma |\Delta_\theta^\varepsilon| \geq \eta) \\
&= P(\sigma_N^\varepsilon < \tau) + P(|\Delta_{t_0}^\varepsilon| > \gamma, \ \sigma_N^\varepsilon \geq t_0) + P_1 + P_2
\end{aligned}$$

where P_1 and P_2 are the third and the fourth terms on the left-hand side of the last inequality.

By the Chebyshev inequality and Lemma 2.1.5

$$\begin{aligned}
P_1 &\leq \delta^{-2} E I_{\{|\Delta_{t_0}^\varepsilon| \leq \gamma, \ \sigma_N^\varepsilon \geq t_0\}} \|\delta^\varepsilon\|_{t_0,\theta}^2 \\
&\leq C\delta^{-2}(E I_{\{\sigma_N^\varepsilon \geq t_0\}} |\delta_{t_0}^\varepsilon|^2 + E I_{|\Delta_{t_0}^\varepsilon| \leq \gamma, \ \{\sigma_N^\varepsilon \geq t_0\}} \|\Delta^\varepsilon\|_{t_0,\theta}^2) \\
&\leq C\delta^{-2}(E|\delta_{t_0}^\varepsilon|^2 + \eta^2) = R_1.
\end{aligned}$$

It remains to show that R_2 is a majorant for P_2. To this aim, we approximate the function φ by a twice continuously differentiable function φ^m in such a way that the inequality $|\varphi^m - \varphi| \leq 1/m$ holds in the compact set $\{(t, x) : \ t \in [0, T], \ |x| \leq N\}$.

Let $\Delta_t^{\varepsilon,m} := y_t^\varepsilon - \varphi^m(x_t^\varepsilon)$. Obviously,

$$|\Delta_t^{\varepsilon,m} - \Delta_t^\varepsilon| \leq 1/m, \quad t \leq \sigma_N^\varepsilon. \tag{2.1.25}$$

According to the Ito formula it follows from (2.1.1) that

$$d\varphi^m(t, x_t^\varepsilon) = b_t^{\varepsilon,m} dt + u_t^{\varepsilon,m} dw_t^x \tag{2.1.26}$$

where

$$b_t^{\varepsilon,m} := \varphi_t^m(x_t^\varepsilon) + \varphi_x^m(x_t^\varepsilon)f(x_t^\varepsilon, y_t^\varepsilon) + \psi(x_t^\varepsilon, y_t^\varepsilon),$$

$$\psi^i(x,y) := \frac{1}{2}\mathrm{tr}\,\varphi_{xx}^{m,i}(x)g(x,y)g^*(x,y),$$

$$u_t^{\varepsilon,m} := \varphi_x^m(x_t^\varepsilon)g(x_t^\varepsilon, y_t^\varepsilon).$$

We set $A_t := F_y(x_t, y_t)$ where x_t and y_t are defined by (2.1.5) and (2.1.6),

$$\alpha_t^{\varepsilon,m} := \varepsilon^{-1}\left(F(x_t^\varepsilon, y_t^\varepsilon) - A_t\Delta_t^{\varepsilon,m}\right) - b_t^{\varepsilon,m}.$$

Subtracting (2.1.26) from (2.1.2) we obtain, after regrouping the terms,

$$d\Delta_t^{\varepsilon,m} = \varepsilon^{-1}A_t\Delta_t^{\varepsilon,m}dt + \alpha_t^{\varepsilon,m}dt + \beta_\varepsilon\varepsilon^{-1/2}G(x_t^\varepsilon, y_t^\varepsilon)dw_t^y - u_t^{\varepsilon,m}dw_t^x.$$

Solving this equation with respect to $\Delta^{\varepsilon,m}$ we get the representation

$$\Delta_t^{\varepsilon,m} = \Phi^\varepsilon(t,t_0)\Delta_{t_0}^{\varepsilon,m} + V_t^1 + \beta_\varepsilon V_t^2 - V_t^3 \qquad (2.1.27)$$

where Φ^ε is the fundamental matrix corresponding to $\varepsilon^{-1}A$, and the processes V^i are the solutions of the following linear equations on the interval $[t_0, T]$:

$$dV_t^1 = \varepsilon^{-1}A_tV_t^1dt + \alpha_t^{\varepsilon,m}dt, \qquad (2.1.28)$$

$$dV_t^2 = \varepsilon^{-1}A_tV_t^2dt + \varepsilon^{-1/2}G(x_t^\varepsilon, y_t^\varepsilon)dw_t^y, \qquad (2.1.29)$$

$$dV_t^3 = \varepsilon^{-1}A_tV_t^3dt + u_t^{\varepsilon,m}dw_t^x, \qquad (2.1.30)$$

with the initial conditions $V_{t_0}^i = 0$.

Clearly,

$$|F(x_t^\varepsilon, y_t^\varepsilon) - A_t\Delta_t^{\varepsilon,m}| \le |F(x_t^\varepsilon, y_t^\varepsilon) - F_y(x_t, y_t)\Delta_t^\varepsilon|$$
$$+ |F_y(x_t, y_t)||\Delta_t^\varepsilon - \Delta_t^{\varepsilon,m}| \qquad (2.1.31)$$

where the second term on the right-hand side for $t \le \sigma_N^\varepsilon$ does not exceed C_N/m.

By the finite increments formula

$$F(x_t^\varepsilon, y_t^\varepsilon) = F(x_t^\varepsilon, \varphi(x_t^\varepsilon) + \Delta_t^\varepsilon) = F(x_t^\varepsilon, \varphi(x_t^\varepsilon)) + \widehat{A}_t\Delta_t^\varepsilon = \widehat{A}_t\Delta_t^\varepsilon$$

where $\widehat{A}_t = \widehat{A}_t^\varepsilon$ is the matrix with the ith column of the form

$$\frac{\partial F}{\partial y_i}(x_t^\varepsilon, \varphi(x_t^\varepsilon) + \vartheta_i\Delta_t^\varepsilon), \quad \vartheta_i = \vartheta_i(x_t^\varepsilon, y_t^\varepsilon) \in [0,1].$$

Thus, the first term on the right-hand side of (2.1.31) for $t \le \sigma_N^\varepsilon$ does not exceed

$$|\widehat{A}_t - F_y(x_t, y_t)||\Delta_t^\varepsilon| \le |\widehat{A}_t - F_y(x_t^\varepsilon, \varphi(x_t^\varepsilon))||\Delta_t^\varepsilon|$$
$$+ |F_y(x_t^\varepsilon, \varphi(x_t^\varepsilon)) - F_y(x_t, y_t)||\Delta_t^\varepsilon|$$
$$\le \rho_N(|\Delta_t^\varepsilon|) + \phi_N(|\delta_t^\varepsilon|)|\Delta_t^\varepsilon|.$$

The function $b_t^{\varepsilon,m}$ for $t \le \sigma_N^\varepsilon$ is bounded by a constant depending on N and m.

The above arguments show that for $t \le \sigma_N^\varepsilon$

$$|\alpha_t^{\varepsilon,m}| \le \varepsilon^{-1}[(\rho_N(|\Delta_t^\varepsilon|) + \phi_N(|\delta_t^\varepsilon|))|\Delta_t^\varepsilon| + C_N/m] + C_{N,m}. \tag{2.1.32}$$

Notice also that for $t \le \sigma_N^\varepsilon$ we have

$$|G(x_t^\varepsilon, y_t^\varepsilon)| \le C_N, \quad |\beta_t^{\varepsilon,m}| \le C_{N,m}.$$

Now, after this preliminary work, it is not difficult to derive the needed majorant for P_2. Since $t \le \sigma_N^\varepsilon$ on the set Γ and hence $|\Delta_\theta^{\varepsilon,m} - \Delta_\theta^\varepsilon| \le 1/m$ we have:

$$P_2 = P(I_\Gamma|\Delta_\theta^\varepsilon| \ge \eta) \le P(I_\Gamma|\Delta_\theta^{\varepsilon,m}| \ge \eta - 1/m)$$
$$\le (\eta - 1/m)^{-1}EI_\Gamma|\Delta_\theta^{\varepsilon,m}|. \tag{2.1.33}$$

It follows from the representation (2.1.27) that

$$EI_\Gamma|\Delta_\theta^{\varepsilon,m}| \le EI_\Gamma|\Phi^\varepsilon(\theta, t_0)||\Delta_{t_0}^{\varepsilon,m}| + EI_\Gamma|V_\theta^1| + \beta EI_\Gamma|V_\theta^2| + EI_\Gamma|V_\theta^3| \tag{2.1.34}$$

Obviously, $|\Phi^\varepsilon(\theta, t_0)|$ is bounded by a certain constant depending on N and we have $|\Delta_{t_0}^{\varepsilon,m}| \le \gamma + 1/m$ on Γ. Thus,

$$EI_\Gamma|\Phi^\varepsilon(\theta, t_0)||\Delta_{t_0}^{\varepsilon,m}| \le C_N(\gamma + 1/m). \tag{2.1.35}$$

The inequality (2.1.32) implies that on Γ we have

$$|\alpha_t^{\varepsilon,m}| \le \varepsilon^{-1}[(\rho_N(\eta) + \phi_N(\delta))\eta + C_N/m] + C_{N,m}.$$

Due to **H.2.1.4** the fundamental matrix for $t \in [t_0, \sigma_N^\varepsilon$ admits the bound

$$|\Phi^\varepsilon(t, t_0)| \le C \exp\{-2\kappa_N \varepsilon^{-1}(t - t_0)\}$$

(see (1.2.2)). Thus, the representation

$$V_\theta^1 = \int_{t_0}^\theta \Phi^\varepsilon(\theta, s)\alpha_s^{\varepsilon,m} ds$$

and above estimates lead to the inequality

$$I_\Gamma|V_\theta^1| \le C_{N,m}\varepsilon + (\rho_N(\eta) + \phi_N(\delta))\eta + C_N/m. \tag{2.1.36}$$

We obtain from Proposition 1.2.4 that for sufficiently small ε

$$EI_\Gamma|V_\theta^2| \le E\|V^{2,N}\|_{t_0,T} \le C_N|\ln\varepsilon|^{1/2}$$

where

$$dV_t^{2,N} = \varepsilon^{-1}A_t^N V_t^{2,N}dt + \varepsilon^{-1/2}G_t^N dw_t^y, \quad V_{t_0}^{2,N} = 0,$$

with

$$A_t^N := A_t I_{\{t\le\sigma_N^\varepsilon\}} - I_{\{t>\sigma_N^\varepsilon\}},$$
$$G_t^N := G(x_{t\wedge\sigma_N^\varepsilon}^\varepsilon, y_{t\wedge\sigma_N^\varepsilon}^\varepsilon).$$

Due to the assumption **H.2.1.5**

$$\beta_\varepsilon EI_\Gamma|V_\theta^2| = C_N\beta_\varepsilon|\ln\varepsilon|^{1/2} \to 0, \quad \varepsilon \to 0. \tag{2.1.37}$$

Applying Proposition 1.2.4 we get

$$EI_\Gamma|V_\theta^3| \le E\|V_{t_0,T}^{3,N}\| \le C_{N,m}\varepsilon^{1/2}(1+|\ln\varepsilon|^{1/2}) \tag{2.1.38}$$

where

$$dV_t^{3,N} = \varepsilon^{-1}A_t^N V_t^{3,N}dt + \varepsilon^{-1/2}u_t^{\varepsilon,m,N}dw_t^y, \quad V_{t_0}^{3,N} = 0,$$

with $u_t^{\varepsilon,m,N} := u_{t\wedge\sigma_N^\varepsilon}^{\varepsilon,m}$.

It follows from $(2.1.33)$–$(2.1.38)$ that $P_2 \le R_2$ and the lemma is proved.
□

2.1.4 Concluding Step

Now we are ready to accomplish the proof of the theorem.

Lemma 2.1.7 *For any* Δ, $\nu > 0$, *and* $N \ge N_0 := 1 + |x^o| + \sup|\tilde{y}_s|$ *there exists* $r_0 = r_0(\eta, \nu, N)$ *such that for* $t_0 = r_0\varepsilon$ *we have*

$$\limsup_{\varepsilon\to 0} P(\|\Delta^\varepsilon\|_{t_0,\sigma_N^\varepsilon\vee t_0} \ge \Delta) \le \nu, \tag{2.1.39}$$

Proof. In order to apply Lemma 2.1.6 we set

$$\delta := \sup\{h : \phi_N(h) \le \nu/4\},$$

$$\eta := \sup\{h : \rho_N(h) \le \nu/4\} \wedge \Delta \wedge \sqrt{\delta^2\nu/(4C_N)},$$

$$\gamma := \nu\eta/4.$$

Let r_0 be equal to the number $r_0(\gamma)$ ensuring the relations (2.1.11) and (2.1.12) in Lemma 2.1.2.

By taking $\limsup_m \limsup_{\varepsilon\to 0}$ of the both sides of (2.1.22) we get that

$$\limsup_{\varepsilon\to 0} P(\|\Delta^\varepsilon\|_{t_0,\sigma_N^\varepsilon\vee t_0} \ge \eta) \le \limsup_{\varepsilon\to 0} P(\sigma_N^\varepsilon < r_0\varepsilon) + \nu = \nu$$

and the result holds. □

Lemma 2.1.8 *For any $\lambda > 0$ and $N \geq N_0$*

$$\lim_{\varepsilon \to 0} P(\|\delta^\varepsilon\|_{\sigma_N^\varepsilon} \geq \lambda) = 0. \tag{2.1.40}$$

Proof. We fix Δ, $\nu > 0$ and choose, in accordance with Lemma 2.1.7, the number r_0 such that (2.1.39) holds with $t_0 = r_0 \varepsilon$. Notice that

$$P(\|\delta^\varepsilon\|_{\sigma_N^\varepsilon} \geq \lambda) \leq P(\sigma_N^\varepsilon < t_0) + P(\|\delta^\varepsilon\|_{t_0} \geq \lambda/2, \ \sigma_N^\varepsilon \geq t_0)$$
$$+ P(\|\delta^\varepsilon\|_{t_0, \sigma_N^\varepsilon} \geq \lambda/2, \ \sigma_N^\varepsilon \geq t_0).$$

The first two terms on the right-hand side of this inequality tend to zero by Lemmas 2.1.7 and 2.1.2.

To show that the last term also tends to zero we introduce the stopping times

$$\theta_\Delta := \inf\{t \geq t_0 : \ |\Delta_t^\varepsilon| \geq \Delta\} \wedge T,$$
$$\tau_\Delta := (\theta_\Delta \wedge \sigma_N^\varepsilon) \vee t_0,$$

depending, of course, on ε and N. The probability of interest does not exceed

$$P(\theta_\Delta \leq \sigma_N^\varepsilon, \ \sigma_N^\varepsilon \geq t_0) + P(\|\delta^\varepsilon\|_{t_0, \tau_\Delta} \geq \lambda/2, \ \theta_\Delta \geq \sigma_N^\varepsilon \geq t_0)$$
$$\leq P(\|\Delta^\varepsilon\|_{t_0, \sigma_N^\varepsilon \vee t_0} \geq \Delta) + P(I_{\{\theta_\Delta \geq \sigma_N^\varepsilon \geq t_0\}} \|\delta^\varepsilon\|_{t_0, \tau_\Delta} \geq \lambda/2). \tag{2.1.41}$$

By the Chebyshev inequality and Lemma 2.1.5 the second term on the right-hand side of the last inequality is less than or equal to

$$4\lambda^{-2} C_N(E|\delta_{t_0}^\varepsilon|^2 + \Delta^2).$$

Making use of (2.1.39) and Proposition 2.1.4 we get that

$$\limsup_{\varepsilon \to 0} P(\|\delta^\varepsilon\|_{t_0, \sigma_N^\varepsilon} \geq \lambda/2, \ \sigma_N^\varepsilon \geq t_0) \leq \nu + 4 C_N \Delta^2 / \lambda^2.$$

The parameters ν and Δ being arbitrary, this implies the result. $\square$

Lemma 2.1.9 *For the exit time σ_N^ε*

$$\lim_{N \to \infty} \limsup_{\varepsilon \to 0} P(\sigma_N^\varepsilon < T) = 0. \tag{2.1.42}$$

Proof. From the definitions we have that

$$\Delta_t^\varepsilon = y_t^\varepsilon - y_t + \varphi(x_t) - \varphi(x_t^\varepsilon).$$

Since the function φ is locally Lipschitz it follows from Lemmas 2.1.7 and 2.1.8 that for arbitrary $\nu > 0$ and $N > N_0$ there is a number r_0 (depending on ν and N) such that for $t_0 = r_0 \varepsilon$

$$\limsup_{\varepsilon \to 0} P(\|y^\varepsilon - y\|_{t_0, \sigma_N^\varepsilon \vee t_0} > 1/2) \leq \nu. \tag{2.1.43}$$

Obviously,

$$P(\sigma_N^\varepsilon < T) \leq P(\tau_N < T) + P(\tau_N^\varepsilon < T)$$
$$\leq 2P(\tau_N < T) + P(\tau_N^\varepsilon < T \wedge \tau_N). \qquad (2.1.44)$$

Let

$$\Gamma := \{t_0 \leq \tau_N^\varepsilon < T \wedge \tau_N, \; \|\delta^\varepsilon\|_{t_0,\sigma_N^\varepsilon \vee t_0} \leq 1/2, \; \|y^\varepsilon - y\|_{t_0,\sigma_N^\varepsilon \vee t_0} \leq 1/2\}.$$

It is easy to see that $\Gamma \subseteq \{\tau_{N-1} \leq \tau_N^\varepsilon < T\} \subseteq \{\tau_{N-1} < T\}$. Therefore,

$$P(\tau_N^\varepsilon < T \wedge \tau_N) \leq P(\tau_N^\varepsilon < t_0) + P(\tau_{N-1} < T) + P(\|\delta^\varepsilon\|_{t_0,\sigma_N^\varepsilon \vee t_0} > 1/2)$$
$$+ P(\|y^\varepsilon - y\|_{t_0,\sigma_N^\varepsilon \vee t_0} > 1/2). \qquad (2.1.45)$$

It follows from (2.1.43)–(2.1.45) and Lemma 2.1.8 that

$$\limsup_{\varepsilon \to 0} P(\sigma_N^\varepsilon < T) \leq 2P(\tau_N < T) + P(\tau_{N-1} < T) + \nu.$$

Since ν is arbitrary and $P(\tau_N < T) \to 0$ as $N \to \infty$, the above bound gives the required result. $\square$

The assertion (2.1.9) of the theorem is a direct corollary of Lemmas 2.1.8 and 2.1.9. To prove (2.1.10) let us take arbitrary numbers $\Delta, \nu > 0$. By Lemma 2.1.9 there exists $N > N_0$ such that

$$\limsup_{\varepsilon \to 0} P(\sigma_N^\varepsilon < T) \leq \nu.$$

Using Lemma 2.1.7 we choose r_0 such that (2.1.39) holds for $t_0 = r_0 \varepsilon$. Then for $\varepsilon \leq S/r_0$ we have

$$P(\|\Delta^\varepsilon\|_{S,T} \geq \Delta) \leq P(\|\Delta^\varepsilon\|_{t_0,\sigma_N^\varepsilon \vee t_0} \geq \Delta, \; \sigma_N^\varepsilon = T) + P(\sigma_N^\varepsilon < T)$$

and hence

$$\limsup_{\varepsilon \to 0} P(\|\Delta^\varepsilon\|_{S,T} \geq \Delta) \leq 2\nu$$

implying the result. $\square$

2.2 The First-Order Asymptotics for Fast Variables

2.2.1 Basic Hypotheses

In this section we study asymptotic expansions for a fast system with time-independent coefficients. We present here successively results concerning the first-order asymptotics and asymptotic approximations of the higher orders. In this particular case, the time-stretching immediately reduces the singularly perturbed SDE to a regularly perturbed one for which the structure of the asymptotic decomposition is well-known. The only difference from the standard regular problem is that we need to prove, under a stability hypothesis, that the approximation holds on the growing time intervals $[0, T/\varepsilon]$. In such a setting the problem of uniform convergence of the solutions outside the boundary layer to the rest point is also of interest and we present here relevant results as well.

Let us consider the stochastic differential equation

$$\varepsilon dy_t^\varepsilon = F(y_t^\varepsilon)dt + \beta\varepsilon^{1/2}G(y_t^\varepsilon)dw_t, \quad y_0^\varepsilon = y^o, \tag{2.2.1}$$

where the coefficients F and G do not depend on time and satisfy the following four assumptions:

H.2.2.1 The functions F with values in $\mathbf{R}^n$ and G with values in the set of $n \times n$ matrices are continuous, of linear growth, and locally Lipschitz.

H.2.2.2 There is a root $\tilde{y}_\infty$ of the equation $F(y) = 0$.

H.2.2.3 The solution of the problem

$$d\tilde{y}_s = F(\tilde{y}_s)ds, \quad \tilde{y}_0 = y^o, \tag{2.2.2}$$

tends to $\tilde{y}_\infty$ as $s \to \infty$:

$$\lim_{s\to\infty} \tilde{y}_s = \tilde{y}_\infty, \tag{2.2.3}$$

that is the initial value y^o belongs to the domain of the influence of the root $\tilde{y}_\infty$.

H.2.2.4 The derivative F' exists, is a continuous function, and the real parts of all eigenvalues of $F'(\tilde{y}_\infty)$ are strictly negative:

$$\operatorname{Re}\lambda(F'(\tilde{y}_\infty)) < -\gamma < 0.$$

We shall use also the hypothesis **H.2.1.5** on the asymptotics of the small parameter, namely, that $\beta_\varepsilon = o(1/\sqrt{|\ln\varepsilon|})$ as $\varepsilon \to 0$.

As usual, it is convenient to assume that $\beta_\varepsilon \in [0, 1]$.

Notice that **H.2.2.4** is a bit weaker than the corresponding version of **H.2.1.4**.

2.2.2 The First-Order Correction

In the stretched time the solution of (2.2.1) approaches the solution $\tilde{y}$ of the reduced equation (2.2.2) uniformly even on the growing intervals. We look now for a correction term needed to obtain the rate of convergence faster than β.

Theorem 2.2.1 *Suppose that the conditions* **H.2.2.1**–**H.2.2.5** *are fulfilled and, moreover, F' is locally Lipschitz. Let $\tilde{y}$ be the solution of (2.2.2) and let $\tilde{y}^1$ be the solution of the following SDE:*

$$dy\tilde{}_s^1 = F'(\tilde{y}_s)\tilde{y}_s^1 ds + G(\tilde{y}_s)d\tilde{w}_s, \quad \tilde{y}_0^1 = 0, \tag{2.2.4}$$

where the Wiener process $\tilde{w}_s := \varepsilon^{-1/2}w_{s\varepsilon}$.

Then

$$y_t^\varepsilon = \tilde{y}_{t/\varepsilon} + \beta\tilde{y}_{t/\varepsilon}^1 + \beta_\varepsilon\Delta_t^1 \tag{2.2.5}$$

where

$$P\text{-}\lim_{\varepsilon\to 0}||\Delta^1||_T = 0. \tag{2.2.6}$$

Notice that in the notations for y^1 and Δ^1 we omit dependence on ε.

Proof. Let us change the time parameter in (2.2.1) putting $s := t/\varepsilon$ and $\tilde{y}_s^\varepsilon := y_{s\varepsilon}^\varepsilon$. Then $\tilde{y}^\varepsilon$ satisfies the equation

$$d\tilde{y}_s^\varepsilon = F(\tilde{y}_s^\varepsilon)ds + \beta G(\tilde{y}_s^\varepsilon)d\tilde{w}_s, \quad \tilde{y}_0^\varepsilon = y^\circ. \tag{2.2.7}$$

We set $\tilde{z}_s^\varepsilon := \tilde{y}_s + \beta\tilde{y}_s^1$ and define

$$\tilde{\Delta}_s^1 := \beta^{-1}(\tilde{y}_s^\varepsilon - \tilde{z}_s^\varepsilon) = \beta^{-1}(\tilde{y}_s^\varepsilon - \tilde{y}_s - \beta\tilde{y}_s^1). \tag{2.2.8}$$

Clearly, $\Delta_t^1 = \tilde{\Delta}_{t/\varepsilon}^1$ and the relation (2.2.6) is equivalent to the uniform convergence in probability as $\varepsilon \to 0$ of the process $\tilde{\Delta}^1$ to zero on the increasing time intervals $[0, T/\varepsilon]$. In other words, we have to check that

$$P\text{-}\lim_{\varepsilon\to 0}||\tilde{\Delta}^1||_{T/\varepsilon} = 0. \tag{2.2.9}$$

By the finite increments formula we have

$$F(\tilde{y}_s^\varepsilon) = F(\tilde{z}_s^\varepsilon + \beta\tilde{\Delta}_s^1) = F(\tilde{z}_s^\varepsilon) + \beta A_s^\varepsilon\tilde{\Delta}_s^1 \tag{2.2.10}$$

where A_s^ε is the matrix whose ith row has the form

$$\frac{\partial F}{\partial y_i}(\tilde{z}_s^\varepsilon + \beta\theta_s^i\tilde{\Delta}_s^1), \quad \theta_s^i \in [0, 1].$$

Put $A_s := F'(\tilde{y}_s)$. It follows from (2.2.8), (2.2.2), (2.2.4), (2.2.7), and (2.2.10) that $\tilde{\Delta}_s^\varepsilon$ can be represented as the solution of the following linear SDE with zero initial condition:

$$d\tilde{\Delta}_s^1 = A_s\tilde{\Delta}_s^1 ds + (A_s^\varepsilon - A_s)\tilde{\Delta}_s^1 ds + \beta^{-1}[F(\tilde{z}_s^\varepsilon) - F(\tilde{y}_s) - \beta F'(\tilde{y}_s)]ds$$
$$+ [G(\tilde{y}_s^\varepsilon) - G(\tilde{y}_s)]dw_s. \tag{2.2.11}$$

Recall that by assumption $\beta_\varepsilon = \alpha_\varepsilon/\sqrt{|\ln\varepsilon|}$ where $\alpha_\varepsilon = o(1)$. We always may assume that α is strictly positive. Let us introduce the stopping time $\rho := \tau_\mu^\varepsilon \wedge \sigma^\varepsilon$, depending on ε and $\mu \in]0, 1]$, where

$$\tau_\mu^\varepsilon := \inf\{s : |\tilde{\Delta}_s^1| \geq \mu\} \wedge (T/\varepsilon),$$
$$\sigma^\varepsilon := \inf\{s : |\tilde{y}_s^1| \geq \sqrt{|\ln\varepsilon|}/\alpha_\varepsilon\} \wedge (T/\varepsilon).$$

Put $N := ||\tilde{y}||_\infty$. For $s \le \rho$ we have, obviously, that

$$|\tilde{z}_s^\varepsilon| \le |\tilde{y}_s| + \beta|\tilde{y}_s^1| \le N + 1, \quad |\tilde{y}_s^\varepsilon| \le |\tilde{z}_s^\varepsilon| + \beta|\tilde{\Delta}_s^1| \le N + 2.$$

Using the local Lipschitz condition for F' and G and the relations (2.2.8) we obtain easily that there exists a constant C such that for all $s \le \rho$

$$|(A_s^\varepsilon - A_s)\tilde{\Delta}_s^1| \le C(|\tilde{z}_s^\varepsilon - \tilde{y}_s| + \beta|\tilde{\Delta}_s^1|) \le C(\beta|\tilde{y}_s^1| + \beta),$$

$$|\beta^{-1}[F(\tilde{z}_s^\varepsilon) - F(\tilde{y}_s) - \beta F'(\tilde{y}_s)]| \le C\beta|\tilde{y}_s^1|,$$

$$|G(\tilde{y}_s^\varepsilon) - G(\tilde{y}_s)| \le C|\tilde{y}_s^\varepsilon - \tilde{y}_s| \le C(\beta|\tilde{y}_s^1| + \beta|\tilde{\Delta}_s^1|).$$

We infer from here that the corresponding terms of the equation (2.2.11) on $[0, \rho]$ allow as a majorant a function which is $o\left(1/\sqrt{|\ln \varepsilon|}\right)$ as $\varepsilon \to 0$. Since A_s converges as $s \to \infty$ to the constant matrix $A_\infty := F'(\tilde{y}_\infty)$ satisfying **H.2.2.4**, it follows that for A the hypothesis **H.1.2.1** on exponential decay of the fundamental matrix is fulfilled. Applying Theorem 1.2.5 we get that

$$P\text{-}\lim_{\varepsilon \to 0} ||\tilde{\Delta}^1||_\rho = 0. \tag{2.2.12}$$

It follows from (1.2.14) that

$$E||\tilde{y}^1||_{T/\varepsilon} \to 0, \qquad \varepsilon \to 0. \tag{2.2.13}$$

Obviously,

$$P(||\tilde{\Delta}^1||_{T/\varepsilon} \ge \mu) \le P(||\tilde{\Delta}^1||_\rho \ge \mu) + P(||\tilde{y}^1||_{T/\varepsilon} \ge \sqrt{|\ln \varepsilon|/\alpha_\varepsilon})$$

and (2.2.9) follows from (2.2.12) and (2.2.13). $\square$

2.2.3 The First-Order Approximation of the Rest Point

The stochastic Tikhonov theorem, being applied to the fast equation (2.2.1) with time-invariant coefficients, asserts that x_t^ε converges in probability to $\tilde{y}_\infty$ uniformly on any interval $[S, T]$ where $S \in]0, T]$. Now we are able to say more about this convergence assuming that the rate of convergence of β to zero is not very fast. We shall use the hypothesis

H.2.2.5 The function $\beta_\varepsilon > 0$ is such that $\beta_\varepsilon = o\left(1/\sqrt{|\ln \varepsilon|}\right)$ and $\beta_\varepsilon^{-p} e^{-1/\varepsilon} \to 0$ for every positive p as $\varepsilon \to 0$.

Theorem 2.2.2 *Suppose that the conditions* **H.2.2.1** – **H.2.2.5** *holds and F' is locally Lipschitz. Let $\tilde{y}$ be the solution of (2.2.2) and let $\hat{y}^1$ be the solution of the following linear SDE with constant coefficients*

$$d\hat{y}_s^1 = A_\infty \hat{y}_s^1 ds + G_\infty d\tilde{w}_s, \quad \tilde{y}_0^1 = 0, \tag{2.2.14}$$

where the Wiener process $\tilde{w}_s := \varepsilon^{-1/2} w_{s\varepsilon}$, $A_\infty := F'(\tilde{y}_\infty)$, $G_\infty := G(\tilde{y}_\infty)$.

Then

$$y_t^\varepsilon = \tilde{y}_\infty + \beta_\varepsilon \widehat{y}_{t/\varepsilon}^1 + \beta_\varepsilon \widehat{\Delta}_t^\varepsilon \qquad (2.2.15)$$

where

$$P\text{-}\lim_{\varepsilon \to 0} \|\widehat{\Delta}^\varepsilon\|_{S,T} = 0, \qquad (2.2.16)$$

S is any point in $]0, T]$.

Proof. It is well-known that **H.2.2.4** implies the exponential rate of convergence to the rest point in (2.2.3). A more formal statement that we need is the following:

Lemma 2.2.3 *Let* $\tilde{R}_s := \tilde{y}_s - \tilde{y}_\infty$. *Then for an arbitrary point* $S \in]0, T]$ *there are positive constants* c_1 *and* c_2 *such that for any* $\varepsilon > 0$

$$\|\tilde{R}\|_{S/\varepsilon,\infty} \leq c_1 e^{-c_2/\varepsilon}. \qquad (2.2.17)$$

Proof. It follows from **H.2.2.4** and the finite increments formula that

$$d\tilde{R}_s = \tilde{A}_s \tilde{R}_s ds$$

where $\tilde{A}_s$ is a matrix with the ith row of the form

$$\frac{\partial F}{\partial y_i}(\tilde{y}_\infty + \vartheta_s^i \tilde{R}_s), \quad \vartheta_s^i \in [0, 1].$$

Since $\tilde{A}_s$ tends to a constant stable matrix A_∞, it satisfies the hypothesis **H.2.1.2** (see discussion at the beginning of Section 1.2) and for all $s \geq s_0 > 0$ we have

$$|\tilde{R}_s| \leq C |\tilde{R}_{s_0}| e^{-\kappa(s-s_0)}.$$

Thus, if $s_0 \in [0, S/\varepsilon]$ we have, for all $s \geq S/\varepsilon$, that

$$|\tilde{R}_s| \leq C |\tilde{R}_{s_0}| e^{\kappa s_0} e^{-\kappa S/\varepsilon},$$

i.e., (2.2.17) holds for sufficiently small ε. Since $\tilde{R}$ is bounded, we can choose such a constant c_1 that the inequality (2.2.17) will hold for all $\varepsilon \in]0, 1]$. $\square$

We need also an exponential bound for the deviation of $\widehat{y}_s^1$ from $\tilde{y}_s^1$.

Lemma 2.2.4 *Put* $\tilde{R}_s^1 := \widehat{y}_s^1 - \tilde{y}_s^1$. *Then for any* $p \geq 1$ *there exist positive constants* γ_1 *and* γ_2 *such that*

$$E\|\tilde{R}^1\|_{S/\varepsilon,T/\varepsilon}^p \leq \gamma_1 e^{-\gamma_2/\varepsilon}. \qquad (2.2.18)$$

Proof. It follows from (2.2.4) and (2.2.15) that

$$d\tilde{R}_s^1 = A_\infty \tilde{R}_s^1 ds + V_s \tilde{y}_s^1 ds + g_s d\tilde{w}_s, \quad \tilde{R}_0^1 = 0, \qquad (2.2.19)$$

where

$$V_s := F'(\tilde{y}_s) - F'(\tilde{y}_\infty),$$
$$g_s := G(\tilde{y}_s) - G(\tilde{y}_\infty).$$

Since the functions F' and G are locally Lipschitz, $||\tilde{y}||_\infty < \infty$, there exists a constant C such that

$$|V_s| + |g_s| \leq C|\tilde{R}_s|. \qquad (2.2.20)$$

Fix a number $t_0 \in \,]0, S[$. It follows from (2.2.19) that for $s \geq t_0/\varepsilon$ we have the representation $\tilde{R}^1_s = X^1_s + X^2_s + X^3_s$ where

$$X^1_s := e^{(s-t_0/\varepsilon)A_\infty}\tilde{R}^1_{t_0/\varepsilon}, \qquad (2.2.21)$$

$$X^2_s := \int_{t_0/\varepsilon}^s e^{(s-u)A_\infty}V_u\tilde{y}^1_u du, \qquad (2.2.22)$$

$$X^3_s := \int_{t_0/\varepsilon}^s e^{(s-u)A_\infty}g_u d\tilde{w}_u. \qquad (2.2.23)$$

Since the real parts of eigenvalues of A_∞ are strictly less than $-\gamma$, by virtue of the condition **H.2.2.4** we have for any $s \geq u$ that

$$|e^{(s-u)A_\infty}| \leq Ce^{-\gamma(s-u)}$$

for some constant C. Therefore, we have the bound

$$E||X^1||^p_{S/\varepsilon,T/\varepsilon} \leq Ce^{-p(S-t_0)/\varepsilon}|\ln \varepsilon|^{p/2}. \qquad (2.2.24)$$

It follows from Proposition 1.2.4 that

$$E||\tilde{y}^1||^p_{T/\varepsilon} \leq C|\ln \varepsilon|^{p/2}.$$

By virtue of (2.2.20) and Lemma 2.2.3

$$||g||_{t_0/\varepsilon,T/\varepsilon} \leq Ce^{-c/\varepsilon} \qquad (2.2.25)$$

and we have from (2.2.22) that

$$E||X^2||^p_{S/\varepsilon,T/\varepsilon} \leq Ce^{-c/\varepsilon}|\ln \varepsilon|^{p/2} \qquad (2.2.26)$$

(the constants here, certainly, depend on p). To estimate the moment of the norm of the process X^3 satisfying the linear SDE

$$dX^3_s := A_\infty X^3_s ds + g_s d\tilde{w}_s, \quad X^3_{t_0/\varepsilon} = 0,$$

we use again Theorem 1.2.4 (with M_G given by the right-hand side of (2.2.25)) which yields the inequality

$$E||X^3||^p_{S/\varepsilon,T/\varepsilon} \leq Ce^{-cp/\varepsilon}|\ln \varepsilon|^{p/2}. \qquad (2.2.27)$$

The bounds (2.2.24), (2.2.26), and (2.2.27) imply that the inequality (2.2.18) holds for all $p \geq 1$. □

To accomplish the proof of the theorem it is sufficient to write that

$$\widehat{\Delta}_t^\varepsilon = \beta^{-1}(y_t^\varepsilon - \tilde{y}_\infty - \beta\widehat{y}_{t/\varepsilon}^1) = \tilde{\Delta}_t^\varepsilon + \beta^{-1}\tilde{R}_{t/\varepsilon} - \tilde{R}_{t/\varepsilon}^1$$

and apply Theorem 2.2.1 and the exponential bounds given by the two previous lemmas. □

2.2.4 Normal Approximation Result

As an easy corollary of Theorem 2.2.2 we have

Theorem 2.2.5 *Suppose that the conditions* **H.2.2.1**–**H.2.2.5** *holds and F' is locally Lipschitz. Then for any m points t_i such that*

$$0 < t_1 < \ldots < t_m < T,$$

the distribution of the random vector $\beta_\varepsilon^{-1}(y_{t_1}^\varepsilon - \tilde{y}_\infty, \ldots, y_{t_m}^\varepsilon - \tilde{y}_\infty)$ converges weakly to the distribution of the random vector $(\xi_1, \ldots, \xi_m)$ whose components are independent identically distributed n-dimensional Gaussian random variables with zero mean and covariance matrix

$$\Gamma := E\xi_1\xi_1^\star = \int_0^\infty e^{sA_\infty}G_\infty G_\infty^\star e^{sA_\infty^\star}ds.$$

Proof. Indeed, Theorem 2.2.2 reduces the problem to the study of the weak convergence of the Gaussian vectors

$$\beta_\varepsilon^{-1}(\tilde{y}_{t_1/\varepsilon}^1, \ldots, \tilde{y}_{t_{m/\varepsilon}}^1).$$

It follows from the representation

$$\tilde{y}_s^1 := \int_0^s e^{(s-u)A_\infty}G_\infty d\tilde{w}_u$$

that

$$\begin{aligned}
\Gamma_s := E\tilde{y}_s^1\tilde{y}_s^{1\star} &= \int_0^s e^{(s-u)A_\infty}G_\infty G_\infty^\star e^{(s-u)A_\infty^\star}du \\
&= \int_{-s}^0 e^{-uA_\infty}G_\infty G_\infty^\star e^{-uA_\infty^\star}du \\
&\to \int_\infty^0 e^{-uA_\infty}G_\infty G_\infty^\star e^{-uA_\infty^\star}du = \Gamma.
\end{aligned}$$

The covariance function of $\tilde{y}^1$ has the form

$$K(s,t) = E\tilde{y}_s^1\tilde{y}_t^{1\star} = \Gamma e^{(t-s)A_\infty}, \quad 0 \leq s \leq t.$$

Obviously, $K(s,t) \to 0$ when $t - s \to \infty$ implying the assertion on independence of the random variables ξ_i. □

2.3 Higher-Order Expansions

We continue to study the model with only fast variables and time-invariant coefficients aiming to construct asymptotic expansions for the solution of the SDE (2.2.1).

Our hypothesis on the asymptotics of β will be

H.2.3.1 The function $\beta_\varepsilon = o(\varepsilon^\delta)$ as $\varepsilon \to 0$ where $\delta > 0$.

At first, we extend Theorem 2.2.1 and find an expansion in the power series in β

$$y_t^\varepsilon = \tilde{y}_s + \beta \tilde{y}_s^1 + \beta^2 \tilde{y}_s^2 + \ldots + \beta^k \tilde{y}_s^k + \ldots \tag{2.3.1}$$

where $s := t/\varepsilon$ and $\tilde{y}_s$ is the solution of (2.2.2). We hope that the use of superscripts to enumerate coefficients (which are vector-valued processes) does not lead to ambiguity.

2.3.1 Formal Expansions

In order to explain the structure of the terms of the expansion we present some simple manipulations concerning a composition of power series. Let

$$Z(\beta) := \sum_{q \geq 0} c_q \beta^q. \tag{2.3.2}$$

Assuming that F and G are infinitely differentiable, we can expand $F(Z(\beta))$ and $G(Z(\beta))$ in power series of β.

The higher-order derivatives of the function $f(\beta) := F(Z(\beta))$ are given by

$$D^q f(\beta) = \sum_{j=1}^{q} \sum A_{i_1,\ldots,i_j}^q F^{(j)}(Z(\beta))(D^{i_1} Z(\beta), \ldots, D^{i_j} Z(\beta))$$

where the interior sum is taken over all integers $i_1, \ldots, i_j$, such that

$$i_1 + i_2 + \ldots + i_j = q.$$

The derivative $F^{(j)}(Z(\beta))$ is a j-linear form on $\mathbf{R}^n$ with values in $\mathbf{R}^n$. It follows that

$$F(Z(\beta)) = \sum_{q \geq 0} F_q(c_0, \ldots, c_q)\beta^q \tag{2.3.3}$$

with $F_0(c_0) = F(c_0)$, $F_1(c_0, c_1) = F'(c_0)c_1$,

$$F_2(c_0, c_1) = F'(c_0)c_1 + \frac{1}{2}F''(c_0)(c_1, c_1),$$

etc. In general, for $q \geq 2$ we have

$$F_q(c_0, \ldots, c_q) = F'(c_0)c_q + f_q(c_0, \ldots, c_{q-1}), \qquad (2.3.4)$$

where

$$f_q(c_0, \ldots, c_{q-1}) := \sum_{j=2}^{q} \sum C_{i_1, \ldots, i_j}^q F^{(j)}(c_0)(c_{i_1}, \ldots, c_{i_j}) \qquad (2.3.5)$$

(in Chapter 4 we shall use for f_q the notation $R^{F,q}$).

It is important to note that F_q and f_q are polynomial functions of $c_1, \ldots, c_q$ of order not greater than q with coefficients depending on c_0 (which are partial derivatives of order not greater than q calculated at c_0). This observation will be used intensively.

Other useful properties: for any integer i we have

$$F_{2i}(c_0, 0, c_1, 0, \ldots, c_{2i-2}) = F_i(c_0, c_1, \ldots, c_{2i})$$

$$F_{2i-1}(c_0, 0, c_1, 0, \ldots, c_{2i-2}, 0) = f_{2i-1}(c_0, 0, c_1, 0, \ldots, c_{2i-2}) = 0.$$

Indeed, if all odd coefficients in (2.3.2) are zero, then the expansion (2.3.2) contains only even powers of β.

Though explicit expressions do not matter in further development, we recall that in the one-dimensional case there is the well-known De Bruno formula for F_q:

$$F_q(c_0, \ldots, c_q) = \sum_{j=1}^{q} \sum \frac{1}{i_1! \ldots i_q!} F^{(j)}(c_0) c_1^{i_1} \ldots c_q^{i_q},$$

where the interior sum is taken over all nonnegative integers $i_1, \ldots, i_q$ such that

$$i_1 + i_2 + \ldots + i_q = j,$$
$$i_1 + 2i_2 + \ldots + qi_q = q.$$

Analogously, we can write the asymptotic expansion for G:

$$G(Z(\beta)) = \sum_{q \geq 0} G_q(c_0, \ldots, c_q)\beta^q. \qquad (2.3.6)$$

Let us substitute the expansion for $\tilde{y}_r^\varepsilon$ into the equation

$$d\tilde{y}_s^\varepsilon = F(\tilde{y}_s^\varepsilon)ds + \beta G(\tilde{y}_s^\varepsilon)d\tilde{w}_s, \quad \tilde{y}_0^\varepsilon = y^o,$$

and make use of (2.3.3) and (2.3.6). We have:

$$\sum_{q \geq 0} \beta^q d\tilde{y}_s^q = \sum_{q \geq 0} \beta^q F_q(\tilde{y}_s^0, \ldots, \tilde{y}_s^q)ds + \sum_{q \geq 0} \beta^{q+1} G_q(\tilde{y}_s^0, \ldots, \tilde{y}_s^q)d\tilde{w}_s.$$

The coefficients at the same powers of β in this formal expansion should be equal. This implies that for $q = 0$

$$d\tilde{y}_s^0 = F(\tilde{y}_s^0)ds, \quad \tilde{y}_0^0 = y^o,$$

i.e., $\tilde{y}^0 = \tilde{y}$.

For $q \geq 1$ we get the SDE

$$d\tilde{y}_s^q = F_q(\tilde{y}_s, \tilde{y}_s^1, \ldots, \tilde{y}_s^q)ds + G_{q-1}(\tilde{y}_s, \tilde{y}_s^1, \ldots, \tilde{y}_s^{q-1})d\tilde{w}_s \qquad (2.3.7)$$

with the initial condition $\tilde{y}_0^q = 0$.

In particular, for $q = 1$,

$$d\tilde{y}_s^1 = F_1(\tilde{y}_s, \tilde{y}_s^1)ds + G(\tilde{y}_s)d\tilde{w}_s, \quad \tilde{y}_0^1 = 0,$$

and this SDE coincides with (2.2.4) because $F_1(\tilde{y}_s, \tilde{y}_s^1) = F'(\tilde{y}_s)\tilde{y}_s^1$.

The structure of the functions F_q and G_q (see (2.3.4)) enables us to solve the system (2.3.7) in a recurrent way. When the solutions of the first q equations are found, the $(q+1)$th equation of the system happens to be the linear SDE

$$d\tilde{y}_s^q = (F'(\tilde{y}_s)\tilde{y}_s^q + \tilde{f}_s)ds + \tilde{G}_{q-1}d\tilde{w}_s, \quad \tilde{y}_0^q = 0, \qquad (2.3.8)$$

with respect to $\tilde{y}_s^q$, where due to our assumptions $F'(\tilde{y}_s)$ tends to the matrix A_∞ which has all its eigenvalues in the left half-plane. The coefficients

$$\tilde{f}_q := f_q(\tilde{y}_s, \ldots, \tilde{y}_s^{q-1}), \qquad \tilde{G}_{q-1} := G_{q-1}(\tilde{y}_s, \ldots, \tilde{y}_s^{q-1}) \qquad (2.3.9)$$

are polynomials in $\tilde{y}_s^1, \ldots, \tilde{y}_s^{q-1}$, whose degrees are q and $q-1$. The coefficients of these polynomials depend on $\tilde{y}_s$.

The above structure of the system allows us to extend easily the approach of the previous section to obtain results for higher-order approximations.

2.3.2 Convergence of the Remainder

Now we formulate a generalization of Theorem 2.2.1.

Theorem 2.3.1 *Suppose* **H.2.2.1**–**H.2.2.4** *and* **H.2.3.1** *be fulfilled. Assume that F and G have k and $k-1$ continuous derivatives, respectively, and $F^{(k)}$ and $G^{(k-1)}$ are locally Lipschitz. Then*

$$y_t^\varepsilon = \tilde{y}_s + \beta\tilde{y}_s^1 + \ldots + \beta^k\tilde{y}_s^k + \beta^k\Delta_t^k, \qquad (2.3.10)$$

where $s = t/\varepsilon$, $\tilde{y}$ is the solution of (2.2.2), $\tilde{y}^q$, $1 \leq q \leq r$, are the solutions of equations (2.3.7) with coefficients given by (2.3.3), (2.3.6), and where the remainder term satisfies

$$P\text{-}\lim_{\varepsilon \to \infty} \|\Delta^k\|_T = 0. \qquad (2.3.11)$$

The reasoning follows the same line as in the proof of Theorem 2.2.1. However, we need some auxiliary results.

Lemma 2.3.2 *Under the assumptions of Theorem 2.3.1 for every $m \in \mathbf{N}$ and $1 \leq q \leq k$ we have*

$$\sup_{s \geq 0} E|\tilde{y}_s^q|^{2m} < \infty. \tag{2.3.12}$$

Proof. We use the induction in q. For $q = 1$ the process $\tilde{y}^1$ is the solution of an equation of the same type as (1.2.22) with $h = 0$ and the deterministic function G. Thus, (2.3.12) holds by virtue of Lemma 1.2.6. Suppose that (2.3.12) holds for all $q \leq M$. Since in the general case the process $\tilde{y}^{M+1}$ also satisfies an equation of the same type as (1.2.22) with the polynomial coefficients h and G of the variables $\tilde{y}^1, \ldots, \tilde{y}^M$, by the induction hypothesis all assumptions of Lemma 1.2.6 are fulfilled and (2.3.12) holds for $q = M + 1$. □

Lemma 2.3.3 *Under the assumptions of Theorem 2.3.1 for every $m \in \mathbf{N}$ and $1 \leq q \leq k$ there are constants $C_0(q, m)$ and $C_1(q, m)$ such that*

$$E||\tilde{y}^q||_T^{2m} \leq C_0(q, m) + C_1(q, m)T. \tag{2.3.13}$$

Proof. The result follows from Lemma 1.2.6 since $\tilde{y}^q$ is the solution of the equation (2.3.8) with coefficients $\tilde{f}_q$ and $\tilde{G}_{q-1}$ satisfying, by virtue of the previous lemma, the condition

$$\sup_{t \geq 0} E|\tilde{f}_q(t)|^{2m} < \infty, \quad \sup_{t \geq 0} E|\tilde{G}_{q-1}(t)|^{2m} < \infty$$

for all $m \in \mathbf{N}$. □

Lemma 2.3.4 *Under the assumptions of Theorem 2.3.1 for any positive α and $1 \leq q \leq k$*

$$\lim_{\varepsilon \to 0} P(||\tilde{y}^q||_{T/\varepsilon} \geq \varepsilon^{-\alpha}) = 0. \tag{2.3.14}$$

Proof. Let m be an integer greater than $1/(2\alpha)$. By the Chebyshev inequality and (2.3.13) we have

$$P(||\tilde{y}^q||_{T/\varepsilon} \geq \varepsilon^{-\alpha}) \leq \varepsilon^{2m\alpha}T||\tilde{y}^q||_{T/\varepsilon}^{2m}$$
$$\leq (\varepsilon^{2m\alpha}C_0(q, m) + \varepsilon^{2m\alpha-1}C_1(q, m)T) \to 0,$$

as $\varepsilon \to 0$. □

Proof of Theorem 2.3.1. Put

$$\tilde{z}_s := \tilde{y}_s + \beta \tilde{y}_s^1 + \ldots + \beta^k \tilde{y}_s^k, \tag{2.3.15}$$
$$\tilde{\Delta}_s^k := \beta^{-k}(\tilde{y}_s^\varepsilon - \tilde{z}_s). \tag{2.3.16}$$

We have to show that

$$P\text{-}\lim_{\varepsilon \to 0} ||\tilde{\Delta}^k||_{T/\varepsilon} = 0. \tag{2.3.17}$$

After regrouping terms, we get from the finite increments formula that

$$
\begin{aligned}
d\tilde{y}_s^\varepsilon &= (F(\tilde{y}_s^\varepsilon) - F(\tilde{z}_s))ds + \beta(G(\tilde{y}_s^\varepsilon) - G(\tilde{z}_s))d\tilde{w}_s + F(\tilde{z}_s)ds + \beta G(\tilde{z}_s)d\tilde{w}_s \\
&= \beta^k A_s^\varepsilon \tilde{\Delta}_s^{(k)} ds + \beta^{k+1} B_s^\varepsilon \tilde{\Delta}_s^k d\tilde{w}_s + F(\tilde{z}_s)ds + \beta G(\tilde{z}_s)d\tilde{w}_s, \tag{2.3.18}
\end{aligned}
$$

where $\tilde{y}_0^\varepsilon = y^o$ and A_s^ε, B_s^ε are matrices with the ith rows, respectively,

$$\frac{\partial F}{\partial y_i}(\tilde{z}_s + \vartheta_s^i(\tilde{y} - \tilde{z}_s)), \quad \frac{\partial G}{\partial y_i}(\tilde{z}_s + \eta_i^s(\tilde{y} - \tilde{z}_s)),$$

$\vartheta_s^i,\ \eta_s^i \in [0,1]$. From (2.3.15), (2.3.18), and (2.3.7) it follows that

$$
\begin{aligned}
d\tilde{\Delta}_s^k &= A_s \tilde{\Delta}_s^k ds + (A_s^\varepsilon - A_s)\tilde{\Delta}_s^k ds + \beta B_s^\varepsilon \tilde{\Delta}_s^k d\tilde{w}_s \\
&\quad + \phi_s ds + \psi_s d\tilde{w}_s, \quad \tilde{\Delta}_0^k = 0, \tag{2.3.19}
\end{aligned}
$$

where

$$\phi_s := \beta^{-k}\left(F(\tilde{z}_s) - \sum_{q=0}^k F_q(\tilde{y}_s, \ldots, \tilde{y}_s^q)\beta^q\right), \tag{2.3.20}$$

$$\psi_s := \beta^{-k+1}\left(G(\tilde{z}_s) - \sum_{q=0}^{k-1} G_q(\tilde{y}_s, \ldots, \tilde{y}_s^q)\beta^q\right). \tag{2.3.21}$$

Fix $\alpha \in \,]0, 1/(k+2)[$ and $\mu \in]0,1]$ and consider the stopping time $\rho := \tau^\varepsilon \wedge \sigma^\varepsilon$ where

$$
\begin{aligned}
\tau^\varepsilon &:= \inf\{s:\ |\tilde{\Delta}_s^k| \geq \mu\} \wedge (T/\varepsilon), \\
\sigma^\varepsilon &:= \inf\{s:\ |\tilde{y}_s^1| + \ldots + |\tilde{y}_s^k| \geq \beta^{-\alpha}\} \wedge (T/\varepsilon).
\end{aligned}
$$

Put $N := ||\tilde{y}||_\infty$. For $s \leq \rho$ we have:

$$
\begin{aligned}
|\tilde{z}_s| &\leq |\tilde{y}_s| + \beta(|\tilde{y}_s^1| + \ldots + |\tilde{y}_s^k|) \leq N + 1, \\
|\tilde{y}_s^\varepsilon| &\leq |\tilde{z}_s| + \beta|\tilde{\Delta}_s^k| \leq N + 2.
\end{aligned}
$$

The local Lipschitz condition for F' implies that there exists a constant K such that for all $r \leq \rho$ we have

$$
\begin{aligned}
|(A_s^\varepsilon - A_s)\tilde{\Delta}_s^k| &\leq K(|\tilde{z}_s - \tilde{y}_s| + \beta|\tilde{\Delta}_s^k|) \\
&\leq K[\beta(|\tilde{y}_s^1| + \ldots + |\tilde{y}_s^k|) + \beta] \leq 2K\beta^{1-\alpha}.
\end{aligned}
$$

By virtue of Lemma 2.3.5 (given below after the proof of the theorem) there exists a constant C_N such that for all $r \leq \rho$

$$
\begin{aligned}
|\phi_r| &\leq C_N\beta^{1-(k+2)\alpha}, \\
|\psi_r| &\leq C_N\beta^{1-(k+1)\alpha}.
\end{aligned}
$$

Applying Theorem 1.2.4 to the equation (2.3.19) we get that

$$\lim_{\varepsilon \to 0} E||\tilde{\Delta}^{(k)}||_\rho = 0. \qquad (2.3.22)$$

Obviously,

$$P(||\tilde{\Delta}^k||_{T/\varepsilon} \geq \mu) \leq P(||\tilde{\Delta}^k||_\rho \geq \mu) + P(\sigma^\varepsilon < T/\varepsilon). \qquad (2.3.23)$$

But

$$P(\sigma^\varepsilon < T/\varepsilon) \leq \sum_{q=1}^{k} P(||\tilde{y}^q||_{T/\varepsilon} \geq \beta^{-\alpha}/k) \to 0, \quad \varepsilon \to 0, \qquad (2.3.24)$$

by virtue of Lemma 2.3.4.

The desired assertion (2.3.17) follows from (2.3.22)–(2.3.24). □

Lemma 2.3.5 *Let $F : \mathbf{R}^n \to \mathbf{R}^n$ be a function such that $F^{(k)}$ is locally Lipschitz and let $z : [0, 1] \to \mathbf{R}^n$ be a function such that*

$$z(\beta) := \sum_{q=0}^{k} c_q \beta^q$$

where $|c_0| \leq N$ and $|c_1| + \ldots + |c_k| \leq \beta M$ for some $M > 1$. Define

$$f(\beta) := F(z(\beta)),$$

$$f_k(\beta) := \sum_{q=0}^{k} \frac{1}{q!} f^{(q)}(0) \beta^q$$

Then there exists a constant C such that

$$\beta^{-k}|f(\beta) - f_k(\beta)| \leq C\beta M^{k+2}.$$

Proof. Without loss of generality we consider the case $n = 1$. Applying the Taylor formula of order $k-1$ to $f(\beta)$, one can see that the problem is reduced to a suitable bound for $|f^{(k)}(\theta\beta) - f^{(k)}(0)|$. The formula for the derivative of a composite function shows that we need to estimate expressions of the form

$$F^{(j)}(z(\theta\beta))(D^{i_1}z(\theta\beta), \ldots, D^{i_j}z(\theta\beta)) - F^{(j)}(z(0))(D^{i_1}z(0), \ldots, D^{i_j}z(0))$$

where $i_1 + \ldots + i_j = k$, $j \geq k$, and D^i denotes the derivative of order i. Notice that

$$D^i z(\theta\beta) = D^i z(0) + R_i(\beta) = i!c_1 + R_i(\beta)$$

where $|R_i(\beta)| \leq C\beta M$. Now, taking into account the Lipschitz condition for $F^{(j)}$, we get that the absolute value of the difference of derivatives does not exceed

$$C(|c_0| + \ldots + |c_k|)^{j+1}\beta M \leq C\beta M^{j+2}.$$

This yields the desired bound. □

2.3.3 Expansion Around the Rest Point

The following result generalizes Theorem 2.2.2.

Theorem 2.3.6 *Let* **H.2.2.1**–**H.2.2.4** *and* **H.2.3.1** *be fulfilled. Suppose that the functions F and G have k and $k-1$ continuous derivatives, respectively and, moreover, that $F^{(k)}$ and $G^{(k-1)}$ are locally Lipschitz.*
Then

$$y_s^\varepsilon = \tilde{y}_\infty + \beta \tilde{y}_s^1 + \ldots + \beta^k \tilde{y}_s^k + \beta^k \widehat{\Delta}_t^k, \qquad (2.3.25)$$

where $s = t/\varepsilon$ and the processes $\tilde{y}^q$, $q \le r$, are the solutions of the following SDEs with zero initial conditions:

$$d\tilde{y}_s^1 = A_\infty \tilde{y}_s^1 ds + G_\infty d\tilde{w}_r, \qquad (2.3.26)$$
$$d\tilde{y}_s^q = A_\infty \tilde{y}_s^q ds + f_q(\tilde{y}_\infty, \tilde{y}_s^1, \ldots, \tilde{y}_s^{q-1}) ds$$
$$+ G_{q-1}(\tilde{y}_\infty, \tilde{y}_s^1, \ldots, \tilde{y}_s^{q-1}) d\tilde{w}_s, \qquad (2.3.27)$$

$A_\infty := F'(\tilde{y}_\infty)$, $G_\infty := G(\tilde{y}_\infty)$, G_{q-1} *and* f_q *are given by (2.2.3)–(2.2.6), and for the remainder term we have*

$$P\text{-}\lim_{\varepsilon \to 0} ||\widehat{\Delta}^k||_{S,T} = 0 \qquad (2.3.28)$$

for any $S \in \,]0, T]$.

This theorem is an obvious corollary of Theorem 2.3.1 and the exponential bound for the difference $\tilde{y}^q - \widehat{y}^q$ given by the following result:

Lemma 2.3.7 *Let $\tilde{R}_s^q := \tilde{y}_s^q - \widehat{y}_s^q$. Then for any $S \in \,]0, T]$ and $m \in \mathbf{N}$ there are constants γ_1, γ_2 such that for all $q \le k$ and $\varepsilon \in \,]0, 1]$*

$$E||\tilde{R}^q||_{S/\varepsilon, T/\varepsilon}^m \le \gamma_1 e^{-\gamma_2/\varepsilon}. \qquad (2.3.29)$$

Proof. It follows from (2.3.8), (2.3.9), and (2.3.27) that

$$d\tilde{R}_s^q = A_\infty \tilde{R}_s^q ds + (V_s \tilde{y}_s^q + h_s^q) ds + g_s^q d\tilde{w}_s, \quad \tilde{R}_0^q = 0, \qquad (2.3.30)$$

where

$$V_s := F'(\tilde{y}_s) - F'(\tilde{y}_\infty),$$
$$h_s^q := f_q(\tilde{y}_s, \ldots, \tilde{y}_s^{q-1}) - f_q(\tilde{y}_\infty, \widehat{y}_s^1, \ldots, \widehat{y}_s^{q-1}),$$
$$g_s^q := G_{q-1}(\tilde{y}_s, \ldots, \tilde{y}_s^{q-1}) - G_{q-1}(\tilde{y}_\infty, \widehat{y}_s^1, \ldots, \widehat{y}_s^{q-1}).$$

The existence of the constants γ_1, γ_2 for $q = 1$ has been proved in Lemma 2.2.4. Now we proceed by induction. Assume that there are constants such that (2.3.29) holds for all $q \le M - 1 < k$ and $\varepsilon \in \,]0, 1]$. Let us show that in this case a similar inequality holds also for $q = M$.

Fix $t_0 \in \,]0, S]$. For $r \ge S/\varepsilon$ we have the representation

$$\tilde{R}_s^M = X_s^1 + X_s^2 + X_s^3$$

where

$$X_s^1 := e^{(s-t_0)A_\infty} \tilde{R}_{t_0}^M,$$

$$X_s^2 := \int_{t_0/\varepsilon}^s e^{(s-u)A_\infty} (V_u \tilde{y}_u^M + h_u^M) du,$$

$$X_s^3 := \int_{t_0/\varepsilon}^s e^{(s-u)A_\infty} g_u^M d\tilde{w}_u.$$

As in the proof of Lemma 2.2.4 we have the bound

$$E||X^1||_{S/\varepsilon,T/\varepsilon}^{4m} \le Ce^{-4m(S-t_0)/\varepsilon}. \tag{2.3.31}$$

The induction hypothesis and the structure of f_M and G_{M-1} imply that for any $p \ge 1$ we have

$$E||h^M||_{t_0/\varepsilon,T/\varepsilon}^p \le \gamma_1 e^{-\gamma_2/\varepsilon}, \tag{2.3.32}$$

$$E||g^M||_{t_0/\varepsilon,T/\varepsilon}^p \le \gamma_1 e^{-\gamma_2/\varepsilon}. \tag{2.3.33}$$

The exponential bounds for V (see (2.2.18), (2.2.20)) and h^M, the inequality

$$|e^{(s-u)A_\infty}| \le Ce^{-\gamma(s-u)},$$

and Lemma 1.2.2 imply the exponential bound for $E||X^2||_{S/\varepsilon,T/\varepsilon}^{4m}$.

Since X^3 for $s \ge t_0/\varepsilon$ satisfies the linear SDE

$$dX_s^3 = A_\infty X_s^3 dr + g_s^M d\tilde{w}_s, \quad X_{t_0/\varepsilon}^3 = 0,$$

with the diffusion coefficient admitting the exponential bound (2.3.33), the inequality (1.2.31) of Lemma 1.2.6 yields the exponential bound for $E||X^3||_{S/\varepsilon,T/\varepsilon}^{4m}$.

The above reasoning gives the desired bound (2.3.29) (for m which is a multiple of 4 and, hence, for all $m \in \mathbf{N}$). $\square$

2.4 Stochastic Approximation: Proofs

2.4.1 Asymptotic Expansion for the Output Signal

We are ready now to prove the results announced in Section 0.4 on asymptotic expansions for the time average over $[r_1^\varepsilon, 1]$ of the output signal given either by (0.4.14) or (0.4.23). The crucial step is to get an asymptotic expansion for the output processes themselves, which are described by the SDEs of the same type as considered in the previous two sections of this chapter. However, there are some specific features: the coefficients are time-dependent and have singularities at zero; moreover, the initial conditions are given at "running" points, converging to zero as $\varepsilon \to 0$. Fortunately, these complications are not serious obstacles to our analysis and the proofs given below follow the same line of ideas. Due to the particular structure of $\widehat{y}_1^\varepsilon$, the boundary layer behavior of y^ε is not important here: the average is taken outside the boundary layer. Conventionally, one may think that the right extremity of the latter is somewhere within the interval $[r_0, r_1]$ with a suitably chosen $r_0 = r_0^\varepsilon$. First of all, we need the property

$$\lim_{\varepsilon \to 0} r_0 T(\varepsilon) = \infty \tag{2.4.1}$$

which ensures, according to Proposition 0.4.1, that

$$\lim_{\varepsilon \to 0} y_{r_0}^\varepsilon = \theta_* \quad \text{a.s.} \tag{2.4.2}$$

This is the only place where the hypothesis **H.0.4.1** is used. We shall continue by studying the asymptotic behavior on the interval $[r_0, 1]$ of the solutions of equations

$$\varepsilon dy_r^\varepsilon = F(y_r^\varepsilon)\gamma_r dr + \beta \varepsilon^{1/2}\gamma_r d\tilde{w}_r \tag{2.4.3}$$

with the initial conditions satisfying (2.4.2). We shall always assume that $F(\theta_*) = 0$ and $\operatorname{Re}\lambda(A) < -2\kappa$ where $A := F'(\theta_*)$ and the function γ_r, which may depend on ε, is positive. The parameter $\beta := \beta^\varepsilon$ must satisfy the hypothesis

H.2.4.1 There exists $\mu > 0$ such that

$$\lim_{\varepsilon \to 0} \beta \varepsilon^{-\mu} = 0.$$

Some further properties of the specification which will be used in our analysis are listed below.

H.2.4.2 There exists $\varepsilon_0 > 0$ and $c > 0$ such that

$$\inf_{r \in [r_0, 1]} \gamma_r^\varepsilon \geq c \qquad \forall \varepsilon \in]0, \varepsilon_0].$$

H.2.4.3 For all $\mu > 0$

$$\lim_{\varepsilon \to 0} \varepsilon^\mu \|\gamma^\varepsilon\|_{r_0, 1} = 0.$$

H.2.4.4 For all $\mu > 0$

$$\lim_{\varepsilon \to 0} \varepsilon^\mu / r_0 = 0.$$

Put

$$\Gamma(t, s) := \int_s^t \gamma_u \, du.$$

H.2.4.5 There exists a function $s_0 = s_0^\varepsilon$ such that $r_0 < s_0 < r_1$ and for all $\alpha > 0$

$$\lim_{\varepsilon \to 0} \beta^{-\alpha} (e^{-\Gamma(s_0, r_0)/\varepsilon} + e^{-\Gamma(r_1, s_0)/\varepsilon}) = 0$$

and

$$\lim_{\varepsilon \to 0} \varepsilon^{-1/2} (\Gamma(s_0, r_0) \wedge \Gamma(r_1, s_0)) = \infty.$$

H.2.4.6 The function $\gamma^\varepsilon \in C^1$ and

$$\lim_{\varepsilon \to 0} \varepsilon^\mu \|\dot\gamma^\varepsilon\|_{r_0, 1} = 0.$$

All these hypotheses are obvious for the first procedure where

$$\gamma_r = r^{-\rho}, \qquad T(\varepsilon) := \varepsilon^{-1/(1-\rho)}, \qquad \beta = \varepsilon^{(1/2)\rho/(1-\rho)},$$

$\rho \in \,]1/2, 1[$, if we choose, e.g.,

$$r_0^\varepsilon := (r_1^\varepsilon)^4 = 1/\ln^4 \varepsilon \tag{2.4.4}$$

and $s_0 := 2r_0$.

For the second model, where $r_1^\varepsilon := 1/\ln_2 T(\varepsilon)$ and the function $T(\varepsilon)$ is implicitly defined for sufficiently small ε by the equality

$$\varepsilon = \frac{1}{e} \frac{\ln_3 T}{\ln T},$$

we may put

$$r_0 := r_0(T(\varepsilon)) \tag{2.4.5}$$

where

$$r_0(T) := (r_1(T))^{\ln_3 T} = (1/\ln_2 T)^{\ln_3 T}. \tag{2.4.6}$$

Notice that for any $m \geq 1$

$$\lim_{T \to \infty} (r_0(T))^m \ln T = \infty. \tag{2.4.7}$$

This implies, in particular, (2.4.1). There is no difficulty to check the hypotheses **H.2.4.1**, **H.2.4.4**, and **H.2.4.6**. To meet **H.2.4.5** we take $s_0 = \sqrt{r_0}$. It is easy to see that

$$\lim_\varepsilon \|\gamma_r - 1/r\|_{r_0, 1} = 0$$

and we have for sufficiently small ε that

$$\Gamma(s_0, r_0) \geq \frac{1}{2} \int_{r_0}^{s_0} r^{-1} dr = -\frac{1}{4} \ln r_0 \to \infty$$

and

$$\Gamma(r_1, s_0) \geq \frac{1}{2} \ln \frac{r_1}{s_0} = \frac{1}{2}[(1/2)\ln_3 T(\varepsilon) - 1] \ln_3 T(\varepsilon) \to \infty$$

as $\varepsilon \to 0$. At last,

$$\beta^{-\alpha} e^{-\Gamma(s_0, r_0)/\varepsilon} \leq \beta^{-\alpha} e^{(1/4\varepsilon) \ln r_0}$$

$$= \left(\frac{T(\varepsilon) \ln_3 T(\varepsilon)}{e \ln T(\varepsilon)} \right)^{\alpha/2} \exp\left\{ -\frac{e}{4} (\ln T(\varepsilon)) \ln_3 T(\varepsilon) \right\} \to 0$$

and, similarly,

$$\beta^{-\alpha} e^{-\Gamma(r_1, s_0)/\varepsilon} \to 0$$

as $\varepsilon \to 0$.

Thus, the needed properties hold for the considered models and we may forget from now on about the stochastic approximation.

First, we obtain the asymptotic expansion of the form

$$y_r^\varepsilon = v_r^\varepsilon + \beta \tilde{y}_r^{1,\varepsilon} + \beta^2 \tilde{y}_r^{2,\varepsilon} + \beta^2 \tilde{\Delta}_r^\varepsilon, \tag{2.4.8}$$

where the function v^ε is the solution of the nonlinear ordinary differential equation

$$\varepsilon dv_r^\varepsilon = F(v_r^\varepsilon) \gamma_r dr, \quad v_{r_0}^\varepsilon = y_{r_0}^\varepsilon, \tag{2.4.9}$$

and the processes $y_r^{1,\varepsilon}$ and $y_r^{2,\varepsilon}$ are given by the linear equations

$$\varepsilon d\tilde{y}_r^{1,\varepsilon} = F'(v_r^\varepsilon) \tilde{y}_r^{1,\varepsilon} \gamma_r dr + \sqrt{\varepsilon} \gamma_r d\tilde{w}_r, \quad \tilde{y}_{r_0}^{1,\varepsilon} = 0, \tag{2.4.10}$$

$$\varepsilon d\tilde{y}_r^{2,\varepsilon} = F'(v_r^\varepsilon) \tilde{y}_r^{2,\varepsilon} \gamma_r dr + \tilde{R}_r^{2,\varepsilon} \gamma_r dr, \quad \tilde{y}_{r_0}^{2,\varepsilon} = 0, \tag{2.4.11}$$

$\tilde{R}_r^{2,\varepsilon} := f_2(v_r^\varepsilon, \tilde{y}_r^{1,\varepsilon})$; in accordance with (2.3.5) the ith component of $f_2(c_0, c_1)$ is

$$f_{2,i}(c_0, c_1) = \frac{1}{2} c_1^\star F_i''(c_0) c_1, \tag{2.4.12}$$

$F_i''(c_0)$ is the Hessian matrix of the ith component of F.

Clearly, (2.4.8) is just the definition of Δ^ε. To show that this process converges to zero uniformly in probability we need some auxiliary results.

Define the process

$$\pi_r^\varepsilon := e^{\kappa \Gamma(r, r_0)/\varepsilon} |v_r^\varepsilon - \theta_*|, \quad r \geq r_0. \tag{2.4.13}$$

Notice that for arbitrary $s, r \in]0, 1]$, $s \leq r$,

$$\varepsilon^{-1} \int_s^r e^{-\Gamma(r, u)/\varepsilon} \gamma_u \, du \leq 1. \tag{2.4.14}$$

This obvious bound will be used frequently in the estimates below.

Lemma 2.4.1 *Assume that $F \in C^1$. Let v^ε a solution of (2.4.9) with the initial condition satisfying (2.4.2). Then*

$$\lim_{\varepsilon \to 0} \|\pi^\varepsilon\|_{r_0, 1} = 0 \quad a.s. \tag{2.4.15}$$

In particular,

$$\lim_{\varepsilon \to 0} \|v^\varepsilon - \theta_*\|_{r_0, 1} = 0 \quad a.s. \tag{2.4.16}$$

Proof. First, check (2.4.16). Put $\Delta_r^\varepsilon := v_r^\varepsilon - \theta_*$. By virtue of (2.4.9)

$$\varepsilon \frac{d\Delta_r^\varepsilon}{dr} = A\Delta_r^\varepsilon \gamma_r + (\hat{A}_r^\varepsilon - A)\Delta_r^\varepsilon \gamma_r, \quad \Delta_{r_0}^\varepsilon = y_{r_0}^\varepsilon - \theta_*,$$

and $\hat{A}_r^\varepsilon$ is the matrix with elements of the form

$$\frac{\partial F_k}{\partial x_i}(\theta_* + \vartheta_r^i \Delta_r^\varepsilon), \quad \vartheta_r^i \in [0,1].$$

By the Cauchy formula on the interval $[r_0, 1]$

$$\Delta_r^\varepsilon = e^{A\Gamma(r,r_0)/\varepsilon}\Delta_{r_0}^\varepsilon + \varepsilon^{-1}\int_{r_0}^r e^{A\Gamma(r,s)/\varepsilon}(\hat{A}_s^\varepsilon - A)\Delta_s^\varepsilon \gamma_s ds. \qquad (2.4.17)$$

Let $L > 0$ be a constant such that

$$|e^{tA}| \le Le^{-2\kappa t} \quad \forall t \ge 0. \qquad (2.4.18)$$

Take arbitrary $\eta > 0$. Choose $\mu \in \,]0, \eta]$ such that

$$\sup_{|x| \le \mu} |F'(\theta_* + x) - F'(\theta_*)| \le \kappa/L$$

and put $\tau_\mu^\varepsilon := \{r \ge r_0 : |\Delta_r^\varepsilon| \ge \mu\} \wedge 1$. We have by virtue of (2.4.2) for all ε less than some ε_0 (depending on ω) the inequality $|\Delta_{r_0}^\varepsilon| < \mu/(2L)$. It follows from (2.4.18) that

$$|\Delta_{\tau_\mu^\varepsilon}^\varepsilon| \le Le^{-2\kappa\Gamma(\tau_\mu^\varepsilon, r_0)/\varepsilon}|\Delta_{r_0}^\varepsilon| + \kappa\mu\varepsilon^{-1}\int_{r_0}^{\tau_\mu^\varepsilon} e^{-2\kappa\Gamma(\tau_\mu^\varepsilon, s)/\varepsilon}\gamma_s\,ds < \mu.$$

Therefore, $\tau_\mu^\varepsilon = 1$ and

$$\|\Delta^\varepsilon\|_{r_0,1} < \mu \le \eta.$$

Thus, the relation (2.4.16) holds. It implies, since F' is a continuous function, that for sufficiently small ε

$$\|\hat{A}^\varepsilon - A\|_{r_0,1} \le \kappa/(2L). \qquad (2.4.19)$$

It follows from the Cauchy formula, (2.4.18), and (2.4.19) that

$$|\pi_r^\varepsilon| \le Le^{-\kappa\Gamma(r,r_0)/\varepsilon}|\Delta_{r_0}^\varepsilon| + L\varepsilon^{-1}\|\hat{A}^\varepsilon - A\|_{r_0,1}\int_{r_0}^r e^{-\kappa\Gamma(r,s)/\varepsilon}|\pi_s^\varepsilon|\gamma_s ds$$

$$\le L|\Delta_{r_0}^\varepsilon| + (1/2)\|\pi^\varepsilon\|_{r_0,1}.$$

Hence,

$$\|\pi^\varepsilon\|_{r_0,1} \le 2L|\Delta_{r_0}^\varepsilon|$$

and we get the result. $\square$

Define the fundamental matrix $\Phi^\varepsilon(t,s)$ by the equation

$$\varepsilon\frac{\partial\Phi^\varepsilon(t,s)}{\partial t} = \gamma_t F'(v_t^\varepsilon)\Phi^\varepsilon(t,s), \quad \Phi^\varepsilon(s,s) = I_n, \tag{2.4.20}$$

where I_n is the identity matrix. Put

$$\phi_t^\varepsilon = \sup_{s\in[r_0,t]} e^{\kappa\Gamma(t,s)/\varepsilon}|\Phi^\varepsilon(t,s)|. \tag{2.4.21}$$

Lemma 2.4.2 *Assume that* $F \in C^1$. *Then*

$$\limsup_{\varepsilon\to 0} \|\phi^\varepsilon\|_{r_0,1} \le 2L. \tag{2.4.22}$$

Proof. The result is a corollary of Lemma 2.4.1 and Proposition A.2.7. □

Lemma 2.4.3 *Let us suppose* $F \in C^2$ *and the hypotheses* **H.2.4.2** *and* **H.2.4.3** *are fulfilled. Then for* $j = 1,2$

$$\lim_{\varepsilon\to 0} P(\|\tilde{y}^{j,\varepsilon}\|_{r_0,1} \ge \varepsilon^{-\mu}) = 0 \quad \forall\mu > 0. \tag{2.4.23}$$

Proof. Introduce the stopping time

$$\sigma^\varepsilon := \inf\{t \ge r_0 : \pi_t^\varepsilon \vee \phi_t^\varepsilon \ge 2L + 1\} \wedge 1 \tag{2.4.24}$$

where L is the constant in (2.4.18). By the above lemmas

$$\lim_{\varepsilon\to 0} P(\sigma^\varepsilon < 1) = 0. \tag{2.4.25}$$

Define also the process ξ^ε with

$$d\xi_r^\varepsilon = A_r^\varepsilon \xi_r^\varepsilon dr + G_r^\varepsilon d\tilde{w}_r, \quad \xi_{r_0}^\varepsilon = 0, \tag{2.4.26}$$

where

$$A_r^\varepsilon := \varepsilon^{-1}\gamma_r F'(v_r^\varepsilon)I_{\{r\le\sigma^\varepsilon\}} - \varepsilon^{-1}\gamma_r\kappa I_n I_{\{r>\sigma^\varepsilon\}},$$
$$G_r^\varepsilon := \varepsilon^{-1/2}\gamma_r I_n.$$

Notice that $\tilde{y}^{1,\varepsilon} = \xi^\varepsilon$ on the interval $[0,\sigma^\varepsilon]$. Let $\tilde{\Phi}^\varepsilon(t,s)$ be given by the equation

$$\frac{\partial\tilde{\Phi}^\varepsilon(t,s)}{\partial t} = A_t^\varepsilon\tilde{\Phi}^\varepsilon(t,s), \quad \tilde{\Phi}^\varepsilon(s,s) = I_n.$$

It follows from the above lemma that for sufficiently small ε

$$|\tilde{\Phi}^\varepsilon(t,s)| \le Ce^{-\kappa\Gamma(t,s)/\varepsilon}, \quad r_0 \le s \le t \le 1,$$

where C is a certain constant. Making use of **H.2.4.2** we infer from here, for some κ_1, our usual exponential bound

$$|\tilde{\Phi}^{\varepsilon}(t,s)| \leq Ce^{-\kappa_1(t-s)/\varepsilon}, \quad r_0 \leq s \leq t \leq 1.$$

Proposition 1.2.4 (applied with $M_A = C\varepsilon^{-1}||\gamma||_{r_0,1}$, $M_G = C\varepsilon^{-1/2}||\gamma||_{r_0,1}$, and $\kappa = \varepsilon^{-1}\kappa_1$) implies that for some constant C and all sufficiently small ε we have

$$E||\xi^{\varepsilon}||_{r_0,1} \leq C||\gamma||_{r_0,1}^3 |\ln \varepsilon|^{1/2}.$$

With this, we get immediately by the Chebyshev inequality that

$$P(|\xi^{\varepsilon}|_{r_0,1} > \varepsilon^{-\mu}) \leq C||\gamma||_{r_0,1}^3 |\ln \varepsilon|^{1/2} \varepsilon^{\mu} \to 0$$

by **H.2.4.3**. The relation (2.4.23) for $j = 1$ follows from here and (2.4.25).

To check (2.4.23) for $j = 2$ we write for $\tilde{y}^{2,\varepsilon}$ the representation

$$\tilde{y}_r^{2,\varepsilon} = \varepsilon^{-1} \int_{r_0}^r \Phi^{\varepsilon}(r,s) \tilde{R}_s^{2,\varepsilon} \gamma_s \, ds.$$

By virtue of (2.4.12) for any c_0 with $|c_0| \leq |\theta_*| + 2L + 1$ we have for some constant C that

$$|f_2(c_0, c_1)| \leq C(1 + |c_1|^2)$$

and, hence,

$$|\tilde{R}_s^{2,\varepsilon}| = |f_2(v_s^{\varepsilon}, \tilde{y}_s^{1,\varepsilon})| \leq C(1 + |\tilde{y}_s^{1,\varepsilon}|^2)$$

when $s \leq \sigma^{\varepsilon}$. Thus, for $r \in [0, \sigma^{\varepsilon}]$,

$$|\tilde{y}_r^{2,\varepsilon}| \leq 2L\varepsilon^{-1} \int_{r_0}^r e^{-\kappa\Gamma(r,s)/\varepsilon} |\tilde{R}_s^{2,\varepsilon}| \gamma_s \, ds \leq C(1 + ||\tilde{y}^{1,\varepsilon}||_{r_0,1}^2)$$

and, using (2.4.23) for $j = 1$, we infer that (2.4.23) holds for $j = 2$ as well. $\square$

Lemma 2.4.4 *Assume that $F \in C^2$ and **H.2.4.2**–**H.2.4.3** holds. Then*

$$P\text{-}\lim_{\varepsilon \to 0} ||\tilde{\Delta}^{\varepsilon}||_{r_0,1} = 0. \tag{2.4.27}$$

Proof. It follows from (2.4.8)–(2.4.11) that

$$\varepsilon d\tilde{\Delta}_r^{\varepsilon} = \beta^{-2} \left(F(y_r^{\varepsilon}) - \sum_{j=0}^{2} \beta^j F_j(v_r^{\varepsilon}, \tilde{y}_r^{1,\varepsilon}, ..., \tilde{y}_r^{j,\varepsilon}) \right) \gamma_r \, dr, \quad \tilde{\Delta}_{r_0}^{k,\varepsilon} = 0, \tag{2.4.28}$$

where $F_j(c_0, c_1, ..., c_j)$ are defined in (2.3.5). Let

$$z_r^{\varepsilon} := v_r^{\varepsilon} + \beta \tilde{y}_r^{1,\varepsilon} + \beta^2 \tilde{y}_r^{2,\varepsilon}.$$

Fix $\mu > 0$ such that $\delta_{\varepsilon} := \beta\varepsilon^{-3\mu} \to 0$ as $\varepsilon \to 0$ and put

$$\tilde{\sigma}_{\mu}^{\varepsilon} := \inf\{r \geq r_0 : |\tilde{y}_r^{1,\varepsilon}| + |\tilde{y}_r^{2,\varepsilon}| > \varepsilon^{-\mu}\} \wedge \sigma^{\varepsilon}, \tag{2.4.29}$$

where the stopping time σ^ε is given by (2.4.24).

It follows from (2.4.23) and (2.4.25) that

$$\lim_{\varepsilon \to 0} P(\tilde{\sigma}_\mu^\varepsilon < 1) = 0. \tag{2.4.30}$$

Obviously, for all $r \leq \tilde{\sigma}_\mu^\varepsilon$ we have the bound

$$|v_r^\varepsilon| + |z_r^\varepsilon| \leq C \tag{2.4.31}$$

for some constant C.

Rewriting the equation (2.4.28) as

$$\varepsilon d\tilde{\Delta}_r^\varepsilon = F'(v_r^\varepsilon)\tilde{\Delta}_r^\varepsilon \gamma_r \, dr + (\eta_r^\varepsilon + \zeta_r^\varepsilon)\gamma_r \, dr, \quad \tilde{\Delta}_{r_0}^\varepsilon = 0,$$

where

$$\eta_r^\varepsilon := \beta^{-2}(F(y_r^\varepsilon) - F(z_r^\varepsilon) - F'(v_r^\varepsilon)\tilde{\Delta}_r^\varepsilon \beta^2),$$

$$\zeta_r^\varepsilon := \beta^{-2}\left(F(z_r^\varepsilon) - \sum_{j=0}^{2} \beta^j F_j(v_r^\varepsilon, \tilde{y}_r^{1,\varepsilon}, \dots, \tilde{y}_r^{j,\varepsilon}) \right),$$

we obtain by the Cauchy formula the representation

$$\tilde{\Delta}_r^{k,\varepsilon} = \varepsilon^{-1} \int_{r_0}^{r} \Phi^\varepsilon(r, s)(\eta_s^\varepsilon + \zeta_s^\varepsilon)\gamma_s \, ds \tag{2.4.32}$$

with $\Phi^\varepsilon(r, s)$ given by (2.4.20).

Fix $\nu \in \,]0, 1[$ and define the stopping time

$$\tau = \tau^\varepsilon := \inf\{r \geq r_0 : |\tilde{\Delta}_r^\varepsilon| > \nu\} \wedge \tilde{\sigma}_\mu^\varepsilon, \tag{2.4.33}$$

Using the finite increments formula, the bound (2.4.31), and local Lipschitz condition for F' we get that for all $r \leq \tau^\varepsilon$

$$|\eta_r^\varepsilon| \leq L(|z_r^\varepsilon - v_r^\varepsilon| + \beta^2|\tilde{\Delta}_r^\varepsilon|)|\tilde{\Delta}_r^\varepsilon| \leq (\beta|\tilde{y}_r^{1,\varepsilon}| + \beta^2|\tilde{y}_r^{2,\varepsilon}| + \beta^2) \leq \delta_\varepsilon. \tag{2.4.34}$$

for sufficiently small $\varepsilon > 0$.

It follows from Lemma 2.3.5 that for any $N > 0$ there exists a constant $L_N > 0$ such that

$$\left| F\left(\sum_{j=0}^{2} \beta^j c_j \right) - \sum_{j=0}^{2} \beta^j F_j(c_0, \dots, c_j) \right| \leq L_N (M\beta)^3$$

for arbitrary vectors c_0, c_1, c_2 with $|c_0| \leq N$ and $|c_1| + |c_2| \leq M$ for some $M \geq 1$.

Taking into account (2.4.34), the definition (2.4.32), and using the last inequality with $N = |\theta_*| + L$ and $M = \varepsilon^{-\mu}$ we get that

$$|\tilde{\zeta}_r^\varepsilon| \le L\beta\varepsilon^{-3\mu} = L\delta_\varepsilon \qquad (2.4.35)$$

for $r \in [r_0, \tilde{\sigma}_\mu^\varepsilon]$.

It follows from (2.4.32), (2.4.22), and the inequalities (2.4.34)–(2.4.35) that

$$|\tilde{\Delta}_\tau^\varepsilon| \le 2L\varepsilon^{-1} \int_{r_0}^\tau e^{-\kappa\Gamma(\tau,s)/\varepsilon}(|\eta_s^\varepsilon| + |\zeta_s^\varepsilon|)\gamma_s \, ds \le C\delta_\varepsilon < \nu$$

for sufficiently small $\varepsilon > 0$. Therefore, we get according to the definition τ in (2.4.33) and using (2.4.35) that for any $\nu > 0$

$$\limsup_{\varepsilon\to 0} P(\|\tilde{\Delta}^\varepsilon\|_{r_0,1} > \nu) \le \limsup_{\varepsilon\to 0} P(\|\tilde{\Delta}^\varepsilon\|_{r_0,\tilde{\sigma}} > \nu) + \limsup_{\varepsilon\to 0} P(\tilde{\sigma}_\mu^\varepsilon < 1)$$

$$\le \limsup_{\varepsilon\to 0} P(|\tilde{\Delta}_\tau^\varepsilon| > \nu) = 0$$

and the proof is complete. $\square$

2.4.2 The Asymptotic Expansion at the Root

We look now for the asymptotic expansion of y^ε of the form

$$y_r^\varepsilon = \theta_* + \beta y_r^{1,\varepsilon} + \beta^2 y_r^{2,\varepsilon} + \beta^2 \Delta_r^\varepsilon, \qquad (2.4.36)$$

where the coefficients are given by the following recursive system of linear equations:

$$\varepsilon dy_r^{1,\varepsilon} = A y_r^{1,\varepsilon}\gamma_r \, dr + \sqrt{\varepsilon}\gamma_r \, d\tilde{w}_r, \quad y_{r_0}^{1,\varepsilon} = 0, \qquad (2.4.37)$$

$$\varepsilon dy_r^{2,\varepsilon} = A y_r^{2,\varepsilon}\gamma_r \, dr + R_r^{2,\varepsilon}\gamma_r \, dr, \quad y_{r_0}^{2,\varepsilon} = 0, \qquad (2.4.38)$$

with

$$R_r^{2,\varepsilon} := f_2(\theta_*, y_r^{1,\varepsilon}) = \frac{1}{2}y_r^{1,\varepsilon\star}F''(\theta_*)y_r^{1,\varepsilon}.$$

The properties of the coefficients $y_r^{j,\varepsilon}$, $j = 1, 2$, are summarized in

Lemma 2.4.5 *Assume that $F \in C^2$ and* **H.2.4.1**–**H.2.4.5** *hold. Then*

$$\lim_{\varepsilon\to 0} P(\|y^{j,\varepsilon}\|_{r_0,1} \ge \varepsilon^{-\mu}) = 0 \quad \forall\mu > 0; \qquad (2.4.39)$$

$$\lim_{\varepsilon\to 0} P(\|\tilde{y}^{j,\varepsilon} - y^{j,\varepsilon}\|_{r_1,1} \ge \beta^\alpha e^{-\alpha/\sqrt{\varepsilon}}) = 0 \quad \forall\alpha > 0. \qquad (2.4.40)$$

Proof. The reasoning used in Lemma 2.4.4 works well to get (2.4.39). To prove (2.4.40) we introduce the stopping time

$$\sigma_\mu^\varepsilon := \inf\left\{r \ge r_0 : \sum_{j=1}^2 (|y_r^{j,\varepsilon}| + |\tilde{y}_r^{j,\varepsilon}|) > \varepsilon^{-\mu}\right\} \wedge \sigma^\varepsilon \qquad (2.4.41)$$

where σ^ε is defined in (2.4.24). The relations (2.4.23) and (2.4.39) imply that

$$\lim_{\varepsilon \to 0} P(\sigma_\mu^\varepsilon < 1) = 0 \qquad \forall \mu > 0. \tag{2.4.42}$$

Put $t_1 := s_0$, $t_2 = r_1$,

$$z_r^{j,\varepsilon} := \tilde{y}_r^{j,\varepsilon} - y_r^{j,\varepsilon}, \quad j = 1, 2.$$

It is sufficient to show that for every $\alpha > 0$

$$\lim_{\varepsilon \to 0} P(\|z^{j,\varepsilon}\|_{t_j,1} > \beta^\alpha e^{-\alpha/\sqrt{\varepsilon}}) = 0. \tag{2.4.43}$$

It follows from (2.4.10) and (2.4.37) that

$$\varepsilon dz_r^{1,\varepsilon} = F'(v_r^\varepsilon)z_r^{1,\varepsilon}\gamma_r \, dr + (F'(v_r^\varepsilon) - A)y_r^{1,\varepsilon}\gamma_r \, dr, \quad z_{r_0}^{1,\varepsilon} = 0,$$

and, hence,

$$z_r^{1,\varepsilon} = \varepsilon^{-1} \int_{r_0}^r \Phi^\varepsilon(r,s)(F'(v_s^\varepsilon) - A)y_s^{1,\varepsilon}\gamma_s \, ds$$

with $\Phi^\varepsilon(r,s)$ is given by (2.4.20). By virtue of the definitions (2.4.24), (2.4.21), and (2.4.13) on $[r_0, \sigma_\mu^\varepsilon]$ the fundamental matrix Φ^ε and the difference $v^\varepsilon - \theta_*$ admit exponential majorants. Since F' is locally Lipschitz, we get that for $r \in [t_1, \sigma_\mu^\varepsilon]$

$$|z_r^{1,\varepsilon}| \le C\varepsilon^{-1} \int_{r_0}^r e^{-\kappa\Gamma(r,s)/\varepsilon}|v_s^\varepsilon - \theta_*|\|y_s^{1,\varepsilon}|\gamma_s \, ds$$
$$\le C\varepsilon^{-(1+\mu)} \int_{r_0}^r e^{-\kappa\Gamma(r,r_0)/\varepsilon}\gamma_s \, ds$$
$$\le C\varepsilon^{-(1+\mu)}e^{-\kappa\Gamma(r,r_0)/\varepsilon}\Gamma(r,r_0)$$
$$\le C\varepsilon^{-\mu}e^{-\kappa\Gamma(r,r_0)/(2\varepsilon)} \le C\varepsilon^{-\mu}e^{-\kappa\Gamma(t_1,r_0)/(2\varepsilon)}$$

(by convention, constants C vary from line to line). Taking into account (2.4.42) and the condition **H.2.4.5** we get (2.4.43) for $j = 1$.

To check (2.4.43) for $j = 2$ we fix arbitrary $\alpha > 0$ and put

$$\sigma_{\mu,\alpha}^\varepsilon := \inf\{r \ge t_1 : |z_r^{1,\varepsilon}| \ge \beta^\alpha e^{-\alpha/\sqrt{\varepsilon}}\} \wedge \sigma_\mu^\varepsilon. \tag{2.4.44}$$

It follows from (2.4.43) for $j = 1$ that

$$\lim_{\varepsilon \to 0} P(\sigma_{\mu,\alpha}^\varepsilon < 1) = 0. \tag{2.4.45}$$

By virtue of (2.4.11) and (2.4.38)

$$\varepsilon dz_r^{2,\varepsilon} = F'(v_r^\varepsilon)z_r^{2,\varepsilon}\gamma_r \, dr + [(F'(v_r^\varepsilon) - A)y_r^{2,\varepsilon} + (\tilde{R}_r^{2,\varepsilon} - R_r^{2,\varepsilon})]\gamma_r \, dr$$

and, hence,

$$z_r^{2,\varepsilon} = \Phi^\varepsilon(r,t_1)z_{t_1}^{2,\varepsilon} + \varepsilon^{-1}\int_{t_1}^r \Phi^\varepsilon(r,s)(F'(v_s^\varepsilon) - A)y_s^{2,\varepsilon}\gamma_s\,ds$$

$$+ \varepsilon^{-1}\int_{t_1}^r \Phi^\varepsilon(r,s)(\tilde{R}_s^{2,\varepsilon} - R_s^{2,\varepsilon})\gamma_s\,ds. \qquad (2.4.46)$$

For $r \in [r_1, \sigma_{\mu,\alpha}^\varepsilon]$ we can estimate the second term on the right-hand side in exactly the same way as for $j = 1$; for the first one we get that

$$|\Phi^\varepsilon(r,t_1)z_{t_1}^{2,\varepsilon}| \le Ce^{-\kappa\Gamma(r,t_1)/\varepsilon}(|\tilde{y}_{t_1}^{2,\varepsilon}| + |y_{t_1}^{2,\varepsilon}|) \le Ce^{-\kappa\Gamma(r_1,t_1)/\varepsilon}\varepsilon^{-2\mu}$$

$$\le C\beta^\alpha e^{-\alpha/\sqrt{\varepsilon}}$$

for sufficiently small $\varepsilon > 0$. There is also no problem to estimate the third term since by virtue of (2.4.12) for $r \in [r_1, \sigma_{\mu,\alpha}^\varepsilon]$ we have

$$|\tilde{R}_r^{2,\varepsilon} - R_r^{2,\varepsilon}| = |f_2(v_r^\varepsilon, \tilde{y}_r^{1,\varepsilon}) - f_2(\theta_*, y_r^{1,\varepsilon})|$$
$$\le C(|v_r^\varepsilon - \theta_*| + |z_r^{1,\varepsilon}|)(1 + |\tilde{y}_r^{1,\varepsilon}|^2 + |y_r^{1,\varepsilon}|^2)$$
$$\le C(e^{-\kappa\Gamma(t_1,r_0)/2\varepsilon} + \beta^\alpha e^{-\alpha/\sqrt{\varepsilon}})\varepsilon^{-2\mu} \le \beta^{\alpha_1}e^{-\alpha_1/\sqrt{\varepsilon}}$$

for arbitrary $\alpha_1 < \alpha$ when $\varepsilon > 0$ is sufficiently small. It follows from the above bounds, due to the relation (2.4.45), that (2.4.43) holds for $j = 2$ (with $\alpha_1 < \alpha$ and hence with every α). $\square$

Lemma 2.4.6 *Assume that $F \in C^2$ and* **H.2.4.1–H.2.4.5** *hold. Then*

$$P\text{-}\lim_{\varepsilon \to 0} \|\Delta^\varepsilon\|_{r_1,1} = 0. \qquad (2.4.47)$$

Lemma 2.4.6 follows from Lemmas 2.4.1, 2.4.4, and 2.4.5 and the condition **H.2.4.4**.

2.4.3 Averaging

To get the assertions of Theorem 0.4.3 it remains to integrate the asymptotic expansion (2.4.36). In particular, the Gaussian random variable ξ^ε in (0.4.20) is simply

$$\xi^\varepsilon := \frac{1}{1 - r_1}\frac{1}{\sqrt{\varepsilon}}\int_{r_1}^1 y_r^{1,\varepsilon}\,dr.$$

Due to (2.4.37) we have $E\xi^\varepsilon = 0$. The claim on asymptotic normality of ξ^ε follows from the first assertion of Lemma 2.4.7 below on properties of the Gaussian process $y^{1,\varepsilon}$ given by the linear equation (2.4.37) where A is a stable matrix, $\gamma^\varepsilon \in C^1$, and the conditions **H.2.4.2**, **H.2.4.3** and **H.2.4.6** are fulfilled. Finally, Lemma 2.4.8 on the limiting behavior of the integral of $y^{2,\varepsilon}$ given by (2.4.38) concludes the proof.

Lemma 2.4.7 *We have:*

$$\varepsilon^{-1/2} \int_{r_1}^1 y_r^{1,\varepsilon}\, dr = -A^{-1}\tilde{w}_1 + \zeta^\varepsilon, \qquad (2.4.48)$$

where $\zeta^\varepsilon \to 0$ in L^2 as $\varepsilon \to 0$;

$$\lim_{\varepsilon \to 0} E \left| \int_{r_1}^1 y_r^{1,\varepsilon} y_r^{1,\varepsilon\star}\, dr - Q \int_{r_1}^1 \gamma_r\, dr \right| = 0, \qquad (2.4.49)$$

where the matrix Q is defined by the equation

$$AQ + QA^\star + I = 0. \qquad (2.4.50)$$

Proof. Let $\vartheta := 1/\gamma$. It follows from (2.4.37) by the Ito formula that

$$d(\vartheta_r y_r^{1,\varepsilon}) = \varepsilon^{-1/2} d\tilde{w}_r + \varepsilon^{-1} A y_r^{1,\varepsilon}\, dr + \dot{\vartheta}_r y_r^{1,\varepsilon}\, dr$$

and, hence,

$$\varepsilon^{-1/2} \int_{r_1}^1 y_r^{1,\varepsilon}\, dr = -A^{-1}(\tilde{w}_1 - \tilde{w}_{r_1})$$

$$+ \varepsilon^{1/2} A^{-1} \left(\vartheta_1 y_1^{1,\varepsilon} - \vartheta_{r_1} y_{r_1}^{1,\varepsilon} - \int_{r_1}^1 \dot{\vartheta}_r y_r^{1,\varepsilon}\, dr \right).$$

Applying the Cauchy formula to (2.4.37) and estimating with the help of the exponential bound (2.4.18), we infer that for all $r \in [r_0, 1]$

$$E|y_r^{1,\varepsilon}|^2 \leq \varepsilon^{-1} \int_{r_0}^r \gamma_s^2 |e^{A\Gamma(r,s)/\varepsilon}|^2\, ds$$

$$\leq C\varepsilon^{-1} \int_{r_0}^r \gamma_s^2 e^{-2\kappa\Gamma(r,s)/\varepsilon}\, ds \leq C\|\gamma\|_{r_0,1} \qquad (2.4.51)$$

and the first claim of the lemma follows by virtue of **H.2.4.3** and **H.2.4.6**.

Put $\eta_r^\varepsilon := y_r^{1,\varepsilon} y_r^{1,\varepsilon\star}$. By the Ito formula

$$d\eta_r^\varepsilon = \varepsilon^{-1}(A\eta_r^\varepsilon + \eta_r^\varepsilon A^\star + \gamma_r I_n)\gamma_r\, dr + \varepsilon^{-1/2}\gamma_r(d\tilde{w}_r y_r^{1,\varepsilon\star} + y_r^{1,\varepsilon} d\tilde{w}_r^\star),$$

and, hence,

$$\vartheta_1 \eta_1^\varepsilon - \vartheta_{r_1} \eta_{r_1}^\varepsilon = \int_{r_1}^1 \eta_r^\varepsilon \dot{\vartheta}_r\, dr + \varepsilon^{-1} A \int_{r_1}^1 \eta_r^\varepsilon\, dr + \varepsilon^{-1} \int_{r_1}^1 \eta_r^\varepsilon\, dr A^\star$$

$$+ \varepsilon^{-1} I_n \int_{r_1}^1 \gamma_r\, dr + \varepsilon^{-1/2} \int_{r_1}^1 (y_r^{1,\varepsilon} d\tilde{w}_r^\star + d\tilde{w}_r y_r^{1,\varepsilon\star})$$

implying (2.4.49). $\square$

Put $h := -A^{-1} h_0$ where the components of the vector h_0 are

$$h_{0i} := \frac{1}{2} \text{tr}\, F_i''(\theta_*) Q. \qquad (2.4.52)$$

Lemma 2.4.8 Let $y^{2,\varepsilon}$ be given by (2.4.38). Then

$$\lim_{\varepsilon \to 0} E \left| \int_{r_1}^1 y_r^{2,\varepsilon} \, dr - h \int_{r_1}^1 \gamma_r \, dr \right| = 0. \qquad (2.4.53)$$

Proof. For $r \geq r_1$ we have by the Cauchy formula that

$$y_r^{2,\varepsilon} = \frac{1}{\varepsilon} \int_{r_0}^r e^{A\Gamma(r,s)/\varepsilon} R_s^{2,\varepsilon} \gamma_s \, ds = e^{A\Gamma(r,r_1)/\varepsilon} y_{r_1}^{2,\varepsilon} + \xi_r^\varepsilon$$

where

$$\xi_r^\varepsilon := \frac{1}{\varepsilon} \int_{r_1}^r e^{A\Gamma(r,s)/\varepsilon} R_s^{2,\varepsilon} \gamma_s \, ds.$$

In virtue of (2.4.51)

$$E|y_{r_1}^{2,\varepsilon}| \leq C ||\gamma||_{r_0,1}$$

and, hence,

$$E \left| \int_{r_1}^1 e^{A\Gamma(r,r_1)/\varepsilon} \, dr \, y_{r_1}^{2,\varepsilon} \right| \leq C ||\gamma||_{r_0,1} \varepsilon \to 0, \quad \varepsilon \to 0.$$

Therefore, to prove (2.4.53) it remains to show that

$$\lim_{\varepsilon \to 0} E \left| \int_{r_1}^1 \xi_r^\varepsilon \, dr - h \int_{r_1}^1 \gamma_r \, dr \right| = 0. \qquad (2.4.54)$$

Indeed,

$$\int_{r_1}^1 \left(\frac{1}{\varepsilon} \int_{r_1}^r e^{A\Gamma(r,s)/\varepsilon} R_s^{2,\varepsilon} \gamma_s \, ds \right) dr = \int_{r_1}^1 \left(\frac{1}{\varepsilon} \int_s^1 e^{A\Gamma(r,s)/\varepsilon} \, dr \right) R_s^{2,\varepsilon} \gamma_s \, ds.$$

Notice that

$$\frac{1}{\varepsilon} \int_s^1 e^{A\Gamma(r,s)/\varepsilon} \, dr = -A^{-1} \gamma_s^{-1} + \vartheta_1 A^{-1} e^{A\Gamma(1,s)/\varepsilon} - A^{-1} g_s^\varepsilon$$

with

$$g_s^\varepsilon := \int_s^1 \dot{\vartheta}_r e^{A\Gamma(r,s)/\varepsilon} \, dr.$$

Using (2.4.51) we obtain that

$$E \left| \int_{r_1}^1 A^{-1} e^{A\Gamma(1,s)/\varepsilon} R_s^{2,\varepsilon} \gamma_s \, ds \right| \leq C ||\gamma||_{r_0,1} \int_{r_1}^1 e^{-\kappa\Gamma(1,s)/\varepsilon} \gamma_s \, ds$$

$$\leq C ||\gamma||_{r_0,1} \varepsilon \to 0.$$

Similarly,

$$E \int_{r_1}^1 |g_s^\varepsilon| |R_s^{2,\varepsilon}|\gamma_s \, ds \le C||\gamma||_{r_0,1} \int_{r_1}^1 |g_s^\varepsilon|\gamma_s \, ds \le C||\dot{\vartheta}||_{r_0,1}||\gamma||_{r_0,1}^2 \varepsilon.$$

To get (2.4.54) it remains to observe that for each component of $R^{2,\varepsilon}$

$$\int_{r_1}^1 R_{ir}^{2,\varepsilon} \, dr = \frac{1}{2}\text{tr}\, F_i''(\theta_*) \int_{r_1}^1 y_r^{1,\varepsilon} y_r^{1,\varepsilon\star} \, dr \to \frac{1}{2}\text{tr}\, F_i''(\theta_*)Q \int_{r_1}^1 \gamma_r \, dr$$

in L^1 by virtue of Lemma 2.4.8. $\square$

Theorems 0.4.4 and 0.4.5 follow from the representation 2.4.36) and Lemmas 2.4.7 and 2.4.8.

2.4.4 Proof of Theorem 0.4.6

We continue to work with the general model given by (2.4.3) assuming **H.2.4.1–H.2.4.6** and adding the assumption

$$\lim_{\varepsilon \to 0} E|y_{r_0}^\varepsilon - \theta_*|^2 = 0, \tag{2.4.55}$$

fulfilled by virtue of Proposition 0.4.1.

Now we can strengthen Lemma 2.4.1 by claiming that $E||\pi^\varepsilon||_{r_0,1} \to 0$ as $\varepsilon \to 0$. Indeed let v be given by (2.4.9). Since $F(\theta_*) = 0$, we get, using **H.0.4.3** that

$$\frac{d}{dr}|v_r^\varepsilon - \theta_*|^2 = 2(v_r^\varepsilon - \theta_*)^* F(v_r^\varepsilon)\gamma_r$$

$$= 2(v_r^\varepsilon - \theta_*)^* A(v_r^\varepsilon, \theta_*)(v_r^\varepsilon - \theta_*)\gamma_r \le -2\kappa|v_r^\varepsilon - \theta_*|^2\gamma_r.$$

By the Gronwall–Bellman lemma, for every $r \in [r_0, 1]$

$$|v_r^\varepsilon - \theta_*|^2 \le |y_{r_0}^\varepsilon - \theta_*|^2 e^{-2\kappa\Gamma(r,r_0)/\varepsilon} \tag{2.4.56}$$

and hence

$$E|v_r^\varepsilon - \theta_*|^2 \le e^{-2\kappa\Gamma(r,r_0)/\varepsilon} E|y_{r_0}^\varepsilon - \theta_*|^2. \tag{2.4.57}$$

In the same way we obtain the exponential bound for the fundamental matrix $\Phi^\varepsilon(t,s)$ given by (2.4.20)

$$|\Phi^\varepsilon(t,s)|^2 \le ne^{-2\kappa\Gamma(t,s)/\varepsilon} \tag{2.4.58}$$

which holds for all $s \le t$ from the interval $[r_0, 1]$; cf. (1.2.2).

Lemma 2.4.9 *Suppose that the assumptions of Theorem 0.4.6 as well as the conditions* **H.2.4.1–H.2.4.6** *are fulfilled. Then for any $\alpha > 0$*

$$\lim_{\varepsilon \to 0} \sup_{r \in [s_0,1]} \beta^{-\alpha} E|\tilde{y}_r^{1,\varepsilon} - y_r^{1,\varepsilon}|^2 = 0 \tag{2.4.59}$$

and

$$\lim_{\varepsilon \to 0} \sup_{r \in [r_1,1]} E|\tilde{\Delta}_r^\varepsilon| = 0, \tag{2.4.60}$$

where $\tilde{y}^{1,\varepsilon}$, $y^{1,\varepsilon}$, and $\tilde{\Delta}^\varepsilon$ are defined by (2.4.10), (2.4.37), and (2.4.8).

Proof. Taking into account the Lipschitz condition for F' and (2.4.53) we get that for $r \geq s_0$

$$|z_r^{1,\varepsilon}| = \sqrt{n}\varepsilon^{-1}\int_{r_0}^r e^{-\kappa\Gamma(r,s)/\varepsilon}|F'(v_s^\varepsilon) - F'(\theta_*)||y_s^{1,\varepsilon}|\gamma_s\,ds$$

$$\leq C\varepsilon^{-1}\int_{r_0}^r e^{-\kappa\Gamma(r,s)/\varepsilon}|v_s^\varepsilon - \theta_*||y_s^{1,\varepsilon}|\gamma_s\,ds.$$

Thus, by the Cauchy–Schwarz inequality,

$$E|z_r^{1,\varepsilon}|^2 \leq C\varepsilon^{-2}\int_{r_0}^r e^{-\kappa\Gamma(r,s)/\varepsilon}E|v_s^\varepsilon - \theta_*|^2 E|y_s^{1,\varepsilon}|^2\gamma_s\,ds\int_{r_0}^r e^{-\kappa\Gamma(r,s)/\varepsilon}\gamma_s\,ds$$

$$\leq C||\gamma||_{r_0,1}E|y_{r_0}^\varepsilon - \theta_*|^2\varepsilon^{-1}\int_{r_0}^r e^{-2\kappa\Gamma(s,r_0)/\varepsilon}e^{-\kappa\Gamma(r,s)/\varepsilon}\gamma_s\,ds$$

$$\leq Ce^{-\kappa\Gamma(s_0,r_0)/\varepsilon}||\gamma||_{r_0,1}E|y_{r_0}^\varepsilon - \theta_*|^2.$$

Taking into account (2.4.55) and **H.2.4.5**, we obtain (2.4.59).

It follows from (2.4.34) that

$$d\tilde{\Delta}_r = \varepsilon^{-1}\beta^{-2}(F(y_r^\varepsilon) - F(z_r^\varepsilon))\gamma_r\,dr + \varepsilon^{-1}\eta_r^\varepsilon\gamma_r\,dr, \qquad \tilde{\Delta}_{r_0} = 0, \quad (2.4.61)$$

where

$$z_r^\varepsilon = v_r^\varepsilon + \beta\tilde{y}_r^{1,\varepsilon} + \beta^2\tilde{y}_r^{2,\varepsilon},$$
$$\eta_r^\varepsilon = \beta^{-2}(F(z_r^\varepsilon) - F(v_r^\varepsilon) - \beta F_1(v_r^\varepsilon, \tilde{y}_r^{1,\varepsilon}) - \beta^2 F_2(v_r^\varepsilon, \tilde{y}_r^{1,\varepsilon}, \tilde{y}_r^{2,\varepsilon})).$$

Using the Taylor formula we get the inequality

$$|\eta_r^\varepsilon| \leq L\beta(1 + |\tilde{y}_r^{1,\varepsilon}|^3 + |\tilde{y}_r^{2,\varepsilon}|^3).$$

It follows from (2.4.10) and (1.1.2) that for any integer $m \geq 1$

$$\sup_{r\in[r_0,1]} E|\tilde{y}_r^{1,\varepsilon}|^{2m} \leq ||\gamma||_{r_0,1}^{2m}(2m-1)!!/(2\kappa)^m. \qquad (2.4.62)$$

Taking into account (2.4.10), (2.4.11), and the inequalities (2.4.58) and (2.4.62) we deduce that

$$\sup_{r\in[r_0,1]} E|\tilde{y}_r^{2,\varepsilon}|^m \leq ||\gamma||_{r_0,1}^m L^m(2m-1)!!. \qquad (2.4.63)$$

It follows from (2.4.62) and (2.4.63) that

$$\sup_{r\in[r_0,1]} E|\eta_r^\varepsilon| \leq \beta(1 + ||\gamma||_{r_0,1}^3)$$

for some constant $L > 0$. Therefore,

$$\lim_{\varepsilon\to 0}\sup_{r\in[r_0,1]} E|\eta_r^\varepsilon| = 0.$$

Rewrite the equation (2.4.61) as

$$d\tilde{\Delta}_r^\varepsilon = \varepsilon^{-1}\beta^{-1}\hat{A}_r^\varepsilon\tilde{\Delta}_r^\varepsilon\gamma_r\,dr + \varepsilon^{-1}\eta_r^\varepsilon\gamma(r)\,dr, \quad \tilde{\Delta}_{r_0}^\varepsilon = 0,$$

where $\hat{A}_r^\varepsilon = A(y_r^\varepsilon, z_r^\varepsilon)$ with A defined in **H.0.4.3**. Thus, on $[r_0, 1]$ we have the representation

$$\tilde{\Delta}_r^\varepsilon = \varepsilon^{-1}\int_{r_0}^r \hat{\Phi}^\varepsilon(r, s)\eta_s^\varepsilon\gamma_s\,ds$$

with the fundamental matrix given by

$$\varepsilon\frac{\partial\hat{\Phi}^\varepsilon(r, s)}{\partial r} = \hat{A}_r^\varepsilon\hat{\Phi}^\varepsilon(r, s)\gamma_r, \qquad \hat{\Phi}^\varepsilon(s, s) = I.$$

Notice that $\hat{\Phi}^\varepsilon(r, s)$ satisfies the inequality (2.4.58). Therefore,

$$E|\tilde{\Delta}_r^\varepsilon| \leq \sqrt{n}\varepsilon^{-1}\int_{r_0}^r e^{-\kappa\Gamma(r, s)/\varepsilon}E|\eta_s^\varepsilon|\gamma_s\,ds \leq \sqrt{n}\sup_{r_0 \leq r \leq 1}E|\eta_r^\varepsilon|/\kappa$$

and we get (2.4.60). $\square$

Lemma 2.4.10 *Suppose that the assumptions of Theorem 0.4.6 as well as the conditions* **H.2.4.1**–**H.2.4.6** *are fulfilled. Then*

$$\lim_{\varepsilon \to 0}\sup_{r \in [r_1, 1]}E|\tilde{y}_r^{2,\varepsilon} - y_r^{2,\varepsilon}| = 0 \tag{2.4.64}$$

where $\tilde{y}^{2,\varepsilon}$ and $y^{2,\varepsilon}$ are defined by (2.4.11) and (2.4.38).

Proof. For arbitrary integer $m \geq 1$ we have the moment bound

$$\sup_{r \in [r_0, 1]}E(|\tilde{y}_r^{2,\varepsilon}|^m + |y_r^{2,\varepsilon}|^m) \leq L^m(2m-1)!!\|\gamma\|_{r_0, 1}^m.$$

It follows from (2.4.46) and (2.4.57) that

$$E|\tilde{y}_r^{2,\varepsilon} - y_r^{2,\varepsilon}| \leq L\left[e^{-\kappa\Gamma(r_1, t_1)/\varepsilon}(E|\tilde{y}_{t_1}^{2,\varepsilon}| + E|y_{t_1}^{2,\varepsilon}|)\right.$$

$$+ \frac{1}{\varepsilon}\int_{t_1}^r e^{-\kappa\Gamma(s, r_0)/\varepsilon}e^{-\kappa\Gamma(r, s)/\varepsilon}(1 + E|\tilde{y}_s^{1,\varepsilon}|^2 + E|y_s^{1,\varepsilon}|^2)\gamma_s\,ds$$

$$\left.+ \frac{1}{\varepsilon}\int_{t_1}^r e^{-\kappa\Gamma(r, s)/\varepsilon}E|\tilde{y}_s^{1,\varepsilon} - y_s^{1,\varepsilon}|(1 + |\tilde{y}_s^{1,\varepsilon}|^2 + |y_s^{1,\varepsilon}|^2)\gamma_s\,ds\right]$$

where $r \geq r_1$ and $t_1 = s_0$. Making use of **H.2.4.5**, (2.4.59), and (2.4.62) we get the needed assertion. $\square$

The assertions of Theorem 0.4.6 follow from (2.4.8), Lemma 2.4.8, (2.4.59), (2.4.60), and Lemma 2.4.10.

3 Large Deviations

In Sections 2.1 and 2.2 we established Theorems 2.1.1 and 2.2.1 on the accuracy of approximation on a finite time interval $[T_0, T]$ of the solution y^ε of the singularly perturbed SDE

$$\varepsilon dy_t^\varepsilon = F(y_t^\varepsilon)dt + \beta\varepsilon^{1/2}G(y_t^\varepsilon)dw_t, \quad y_0^\varepsilon = y^o, \tag{3.0.1}$$

(which is a "fast" process) by the deterministic function $\tilde{y}_{./\varepsilon}$ where $\tilde{y}$ satisfies the ordinary differential equation

$$d\tilde{y}_s = F(\tilde{y}_s)ds, \quad \tilde{y}_0 = y^o. \tag{3.0.2}$$

Theorems 2.2.2 and 2.3.6 give some idea on the accuracy of approximation of y^ε, outside the boundary layer, by the constant function $\tilde{y}_\infty$ which is the rest point of (3.0.2). The above results are formulated in terms of asymptotic expansions with remaining terms tending to zero in probability. In this chapter we investigate the problem from another point of view. In Section 3.1 we find the logarithmic asymptotics of the deviation probability $P(\sup_{t \leq T} |y_t^\varepsilon - \tilde{y}_{t/\varepsilon}| \geq \eta)$ as $\varepsilon \to 0$ assuming that β is "rapidly" decreasing, namely, that $\beta = o(\varepsilon^{1/2})$. Section 3.2 contains a result on the logarithmic asymptotics of the deviation probability of y^ε from the constant function $\tilde{y}_\infty$; the main feature of Theorem 3.2.1 is that it uses the $L^2[0, T]$-metric and not a uniform metric. It happens that such "non-standard" large deviations can be applied in various problems of statistical estimation and filtering. Some of them will be discussed later, in Chapter 6.

Notice that the time change $s := t/\varepsilon$ transforms (3.0.1) into the "regularly" perturbed SDE

$$d\tilde{y}_s^\varepsilon = F(\tilde{y}_s^\varepsilon)ds + \beta G(\tilde{y}_s^\varepsilon)d\tilde{w}_s, \quad \tilde{y}_0^\varepsilon = 0, \tag{3.0.3}$$

where $\tilde{y}_s^\varepsilon := x_{s\varepsilon}$, $\tilde{w}_s := w_{s\varepsilon}/\sqrt{\varepsilon}$ is a Wiener process. The difference between our model and the classical Wentzell–Freidlin scheme is that we need to study asymptotics of large deviations of the norm of $\tilde{y}^\varepsilon - \tilde{y}$ from zero not on the fixed but on the increasing time intervals $[0, T/\varepsilon]$ and this can be done because of the assumed exponential asymptotic stability of the associated equation (3.0.2).

3.1 Deviations in the Uniform Metric

3.1.1 Formulation of the Result

We need the following assumptions on the coefficients of (3.0.1):

H.3.1.1 The function $F : \mathbf{R}^n \to \mathbf{R}^n$ is continuously differentiable. There exists a function $A(y_1, y_2)$ taking values in the set of $n \times n$ matrices such that $|A| \leq L$ and for all y_1, y_2

$$F(y_1) - F(y_2) = A(y_1, y_2)(y_1 - y_2) \tag{3.1.1}$$

and

$$z^* A(y_1, y_2) z \leq -\gamma z^* z \quad \text{for every} \ \ z \in \mathbf{R}^n \tag{3.1.2}$$

where $\gamma > 0$ is a constant.

It is clear that **H.3.1.1** implies the Lipschitz and linear growth conditions for F. Notice that in the one-dimensional case $A(y_1, y_2) = F'(y_1 + \theta(y_2 - y_1))$ for some $\theta = \theta(y_1, y_2) \in [0, 1]$ and (3.1.2) means simply that $F' \leq -\gamma$.

H.3.1.2 There exists a root $\tilde{y}_\infty$ of the equation $F(y) = 0$.

It is easy to see that under **H.3.1.1** and **H.3.1.2** the solution of the differential equation (3.0.2) tends to $\tilde{y}_\infty$:

$$\lim_{s \to \infty} \tilde{y}_s = \tilde{y}_\infty. \tag{3.1.3}$$

H.3.1.3 The function G taking values in the set of $n \times n$ matrices is Lipschitz, bounded ($|G(y)| \leq L$), and such that the function $B(y) := G(y)G^*(y)$ satisfies the uniform ellipticity condition:

$$z^* B(y) z \geq \gamma z^* z \quad \forall \ y, z \in \mathbf{R}^n.$$

Obviously, the uniform ellipticity of $B(y)$ implies the boundedness of $B^{-1}(y)$. It follows from the identity

$$B^{-1}(y_1) - B^{-1}(y_2) = B^{-1}(y_1)(B(y_2) - B(y_1))B^{-1}(y_2)$$

that **H.3.1.3** ensures the Lipschitz condition for $B^{-1}(y)$. To avoid new notations for constants we shall assume without loss of generality that

$$|B^{-1}(y)| \leq L, \quad |B^{-1}(y_1) - B^{-1}(y_2)| \leq L|y_1 - y_2|. \tag{3.1.4}$$

H.3.1.4 The function $\beta = \beta_\varepsilon$ is strictly positive and $\beta = o(\sqrt{\varepsilon})$.

H.3.1.5 The function $\beta = \beta_\varepsilon$ is strictly positive and $\beta = o(1/\sqrt{|\ln \varepsilon|})$.

In this section we establish a theorem on the logarithmic asymptotics of deviation probabilities when the deviations are measured in the uniform metric under the assumption **H.3.1.4**. However, the important exponential bound of Proposition 3.1.2, used also in Section 3.2, will be proved under the weaker assumption **H.3.1.5**.

We shall denote by $C^a[0,T]$ (resp., by $C_0[0,T]$) the subspace of $C[0,T]$ formed by the absolute continuous functions (resp., by the functions y with $y_0 = 0$). The notations $C_0^a[0,T]$, $C_0(\mathbf{R}_+)$ etc., have obvious meanings.

Let $d_T(x,y) := ||x - y||_T$.

Let us define on $C(\mathbf{R}_+)$ the functional $S_T^\varepsilon(y)$ by the formula

$$S_T^\varepsilon(y) := \frac{1}{2} \int_0^{T/\varepsilon} |B^{-1/2}(y_s)(\dot{y}_s - F(y_s))|^2 ds \qquad (3.1.5)$$

for $y \in C^a[0, T/\varepsilon]$ and put $S_T^\varepsilon(y) = \infty$ for $y \notin C^a[0, T/\varepsilon]$. Similarly,

$$S(y) := \frac{1}{2} \int_0^\infty |B^{-1/2}(y_s)(\dot{y}_s - F(y_s))|^2 ds, \qquad (3.1.6)$$

if $y \in C^a(\mathbf{R}_+)$ and $S(y) = +\infty$ if $y \notin C^a(\mathbf{R}_+)$.

The functional $S_T^\varepsilon(y)$ can be considered also as a functional on $C[0, T/\varepsilon]$.

Let $\tilde{y}$ be the solution of the associated equation (3.0.2). For $\eta > 0$ we put

$$\Gamma(\eta) := \{y \in C_0(\mathbf{R}_+) : \ ||y - \tilde{y}||_\infty > \eta\}. \qquad (3.1.7)$$

That is, $\Gamma(\eta)$ is the exterior of the closed ball of radius η and center $\tilde{y}$ in the space $C(\mathbf{R}_+)$ equipped with the uniform metric.

Theorem 3.1.1 *Assume that* **H.3.1.1**–**H.3.1.4** *hold. Then for any* $\eta > 0$

$$\lim_{\varepsilon \to 0} \beta^2 \ln P(||\tilde{y}^\varepsilon - \tilde{y}||_{T/\varepsilon} > \eta) = -\pi(\eta) \qquad (3.1.8)$$

where

$$\pi(\eta) = \inf_{y \in \Gamma(\eta)} S(y). \qquad (3.1.9)$$

3.1.2 A Lower Exponential Bound for the Non-Exit Probability

The proof of the theorem is based on some exponential bounds. We start with the lower bound for the probability of the event that the trajectory of (3.0.3) remains in the neighborhood of a fixed but arbitrary function $y \in C_0^a(\mathbf{R}_+)$ with $S(y) < \infty$.

Proposition 3.1.2 *Assume that* **H.3.1.1**–**H.3.1.3** *and* **H.3.1.5** *hold. Let* y *be an arbitrary function from* $C_0^a(\mathbf{R}_+)$ *such that* $S(y) < \infty$. *Then for any* ν, $\eta > 0$ *there exists* $\varepsilon_0 > 0$ *such that*

$$P(||\tilde{y}^\varepsilon - y||_{T/\varepsilon} < \eta) \geq \exp\{-\beta^{-2}S_T^\varepsilon(y)(1 + \nu)\} \qquad (3.1.10)$$

for all $\varepsilon \in]0, \varepsilon_0]$.

Proof. Put

$$z_t^\varepsilon := \tilde{y}_t^\varepsilon - y_t, \quad \psi_t := y_t - \int_0^t F(y_s)ds.$$

It follows from (3.0.3) that

$$dz_t^\varepsilon = f(z_t^\varepsilon, t)dt + \dot{\psi}_t dt + \beta g(z_t^\varepsilon, t)d\tilde{W}_t, \quad z_0^\varepsilon = 0, \tag{3.1.11}$$

where

$$f(z, t) := F(z + y_t) - F(y_t), \quad g(z, t) := G(z + y_t).$$

Let us consider the process ξ^ε which is the solution of the stochastic differential equation

$$d\xi_t^\varepsilon = f(\xi_t^\varepsilon, t)dt + \beta g(\xi_t^\varepsilon, t)d\tilde{W}_t, \quad \xi_0^\varepsilon = 0. \tag{3.1.12}$$

Let denote by μ_z^ε, μ_ξ^ε the distributions of the process z^ε, ξ^ε in the space $C[0, T/\varepsilon]$. Notice that $\mu_z^\varepsilon \ll \mu_\xi^\varepsilon$ and the value of the Radon–Nikodym derivative $\rho_{T/\varepsilon} := d\mu_z^\varepsilon/d\mu_\xi^\varepsilon$ at the point ξ^ε is given by

$$\rho_{T/\varepsilon}(\xi^\varepsilon) = \exp\{\beta^{-1}U_{T/\varepsilon}^\varepsilon - (1/2)\beta^{-2}\langle U^\varepsilon\rangle_{T/\varepsilon}\} \tag{3.1.13}$$

where

$$U_t^\varepsilon := \int_0^t \dot{\psi}_s^\star b^{-1}(\xi_s^\varepsilon, s)g(\xi_s^\varepsilon, s)d\tilde{W}_s, \quad \langle U^\varepsilon\rangle_t := \int_0^t \dot{\psi}_s^\star b^{-1}(\xi_s^\varepsilon, s)\dot{\psi}_s ds,$$

$b(u, t) := B(u + y_t)$. Thus, for any measurable bounded function V on $C[0, T/\varepsilon]$ we have

$$EV(z^\varepsilon) = E\rho_{T/\varepsilon}(\xi^\varepsilon)V(\xi^\varepsilon). \tag{3.1.14}$$

By virtue of **H.3.1.1**, $f(\xi_t^\varepsilon, t) = \widehat{A}_t^\varepsilon \xi_t^\varepsilon$ where the matrix $\widehat{A}_t^\varepsilon := A(\xi_t^\varepsilon, y_t)$ is uniformly negative definite. Applying the Chebyshev inequality and the bound (1.1.21), we conclude that

$$\lim_{\varepsilon \to 0} P(\|\xi^\varepsilon\|_{T/\varepsilon} \geq r_\varepsilon) = 0 \tag{3.1.15}$$

where $r_\varepsilon := \sqrt{\beta|\ln \varepsilon|}$. Hence, for $\Gamma_1^\varepsilon := \{\|\xi^\varepsilon\|_{T/\varepsilon} < r_\varepsilon\}$ we have, when ε is sufficiently small, the inequality

$$P(\Gamma_1^\varepsilon) \geq 3/4. \tag{3.1.16}$$

Notice also that (1.1.9) implies the bound

$$E|\xi_t^\varepsilon| \leq C\beta \tag{3.1.17}$$

where C is a constant (depending on L).

Since **H.3.1.3** implies the Lipschitz condition for B^{-1} we have that

$$\dot{\psi}_s^\star b^{-1}(\xi_s^\varepsilon, s)\dot{\psi}_s \leq \dot{\psi}_s^\star B^{-1}(y_s)\dot{\psi}_s(1 + L|\xi_s^\varepsilon|). \tag{3.1.18}$$

The process U^ε is a square integrable martingale with the characteristics $\langle U^\varepsilon \rangle$. It follows from (3.1.17) and (3.1.18) that

$$E|U_{T/\varepsilon}^\varepsilon|^2 = E\langle U^\varepsilon\rangle_{T/\varepsilon} \leq S_T^\varepsilon(y)(1 + CL\beta).$$

For a fixed number $\lambda > 2$ we consider the set

$$\Gamma_2^\varepsilon := \left\{ U_{T/\varepsilon}^\varepsilon \geq -\lambda\sqrt{S_T^\varepsilon(y)}\right\}.$$

It is clear that

$$P(\bar{\Gamma}_2^\varepsilon) \leq P\left(|U_{T/\varepsilon}^\varepsilon| > \lambda\sqrt{S_T^\varepsilon(y)}\right) \leq \frac{E|U_{T/\varepsilon}^\varepsilon|^2}{\lambda^2 S_T^\varepsilon(y)} \leq \frac{1 + L\beta}{\lambda^2} \leq \frac{1}{4} \tag{3.1.19}$$

for sufficiently small ε.

Comparing the last bound with (3.1.16), we conclude that for any ε less than some positive ε_0

$$P(\Gamma_1^\varepsilon \cap \Gamma_2^\varepsilon) \geq 1/2.$$

Notice that on the set $\Gamma_1^\varepsilon \cap \Gamma_2^\varepsilon$

$$\rho_{T/\varepsilon}(\xi^\varepsilon) \geq \exp\left\{-\lambda\beta^{-1}\sqrt{S_T^\varepsilon(y)} - \beta^{-2}S_T^\varepsilon(y)(1 + Lr_\varepsilon)\right\}.$$

Using (3.1.14) and the above bound we have:

$$P(\|z^\varepsilon\|_{T/\varepsilon} < \eta) \geq EI_{\Gamma_1^\varepsilon}\rho_{T/\varepsilon}(\xi^\varepsilon) \geq EI_{\Gamma_1^\varepsilon \cap \Gamma_2^\varepsilon}\rho_{T/\varepsilon}(\xi^\varepsilon)$$

$$\geq \exp\left\{-\beta^{-2}[S_T^\varepsilon(y)(1 + Lr_\varepsilon) + \lambda\beta\sqrt{S_T^\varepsilon(y)}]\right\} P(\Gamma_1^\varepsilon \cap \Gamma_2^\varepsilon)$$

$$\geq \frac{1}{2}\exp\left\{-\beta^{-2}[S_T^\varepsilon(y)(1 + Lr_\varepsilon) + \lambda\beta\sqrt{S_T^\varepsilon(y)}]\right\}. \tag{3.1.20}$$

The assertion of the theorem follows from this in obvious way. $\square$

3.1.3 An Upper Bound for the Probability of Deviation of a Trajectory from the Lebesgue Sets of S_T^ε

Let us consider the Lebesgue set of the functional S_T^ε corresponding to the level h defined by

$$H^\varepsilon(h) := \{y \in C_0(\mathbf{R}_+) : \ S_T^\varepsilon(y) \leq h\} \tag{3.1.21}$$

Proposition 3.1.3 *Assume that* **H.3.1.1–H.3.1.3** *and* **H.3.1.4** *hold. Then for any* α, h, $\nu > 0$ *there exists* $\varepsilon_0 > 0$ *such that*

$$P(d_{T/\varepsilon}(\tilde{y}^\varepsilon, H^\varepsilon(h)) > \alpha) \leq \exp\{-\beta^{-2}(h - \nu)\} \tag{3.1.22}$$

for any positive $\varepsilon \leq \varepsilon_0$.

For the study of bounds like (3.1.22), the following notion is useful:

Definition. We say that the family of nonnegative random variables ξ^ε admits an *exponential majorant of order* $\theta = (\theta_\varepsilon)$ where $\theta_\varepsilon \to 0$ as $\varepsilon \to 0$ if there exists $a > 0$ such that for any $\alpha > 0$

$$P(\xi^\varepsilon > \alpha) \le \exp\{-\theta_\varepsilon^{-1} a\} \tag{3.1.23}$$

for all positive ε which is less than some $\varepsilon_0 = \varepsilon_0(\alpha, a) > 0$. A supremum of all a with the above property we shall call the *threshold*.

Thus, the statement of Proposition 3.1.3 means that for any h the family of random variables $d_{T/\varepsilon}(\tilde{y}^\varepsilon, H^\varepsilon(h))$ admits an exponential majorant of order β^2 with a threshold which is not less than h.

It is easy to verify that if ξ^ε and $\tilde{\xi}^\varepsilon$ admit exponential majorants of orders θ and $\tilde{\theta}$ then $\xi^\varepsilon + \tilde{\xi}^\varepsilon$ admits an exponential majorant of order $\theta \vee \tilde{\theta} := \max\{\theta, \tilde{\theta}\}$.

The proof of Proposition 3.1.3 is based on a very simple idea. Let us consider a piecewise linear function l^ε approximating the trajectory $\tilde{y}^\varepsilon$. It is clear that

$$\begin{aligned}
P(d_{T/\varepsilon}(\tilde{y}^\varepsilon, H^\varepsilon(h)) > \alpha) &\le P(S_T^\varepsilon(l^\varepsilon) > h) \\
&\quad + P(d_{T/\varepsilon}(\tilde{y}^\varepsilon, H^\varepsilon(h)) > \alpha, \ S_T^\varepsilon(l^\varepsilon) \le h) \\
&\le P(S_T^\varepsilon(l^\varepsilon) > h) + P(\|\tilde{y}^\varepsilon - l^\varepsilon\|_{T/\varepsilon} > \alpha)
\end{aligned}$$

and the problem is reduced to prove appropriate exponential bounds for two terms in the right-hand side of the last inequality for a suitably chosen l^ε. This is not very difficult since for piecewise linear functions the value of the functional $S_T^\varepsilon(l^\varepsilon)$ can be calculated in a rather explicit form. To realize this program we need some technical results concerning large deviation probabilities for an approximating discrete model.

Let $X_t^\varepsilon := \tilde{y}_t^\varepsilon - \tilde{y}_\infty$.

Using **H.3.1.1** we represent X_t^ε as the solution of the linear equation

$$dX_t^\varepsilon = A_t^\varepsilon X_t^\varepsilon dt + \beta G_t^\varepsilon d\tilde{W}_t, \quad X_0^\varepsilon = -\tilde{y}_\infty, \tag{3.1.24}$$

where $A_t^\varepsilon := A(\tilde{y}_t^\varepsilon, \tilde{y}_\infty)$, $G_t^\varepsilon := G(\tilde{y}_t^\varepsilon)$. Let us divide the interval $[0, T/\varepsilon]$ into N equal parts by the points $t_k := k\Delta$ with $\Delta := T/(\varepsilon N)$ choosing the length Δ (depending on ε as well as N) to satisfy the relations

$$\lim_{\varepsilon \to 0} \beta^2/(\varepsilon\Delta) = 0, \qquad \lim_{\varepsilon \to 0} \Delta = 0. \tag{3.1.25}$$

This is possible: by virtue of **H.3.1.4** $\beta_\varepsilon = \alpha_\varepsilon \varepsilon^{1/2}$ where $\alpha_\varepsilon \to 0$ as $\varepsilon \to 0$ and we can take, e.g., $N(\varepsilon)$ to be equal to the integer part of $(\alpha_\varepsilon \varepsilon)^{-1}$.

Let us consider the process ξ_t^ε which coincides on the interval $]t_{k-1}, t_k]$ with the solution of the equation

$$d\xi_t^\varepsilon = A_t^\varepsilon \xi_t^\varepsilon dt + \beta G_t^\varepsilon d\tilde{W}_t, \quad \xi_{t_{k-1}+}^\varepsilon = 0.$$

By the Cauchy formula we have for $t \in \,]t_{k-1}, t_k]$ that

$$X_t^\varepsilon = \Phi(t, t_{k-1}) X_{t_{k-1}}^\varepsilon + \xi_t^\varepsilon, \qquad (3.1.26)$$

where the fundamental matrix $\Phi(t, s)$ is the solution of the linear equation

$$\frac{\partial \Phi(t, s)}{\partial t} = A_t^\varepsilon \Phi(t, s), \quad \Phi(s, s) = I.$$

Let $X_k := X_{t_k}^\varepsilon$. It follows from (3.1.26) that

$$X_k = \Lambda_k X_{k-1} + \xi_k \qquad (3.1.27)$$

where

$$\Lambda_k := \Phi(t_k, t_{k-1}), \quad \xi_k := \xi_{t_k}^\varepsilon, \quad X_0 = -\tilde{y}_\infty$$

(certainly, X_k, Λ_k, and ξ_k depend on ε).

Lemma 3.1.4 *Assume that* **H.3.1.1**–**H.3.1.3** *and also* **H.3.1.4** *hold. Let* $\rho := e^{-\gamma \Delta}$. *Then*

$$\sum_{k=1}^{N} |X_k|^2 \leq \frac{2n|\tilde{y}_\infty|^2}{1 - \rho^2} + \frac{2n}{(1 - \rho)^2} \sum_{k=1}^{N} |\xi_k|^2 \qquad (3.1.28)$$

and for any $\lambda < (2L^2 \beta^2 \Delta)^{-1}$ *and* $\theta > 0$

$$P\left(\sum_{k=1}^{N} |X_k|^2 > \theta \right) \leq e^{-\lambda g(\theta)} (1 - 6\lambda L^2 \beta^2 \Delta)^{-N} \qquad (3.1.29)$$

where

$$g(\theta) := \frac{(1 - \rho)^2}{2n} \theta - \frac{1 - \rho}{1 + \rho} |\tilde{y}_\infty|^2.$$

Proof. Making use of the semigroup property of the fundamental matrix Φ we write the solution of (3.1.27) as follows:

$$X_k = \Lambda_k \dots \Lambda_1 X_0 + \sum_{j=1}^{k-1} \Lambda_k \dots \Lambda_{k-j+1} \xi_{k-j} + \xi_k$$

$$= -\Phi(t_k, 0)\tilde{y}_\infty + \sum_{j=0}^{k-1} \Phi(t_k, t_{k-j})\xi_{k-j}.$$

By virtue of (1.2.2) we have $|\Phi(t_k, t_{k-j})| \leq n^{1/2} \rho^j$. Thus,

$$|X_k| \leq n^{1/2} \rho^k |\tilde{y}_\infty| + n^{1/2} \sum_{j=0}^{k-1} \rho^j |\xi_{k-j}|.$$

It follows from the Cauchy–Schwarz inequality that

$$\sum_{k=1}^{N}|X_k|^2 \leq 2n|\tilde{y}_\infty|^2 \sum_{k=1}^{N}\rho^{2k} + 2n\sum_{k=1}^{N}\left(\sum_{j=0}^{k-1}\rho^j\right)\left(\sum_{j=0}^{k-1}\rho^j|\xi_{k-j}|^2\right)$$

$$\leq \frac{2n|\tilde{y}_\infty|^2}{1-\rho^2} + \frac{2n}{(1-\rho)^2}\sum_{k=1}^{N}|\xi_k|^2$$

and (3.1.28) is proved.

The bound (3.1.28) and the exponential Chebyshev inequality imply that

$$P\left(\sum_{k=1}^{N}|X_k|^2 > \theta\right) \leq P\left(\sum_{k=1}^{N}|\xi_k|^2 > g(\theta)\right) \leq e^{-\lambda g(\theta)}E\exp\left\{\lambda\sum_{k=1}^{N}|\xi_k|^2\right\}.$$

Applying (1.1.16) we get (3.1.29). □

Let us introduce the notations:

$$\mu_k := \sup_{t\in]t_{k-1},t_k]}|X_t^\varepsilon - X_{t_{k-1}}^\varepsilon| = \sup_{t\in]t_{k-1},t_k]}|\tilde{y}_t^\varepsilon - \tilde{y}_{t_{k-1}}^\varepsilon|,$$

$$u_k := \sup_{t\in]t_{k-1},t_k]}|\xi_t^\varepsilon|.$$

Lemma 3.1.5 *Assume that* **H.3.1.1–H.3.1.3** *and* **H.3.1.4** *hold and Δ satisfies (3.1.25). Then for any $\kappa \in [0,1[$ the family $\Delta^{-\kappa}\sum_{k=1}^{N}\mu_k^2$ admits an exponential majorant of order $\beta^2\Delta^\nu$ where ν is any number from the interval $[0,1-\kappa[$, i.e., for any $\alpha > 0$ and sufficiently small ε*

$$P\left(\Delta^{-\kappa}\sum_{k=1}^{N}\mu_k^2 > \alpha\right) \leq \exp\{-\beta^{-2}\Delta^{-\nu}\}. \tag{3.1.30}$$

Proof. It follows from (3.1.26) that

$$X_t^\varepsilon - X_{t_{k-1}}^\varepsilon = (I - \Phi(t, t_{k-1}))X_{t_{k-1}}^\varepsilon + \xi_t^\varepsilon, \quad t \in]t_{k-1}, t_k],$$

where I is the unit matrix. Taking into account **H.3.1.1** and the linear equation defining the fundamental matrix Φ, it is easy to deduce from here that

$$\mu_k \leq L\Delta|X_{k-1}| + |u_k|. \tag{3.1.31}$$

Assuming without loss of generality that $L \geq 1$ and estimating the sum of squares of $|X_k|$ by (3.1.28) we get that

$$\sum_{k=1}^{N}\mu_k^2 \leq 2L^2\left(\Delta^2\sum_{k=0}^{N}|X_k|^2 + \sum_{k=1}^{N}u_k^2\right)$$

$$\leq 4L^2 n \left(\frac{|\tilde{y}_\infty|^2 \Delta^2}{1 - \rho^2} + \frac{\Delta^2}{(1-\rho)^2} \sum_{k=1}^{N} |\xi_k|^2 + \sum_{k=1}^{N} u_k^2 \right)$$

$$\leq 4L^2 n (1 \vee |\tilde{y}_\infty|^2) \left(h_\Delta \frac{1-\rho}{1+\rho} + (h_\Delta + 1) \sum_{k=1}^{N} u_k^2 \right)$$

where $h_\Delta := \Delta^2 (1-\rho)^{-2} \to \gamma^{-2}$ as $\varepsilon \to 0$. Since $1 - \rho = O(\Delta)$, we have for sufficiently small ε the bound

$$P \left(\sum_{k=1}^{N} \mu_k^2 > \alpha \Delta^\kappa \right) \leq P \left(\sum_{k=1}^{N} u_k^2 > \alpha a \Delta^\kappa \right),$$

where $1/a = 16 L^2 n (1 \vee |\tilde{y}_\infty|^2)(1 + \gamma^{-2})$. Applying the exponential Chebyshev inequality with the exponent $\lambda = (12 L^2 \beta^2 \Delta)^{-1}$ and the bound (1.1.16) we get that

$$P \left(\sum_{k=1}^{N} u_k^2 > a\alpha \Delta^\kappa \right) \leq \exp\{ -a\alpha(12L^2)^{-1}\beta^{-2}\Delta^{\kappa-1} + N \ln 2 \}$$

$$= \exp\{ -\beta^{-2}\Delta^{-\nu}[a\alpha(12L^2)^{-1}\Delta^{\nu+\kappa-1} - \beta^2 \Delta^\nu N \ln 2] \}$$

$$\leq \exp\{ -\beta^{-2}\Delta^{-\nu} \} \qquad (3.1.32)$$

for sufficiently small ε, because

$$\lim_{\varepsilon \to 0} \beta^2 N = T \lim_{\varepsilon \to 0} \beta^2 / (\varepsilon \Delta) = 0 \qquad (3.1.33)$$

by virtue of (3.1.25). $\square$

We introduce the following abbreviations:

$$g_k(t) := G(\tilde{y}_t^\varepsilon) - G(\tilde{y}_{t_{k-1}}^\varepsilon),$$

$$\chi_k(t) := \Delta^{-1/2} \int_{t_{k-1}}^{t_k} g_t^k d\tilde{W}_t,$$

$$\chi_k := \chi_k(t_k).$$

Lemma 3.1.6 *Assume that* **H.3.1.1–H.3.1.3** *and* **H.3.1.4** *hold and Δ satisfies (3.1.25). Then for any $\eta \in [0, 1/4[$ the family $\beta^2 \Delta^{-\eta} \sum_{k=1}^{N} |\chi_k|^2$ admits an exponential majorant of order $\beta^2 \Delta^{1/4-\eta}$.*

Proof. Let us consider the stopping times

$$\tau_k := \inf\{t \geq t_{k-1} : |g_k(t)| \geq 2L\Delta^{1/8}\} \wedge t_k.$$

Since G is Lipschitz, in the set $\left\{ \sum_{k=1}^{N} \mu_k^2 \leq \Delta^{1/4} \right\}$ we have

$$\sup_{t \in [t_{k-1}, t_k]} |g_k(t)| \leq L\mu_k \leq L\Delta^{1/8},$$

and, hence, $\tau_k = t_k$.

Using the moment inequality for an even power of a stochastic integral with bounded integrand, we get for any $m \geq 1$ that

$$E|\chi_k(\tau_k)|^{2m}|\mathcal{F}_{t_{k-1}}) \leq (4n)^m(2m-1)!!L^{2m}\Delta^{m/4} \leq m!(8nL^2\Delta^{1/4})^m.$$

Thus, for $\lambda \in]0, (8nL^2\Delta^{1/4})^{-1}[$ we have

$$E\exp\{\lambda|\chi_k(\tau_k)|^2\}|\mathcal{F}_{t_{k-1}}) \leq 1 + \sum_{m=1}^{\infty}(8\lambda nL^2\Delta^{1/4})^m = (1 - 8\lambda nL^2\Delta^{1/4})^{-1}.$$

It follows that

$$E\exp\left\{\lambda\sum_{k=1}^{N}|\chi_k(\tau_k)|^2\right\} \leq (1 - 8\lambda nL^2\Delta^{1/4})^{-N}$$

for any $\lambda \in]0, (8nL^2\Delta^{1/4})^{-1}[$.

Using (3.1.30) (with $\kappa = \nu = 1/4$), the exponential Chebyshev inequality with $\lambda = (16nL^2\Delta^{1/4})^{-1}$, and the above bound, we have for sufficiently small ε:

$$P\left(\beta^2\Delta^{-\eta}\sum_{k=1}^{N}|\chi_k|^2 \geq \alpha\right)$$

$$\leq P\left(\sum_{k=1}^{N}\mu_k^2 \geq \Delta^{1/4}\right) + P\left(\sum_{k=1}^{N}|\chi_k(\tau_k)|^2 \geq \Delta^{\eta}\beta^{-2}\right)$$

$$\leq \exp\{-\beta^{-2}\Delta^{-1/4}\} + \exp\{-(16nL^2)^{-1}\beta^{-2}\Delta^{-1/4+\eta} + N\ln 2\}.$$

Since $\beta^2 N \to 0$ by virtue of (3.1.25), the assertions of Lemma 3.1.6 follow from here in obvious way. $\square$

Lemma 3.1.7 *Let ζ^k, $k \leq N$, be independent Gaussian vectors in $\mathbf{R}^n$ with zero means and unit covariance matrix, $N = N(\varepsilon) \to \infty$ as $\varepsilon \to 0$, and*

$$\lim_{\varepsilon \to 0}\beta^2 N = 0. \tag{3.1.34}$$

Then for any $h, \nu > 0$ and sufficiently small ε

$$P\left(\beta^2\sum_{k=1}^{N}|\zeta_k|^2 > 2h\right) \leq \exp\{-\beta^{-2}(h-\nu)\}. \tag{3.1.35}$$

Proof. Using the exponential Chebyshev inequality with $\kappa \in]0, 1/2[$ we have:

$$P\left(\sum_{k=1}^{N}|\zeta_k|^2 \geq 2\beta^{-2}h\right) \leq \exp\{-2\kappa\beta^{-2}\}(E\exp\{\kappa|\zeta_k|^2\})^N$$

$$= \exp\{-2\kappa\beta^{-2}\}(1-2\kappa)^{-nN/2}$$

$$= \exp\{-\beta^{-2}[2\kappa h + (1/2)nN\beta^2\ln(1-2\kappa)]\}. \tag{3.1.36}$$

Since the value of 2κ can be chosen arbitrarily close to unit, the bound (3.1.35) follows from (3.1.36) and the assumption (3.1.34). $\square$

Lemma 3.1.8 *Let*

$$\xi^\varepsilon := \sum_{k=1}^N |\zeta_k^\varepsilon|^2, \quad \tilde{\xi}^\varepsilon := \sum_{k=1}^N |\tilde{\zeta}_k^\varepsilon|^2, \quad \eta^\varepsilon := \sum_{k=1}^N |\zeta_k^\varepsilon||\tilde{\zeta}_k^\varepsilon|, \tag{3.1.37}$$

where the families of random variables ξ^ε and $\tilde{\xi}^\varepsilon$ admit exponential majorants of order θ^ε and $\tilde{\theta}^\varepsilon$. Then η^ε admits an exponential majorant of order $\theta^\varepsilon \vee \tilde{\theta}^\varepsilon$.

Proof. Making use of the Cauchy–Schwarz inequality, for sufficiently small ε we have:

$$P\left(\sum_{k=1}^N |\zeta_k^\varepsilon||\tilde{\zeta}_k^\varepsilon| > \alpha\right) \leq P\left(\sum_{k=1}^N |\zeta_k^\varepsilon|^2 \sum_{k=1}^N |\tilde{\zeta}_k^\varepsilon|^2 > \alpha^2\right)$$

$$\leq P\left(\sum_{k=1}^N |\zeta_k^\varepsilon|^2 > \alpha\right) + P\left(\sum_{k=1}^N |\tilde{\zeta}_k^\varepsilon|^2 > \alpha\right)$$

$$\leq \exp\{-a(\theta^\varepsilon)^{-1}\} + \exp\{-\tilde{a}(\tilde{\theta}^\varepsilon)^{-1}\}$$

$$\leq 2\exp\{-a \wedge \tilde{a}(\theta^\varepsilon \vee \tilde{\theta}^\varepsilon)^{-1}\}$$

$$\leq \exp\{-(1/2)a \wedge \tilde{a}(\theta^\varepsilon \vee \tilde{\theta}^\varepsilon)^{-1}\}$$

and the desired assertion holds. $\square$

Let $l = l^\varepsilon$ be the piecewise linear function with the graph connecting successive points $(t_k, \tilde{y}_{t_k}^\varepsilon)$, i.e.,

$$l := \sum_{k=1}^N l^k I_{[t_{k-1}, t_k[}, \tag{3.1.38}$$

$$l_t^k := (\tilde{y}_{t_k}^\varepsilon - \tilde{y}_{t_{k-1}}^\varepsilon)\frac{t - t_{k-1}}{\Delta} + \tilde{y}_{t_{k-1}}^\varepsilon. \tag{3.1.39}$$

Lemma 3.1.9 *Assume that* **H.3.1.1–H.3.1.3** *and* **H.3.1.4** *hold and Δ satisfies (3.1.25). Then for any $\nu > 0$ and sufficiently small ε*

$$P(S_T^\varepsilon(l) > h) \leq \exp\{-\beta^{-2}(h - \nu)\}. \tag{3.1.40}$$

Proof. Let us consider the random variables

$$\rho_k := \beta\Delta^{-1}G(\tilde{y}_{t_{k-1}}^\varepsilon)(\tilde{W}_{t_k} - \tilde{W}_{t_{k-1}}),$$

$$\chi_k := \Delta^{-1/2}\int_{t_{k-1}}^{t_k}[G(\tilde{y}_t^\varepsilon) - G(\tilde{y}_{t_{k-1}}^\varepsilon)]d\tilde{W}_t,$$

$$\alpha_k := \beta\Delta^{-1/2}\chi_k.$$

For $t \in [t_{k-1}, t_k]$ put

$$D_t^k := \Delta^{-1} \int_{t_{k-1}}^{t_k} F(\tilde{y}_s^\varepsilon) ds - F(l_t^k),$$

$$\zeta_t^k := \rho_k + \alpha_k + D_t^k = \Delta^{-1}(\tilde{y}_{t_k}^\varepsilon - \tilde{y}_{t_{k-1}}^\varepsilon) - F(l_t^k) = \dot{l}_t^k - F(l_t^k),$$

$$\Psi_t^k := B^{-1}(l_t^k) - B^{-1}(\tilde{y}_{t_{k-1}}^\varepsilon).$$

It follows from the definitions that

$$2S_T^\varepsilon(l) = M + r_1 + r_2,$$

where

$$M := \sum_{k=1}^{N} \Delta \rho_k^\star B^{-1}(\tilde{y}_{t_{k-1}}^\varepsilon) \rho_k = \sum_{k=1}^{N} \Delta |B^{-1/2}(\tilde{y}_{t_{k-1}}^\varepsilon) \rho_k|^2$$

$$= \sum_{k=1}^{N} \beta^2 \Delta^{-1} |\tilde{W}_{t_k} - \tilde{W}_{t_{k-1}}|^2,$$

$$r_1 := \sum_{k=1}^{N} \int_{t_{k-1}}^{t_k} \zeta_t^{k\star} \Psi_t^k \zeta_t^k dt,$$

$$r_2 := \sum_{k=1}^{N} \int_{t_{k-1}}^{t_k} \left(|B^{-1/2}(\tilde{y}_{t_{k-1}}^\varepsilon)(\alpha_k + D_t^k)|^2 + 2\zeta_t^{k\star} B^{-1}(\tilde{y}_{t_{k-1}}^\varepsilon)(\alpha_k + D_t^k) \right) dt.$$

Since $G^\star B^{-1} G = I$, we have

$$\beta^2 \Delta \rho_k^\star B^{-1}(\tilde{y}_{t_{k-1}}^\varepsilon) \rho_k = \beta^2 \Delta^{-1} |\tilde{W}_{t_k} - \tilde{W}_{t_{k-1}}|^2,$$

and by virtue of Lemma 3.1.7 for any positive h and ν

$$P(M > 2h) \leq \exp\{-\beta^{-2}(h - \nu/2)\} \tag{3.1.41}$$

when ε is sufficiently small. Thus, to prove the desired assertion we need to check that $|r_i|$ admits exponential majorants of order $\beta^2 \Delta^\kappa$ with $\kappa > 0$.

The Lipschitz condition for B^{-1} and F leads to the bounds

$$||\Psi^k||_{t_{k-1}, t_k} \leq L\mu_k,$$

$$||D^k||_{t_{k-1}, t_k} \leq \Delta^{-1} \int_{t_{k-1}}^{t_k} |F(\tilde{y}_s^\varepsilon) - F(\tilde{y}_{t_{k-1}}^\varepsilon)| ds + ||F(l^k) - F(\tilde{y}_{t_{k-1}}^\varepsilon)||_{t_{k-1}, t_k}$$

$$\leq 2L\mu_k,$$

implying that

$$|r_1| \leq C \sum_{k=1}^{N} (\Delta|\rho_k|^2 + \beta^2|\chi_k|^2 + \Delta|\mu_k|^2)\mu_k, \tag{3.1.42}$$

$$|r_2| \leq C \sum_{k=1}^{N} (\beta^2|\chi_k|^2 + \Delta|\mu_k|^2 + \beta\Delta|\rho_k||\chi_k| + \Delta|\rho_k||\mu_k|). \tag{3.1.43}$$

It follows from (3.1.41) that for any positive κ the family $\Delta^\kappa M$ admits an exponential majorant of order $\beta^2 \Delta^\nu$ for any $\nu \in \,]0, \kappa[$. Since $B^{-1/2}$ is bounded the same property holds for the family $\sum_{k=1}^n \Delta |\rho_k|^2$.

The desired assertion on the existence of the exponential majorants for $|r_i|$ follows from (3.1.42), (3.1.43) and Lemmas 3.1.6–3.1.8. $\square$

Lemma 3.1.10 *Assume that* **H.3.1.1–H.3.1.3** *and* **H.3.1.4** *hold, and* Δ *satisfies (3.1.25). Then for any* $\alpha > 0$, $\nu \in \,]0, 1[$, *and sufficiently small* ε

$$P(\|\tilde{y}^\varepsilon - l^\varepsilon\|_{T/\varepsilon} > \alpha) \leq \exp\{-\beta^{-2}\Delta^{-\nu}\}. \tag{3.1.44}$$

In other words, $\|\tilde{y}^\varepsilon - l^\varepsilon\|_{T/\varepsilon}$ admits an exponential majorant of order $\beta^2 \Delta^\nu$.

Proof. Notice that

$$\|\tilde{y}^\varepsilon - l^\varepsilon\|_{t_{k-1}, t_k} \leq 2 \sup_{t \in [t_{k-1}, t_k]} |\tilde{y}^\varepsilon_t - \tilde{y}^\varepsilon_{t_{k-1}}| = 2\mu_k.$$

Thus,

$$P(\|\tilde{y}^\varepsilon - l^\varepsilon\|_{T/\varepsilon} > \alpha) \leq P\left(\bigcup_{k=1}^N \{\mu_k > \alpha/2\}\right) \leq P\left(\sum_{k=1}^N |\mu_k|^2 > \alpha^2/4\right)$$
$$\leq \exp\{-\beta^{-2}\Delta^{-\nu}\}$$

where the last inequality holds by virtue of Lemma 3.1.5. $\square$

Proof of Proposition 3.1.3. Since

$$P(d_{T/\varepsilon}(\tilde{y}^\varepsilon, H^\varepsilon(h)) > \alpha) \leq P(S_T^\varepsilon(l^\varepsilon) > h) + P(\|\tilde{y}^\varepsilon - l^\varepsilon\|_{T/\varepsilon} > \alpha),$$

the desired assertion follows from the bounds (3.1.34) and (3.1.44). $\square$

3.1.4 Proof of Theorem 3.1.1

First, we show that

$$\liminf_{\varepsilon \to 0} \beta^2 \ln P(\|\tilde{y}^\varepsilon - \tilde{y}\|_{T/\varepsilon} > \eta) \geq -\pi(\eta). \tag{3.1.45}$$

By the definition of $\pi(\eta)$, for any $\sigma > 0$ there exists a function $y^\sigma \in \Gamma(\eta)$ with

$$S(y^\sigma) < \pi(\eta) + \sigma. \tag{3.1.46}$$

But

$$\Gamma(\eta) = \bigcup_{\varepsilon > 0} \Gamma^\varepsilon(\eta),$$

where $\Gamma^\varepsilon(\eta) := \{y \in C_0(\mathbf{R}_+) : \|y - \tilde{y}\|_{T/\varepsilon} > \eta\}$ are open sets in $C_0(\mathbf{R}_+)$ increasing as ε decreases. Obviously, $y^\sigma \in \Gamma^\varepsilon(\eta)$ for all sufficiently small $\varepsilon > 0$. Moreover, there exists $\lambda > 0$ such that

$$B^\varepsilon(\lambda) := \{y \in C_0(\mathbf{R}_+): \ ||y - y^\sigma||_{T/\varepsilon} < \lambda\} \subseteq \Gamma^\varepsilon(\eta)$$

for all sufficiently small ε. Thus,

$$P(||\tilde{y}^\varepsilon - \tilde{y}||_{T/\varepsilon} > \eta) = P(\tilde{y}^\varepsilon \in \Gamma^\varepsilon(\eta))$$
$$\geq P(\tilde{y}^\varepsilon \in B^\varepsilon(\lambda)) = P(||\tilde{y}^\varepsilon - y^\sigma||_{T/\varepsilon} < \lambda). \quad (3.1.47)$$

The inequality (3.1.45) follows from (3.1.47), Proposition 3.1.2, and (3.1.46).

To finish the proof of the theorem it remains to show that

$$\limsup_{\varepsilon \to 0} \beta^2 \ln P(||\tilde{y}^\varepsilon - \tilde{y}||_{T/\varepsilon} > \eta) \leq -\pi(\eta). \quad (3.1.48)$$

We shall consider only the nontrivial case of $\pi(\eta) > 0$. Let

$$\pi_\varepsilon(\eta) := \inf_{y \in \Gamma^\varepsilon(\eta)} S_T^\varepsilon(y).$$

For any function $y \in \Gamma^\varepsilon(\eta)$ there exists a function $\widehat{y} \in \Gamma^\varepsilon(\eta)$ with $S(\widehat{y}) = S_T^\varepsilon(y)$ (one can take as $\widehat{y}$ the continuous function which coincides with y on $[0, T/\varepsilon]$ and goes further along the integral curve of the associated equation (3.0.2)). It follows that

$$\pi_\varepsilon(\eta) = \inf_{y \in \Gamma^\varepsilon(\eta)} S(y). \quad (3.1.49)$$

Since $\Gamma(\eta)$ is the union of the sets $\Gamma^\varepsilon(\eta)$, we have

$$\lim_{\varepsilon \to 0} \pi_\varepsilon(\eta) = \pi(\eta). \quad (3.1.50)$$

Fix $\sigma \in]0, \pi(\eta)[$. Let $h_\varepsilon := \pi_\varepsilon(\eta) - \sigma$,

$$H^\varepsilon := H^\varepsilon(h_\varepsilon) := \{y \in C_0(\mathbf{R}_+): \ S_T^\varepsilon(y) \leq h_\varepsilon\}.$$

The function h_ε as well as the sets H^ε decreases as ε decreases, h_ε is positive and bounded from above by $2\pi(\eta)$ for sufficiently small ε; we shall consider this only from now on.

Let $\rho_\varepsilon := d_{T/\varepsilon}(\Gamma^\varepsilon(\eta), H^\varepsilon)$.

Lemma 3.1.11 *There exists $\rho > 0$ such that $\rho^\varepsilon > \rho$ for all sufficiently small ε.*

The above assertion implies that

$$\Gamma^\varepsilon(\eta) \subseteq \{y \in C(\mathbf{R}_+): \ d_{T/\varepsilon}(y, H^\varepsilon) > \rho\}$$

for sufficiently small ε and, so,

$$P(||\tilde{y}^\varepsilon - \tilde{y}||_{T/\varepsilon} > \eta) \leq P(d_{T/\varepsilon}(\tilde{y}^\varepsilon, H^\varepsilon) > \rho).$$

The desired inequality (3.1.48) follows from here by virtue of Proposition 3.1.3.

To prove Lemma 3.1.11 we need another auxiliary assertion.

Lemma 3.1.12 *There exists a constant $C = C(\eta)$ such that for any $y \in H^\varepsilon$*

$$\|y\|_{T/\varepsilon} \leq C, \tag{3.1.51}$$

$$|y_t - y_s| \leq C\sqrt{t - s} \tag{3.1.52}$$

for all $t, s \in [0, T/\varepsilon]$, $t > s$, $t - s \leq 1$.

Proof. Let $y \in H^\varepsilon$ and $\xi_t := B^{-1/2}(y_t)(\dot{y}_t - F(y_t))$. By the definition of the sets H^ε

$$\int_0^{T/\varepsilon} |\xi_t|^2 dt \leq 2h_\varepsilon \leq 4\pi(\eta). \tag{3.1.53}$$

The assumption **H.3.1.1** implies that $F(y_t) = A(y_t, \tilde{y}_\infty)(y_t - \tilde{y}_\infty)$. Plugging this expression into the definition of ξ_t we get that

$$\dot{y}_t = A(y_t, \tilde{y}_\infty)(y_t - \tilde{y}_\infty) + B^{1/2}(y_t)\xi_t.$$

By the Cauchy formula

$$y_t - \tilde{y}_\infty = -\Phi(t, 0)\tilde{y}_\infty + \int_0^t \Phi(t, s)B^{1/2}(y_s)\xi_s ds \tag{3.1.54}$$

where the fundamental matrix $\Phi(t, s)$ is the solution of the equation

$$\frac{\partial \Phi(t, s)}{\partial t} = A(y_t, \tilde{y}_\infty)\Phi(t, s), \quad \Phi(s, s) = I. \tag{3.1.55}$$

According to (1.2.2) it admits the bound

$$|\Phi(t, s)| \leq \sqrt{n}e^{-\gamma(t-s)}. \tag{3.1.56}$$

It follows from (3.1.53), (3.1.54), (3.1.56), the assumption **H.3.1.3**, and the Cauchy–Schwarz inequality that for $t \in [0, T/\varepsilon]$

$$|y_t| \leq (\sqrt{n} + 1)|\tilde{y}_\infty| + L\sqrt{n}\int_0^t e^{-\gamma(t-s)}|\xi_s|ds$$

$$\leq (\sqrt{n} + 1)|\tilde{y}_\infty| + 2L\gamma^{-1}\sqrt{n}\sqrt{\pi(\eta)}$$

implying (3.1.51).

Using (3.1.53), we deduce in a similar way that

$$|y_t - y_s| \leq \int_s^t |\dot{y}_u|du \leq \int_s^t |F(y_u)|du + \int_s^t |B^{1/2}(y_u)\xi_u|du$$

$$\leq (t - s)\sup_{|x| \leq C} |F(x)| + 2L\sqrt{(t - s)\pi(\eta)}.$$

Obviously, the bound (3.1.52) follows from here. Lemma 3.1.12 is proved. □

Proof of Lemma 3.1.11. Assume that the assertion of the lemma fails. Then there exist a sequence of positive numbers ε_n decreasing to zero and sequences of functions $x^n \in H^n$ and $y^n \in \Gamma^n(\eta)$ such that $\lambda^n := ||x^n - y^n||_{T_n}$ tends to zero. We use here the simplified notations H^n, h_n, $\Gamma^n(\eta)$, and T_n instead H^{ε_n}, h_{ε_n}, $\Gamma^{\varepsilon_n}(\eta)$, and T/ε_n.

The condition $y^n \in \Gamma^n(\eta)$ implies that there exists a point $t_n \in [0, T_n]$ such that

$$|y_{t_n}^n - \tilde{y}_{t_n}| > \eta. \tag{3.1.57}$$

Let (n') be a subsequence such that $t_{n'} \to a$ where $a \le T$. All x^n belong to the set H^1. The set of the restrictions of functions from H^ε to the interval $[0, T/\varepsilon]$ is the set $\{x \in C_0[0, T/\varepsilon] : S_T^\varepsilon(x) \le h_\varepsilon\}$ which is a compact in $C_0[0, T/\varepsilon]$ (see [27]). Thus, we can extract from (n') a subsequence (n'') such that $x^{n''}$ (hence, $y^{n''}$) converges uniformly on $[0, T/\varepsilon_1]$ to some function $\bar{y}$. Taking in (3.1.57) a partial limit along the subsequence (n'') we obtain the inequality $|\bar{y}_a - \tilde{y}_a| \ge \eta$ which can hold only if $a > 0$. Thus,

$$\liminf_{n \to \infty} t_n > 0. \tag{3.1.58}$$

Let $\Delta_n := \lambda_n^{1/2}$, $r_n := t_n - \Delta_n$. It follows from (3.1.58) that $r_n > 0$ for sufficiently large n; we shall consider this only. Let us introduce the function

$$z_t^n := x_t^n I_{[0,r_n[}(t) + \left(\frac{y_{t_n}^n - x_{r_n}^n}{\Delta_n}(t - r_n) + x_{r_n}^n \right) I_{[r_n, t_n[}(t) + \tilde{z}_t^n I_{[t_n, \infty[}(t), \tag{3.1.59}$$

where $\tilde{z}^n$ is the integral curve of the associated equation (3.0.2) such that $\tilde{z}_{t_n}^n = y_{t_n}^n$. Clearly, $z^n \in \Gamma^n(\eta)$.

In the following calculations we shall use the notations

$$D(x_t^n) := \dot{x}_t^n - F(x_t^n), \quad \xi_t^n := B^{-1/2}(x_t^n)D(x_t^n). \tag{3.1.60}$$

Taking into account that $x^n \in H^n$ and $t_n \le T_n$ we have:

$$2h_n \ge \int_0^{t_n} |\xi_t^n|^2 dt = \int_0^{r_n} |\xi_t^n|^2 dt + \int_{r_n}^{t_n} |B^{-1/2}(x_{r_n}^n)D(x_t^n)|^2 dt + R_1^n \tag{3.1.61}$$

where

$$R_1^n := \int_{r_n}^{t_n} |B^{-1/2}(x_t^n)D(x_t^n)|^2 dt - \int_{r_n}^{t_n} |B^{-1/2}(x_{r_n}^n)D(x_t^n)|^2 dt$$

$$= \int_{r_n}^{t_n} D(x_t^n)^\star (B^{-1}(x_t^n) - B^{-1}(x_{r_n}^n))D(x_t^n)dt.$$

Using the Lipschitz condition for B^{-1} and applying the inequality (3.1.52) to the increments of x^n we get that

$$|R_1^n| \leq L \int_{r_n}^{t_n} |x_t^n - x_{r_n}^n| |D(x_t^n)|^2 dt \leq \Delta_n^{1/2} L \int_0^{t_n} |D(x_t^n)|^2 dt \to 0 \quad (3.1.62)$$

(integrals on the right-hand side of the last inequality are bounded due to (3.1.60), (3.1.61) and the boundedness of B).

Let

$$D_1^n := \Delta_n^{-1} \left(x_{t_n}^n - x_{r_n}^n - \int_{r_n}^{t_n} F(x_t^n) dt \right),$$

It follows from the Cauchy–Schwarz inequality that

$$\int_{r_n}^{t_n} |B^{-1/2}(x_{r_n}^n) D(x_t^n)|^2 dt \geq \Delta_n^{-1} \left| \int_{r_n}^{t_n} B^{-1/2}(x_{r_n}^n) D(x_t^n) dt \right|^2$$

$$= \Delta_n |B^{-1/2}(x_{r_n}^n) D_1^n|^2$$

$$= \int_{r_n}^{t_n} |B^{-1/2}(z_t^n) D(z_t^n)|^2 dt + R_2^n + R_3^n \quad (3.1.63)$$

where

$$R_2^n := \int_{r_n}^{t_n} D(z_t^n)^\star (B^{-1}(x_{r_n}^n) - B^{-1}(z_t^n)) D(z_t^n) dt,$$

$$R_3^n := \int_{r_n}^{t_n} (|B^{-1/2}(x_{r_n}^n) D_1^n|^2 - |B^{-1/2}(x_{r_n}^n) D(z_t^n)|^2) dt.$$

Using the Lipschitz condition for B^{-1} and the inequality (3.1.52) applied to x^n, we get that for $t \in [r_n, t_n]$

$$|B^{-1}(x_{r_n}^n) - B^{-1}(z_t^n)| \leq L|x_{r_n}^n - z_t^n| \leq L|y_{t_n}^n - x_{r_n}^n|$$

$$\leq L|y_{t_n}^n - x_{t_n}^n| + L|x_{t_n}^n - x_{r_n}^n| \quad (3.1.64)$$

$$\leq L\lambda_n + LC\Delta_n^{1/2}.$$

Applying again Lemma 3.1.12, we have on the interval $[r_n, t_n]$:

$$|D(z_t^n)| = |\Delta_n^{-1}(y_{t_n}^n - x_{r_n}^n) - F(z_t^n)|$$

$$\leq \Delta_n^{-1}|y_{t_n}^n - x_{t_n}^n| + \Delta_n^{-1}|x_{t_n}^n - x_{r_n}^n| + |F(z_t^n)|$$

$$\leq C\lambda_n^{1/2} + C\Delta_n^{-1/2} + \sup_{|x| \leq C} |F(x)|.$$

Thus, there exists a constant c such that

$$|D(z_t^n)| \leq c\Delta_n^{-1/2}. \quad (3.1.65)$$

Similar arguments leads to the inequality

$$|D_1^n| \leq c\Delta_n^{-1/2}. \quad (3.1.66)$$

Substituting the bounds (3.1.65) and (3.1.65) into the expression for R_2^n we get that

$$|R_2^n| \le \int_{r_n}^{t_n} |D(z_t^n)|^2 |B^{-1}(x_{r_n}^n) - B^{-1}(z_t^n)|dt \le c^2 L(\lambda_n + C\Delta_n^{1/2}) \to 0.$$
(3.1.67)

The integrand in the definition of R_3^n we can represent as the product $K_t^n M_t^n$ where

$$K_t^n := |B^{-1/2}(x_{r_n}^n)D_1^n| - |B^{-1/2}(x_{r_n}^n)D(z_t^n)|,$$
$$M_t^n := |B^{-1/2}(x_{r_n}^n)D_1^n| + |B^{-1/2}(x_{r_n}^n)D(z_t^n)|.$$

It is easily seen that on the interval $[r_n, t_n]$

$$|K_t^n| \le |B^{-1/2}(x_{r_n}^n)(D_1^n - D(z_t^n))|$$

$$\le L\Delta_n^{-1}|y_{t_n}^n - x_{t_n}^n| + |F(x_{r_n}^n) - F(z_t^n)| + \Delta_n^{-1}\int_{r_n}^{t_n}|F(x_{r_n}^n) - F(x_t^n)|dt$$

$$\le L(\lambda_n^{1/2} + \lambda_n + \Delta_n^{1/2}).$$
(3.1.68)

It follows from (3.1.65), (3.1.68), and the boundedness of $B^{-1/2}$ that

$$|M_t^n| \le c_1 \Delta_n^{-1/2}.$$
(3.1.69)

The inequalities (3.1.68) and (3.1.69) show that the function $|K_t^n M_t^n|$ is bounded on the interval $[r_n, t_n]$. Hence, $R_3^n \to 0$.

It follows from (3.1.61) and (3.1.63) that

$$2h_n \ge \int_0^{r_n} |\xi_t^n|^2 dt + \int_{r_n}^{t_n} |B^{-1/2}(z_t^n)D(z_t^n)|^2 dt + R_1^n + R_2^n + R_3^n$$
$$= 2S(z_n) + R_1^n + R_2^n + R_3^n$$
$$\ge 2\pi_n(\eta) + R_1^n + R_2^n + R_3^n$$

where the last inequality holds since $z_n \in \Gamma^n(\eta)$. Taking a limit as $n \to \infty$ and using (3.1.50) we get the inequality $2\pi(\eta) - 2\sigma \ge 2\pi(\eta)$ which contradicts to the assumption that σ is positive.

Lemma 3.1.11 is proved and the proof of Theorem 3.1.1 is completed. $\square$

3.1.5 Example: the Ornstein–Uhlenbeck Process

Proposition 3.1.13 *Let y^ε be the Ornstein–Uhlenbeck procces satisfying the linear SDE with constant coefficients*

$$\varepsilon dy_t^\varepsilon = Ay_t^\varepsilon dt + \beta\varepsilon^{1/2}Gdw_t, \quad y_0^\varepsilon = y^o,$$
(3.1.70)

where the matrix A is symmetric and stable, the matrix $B := GG^\star$ is nondegenerated and commuting with A, and $0 < \beta = O(\varepsilon^{1/2})$. Then

$$\lim_{\varepsilon \to 0} \beta^2 \ln P(\|\tilde{y}^\varepsilon - \tilde{y}\|_{T/\varepsilon} > \eta) = -\eta^2 \lambda_{\min}(-AB^{-1}).$$
(3.1.71)

Proof. By virtue of Theorem 3.1.1 it remains to check that

$$\pi(\eta) := \frac{1}{2} \inf_{y \in \Gamma(\eta)} \int_0^\infty |B^{-1/2}(\dot{y}_s - Ay_s)|^2 ds = \eta^2 \lambda_{\min}(-AB^{-1}) \qquad (3.1.72)$$

where $\Gamma(\eta)$ is the set of all absolute continuous functions y such that $y_0 = 0$ and $||y - \tilde{y}||_\infty > \eta$ for $\tilde{y}_t = e^{At}y^\circ$.

We give the arguments in the scalar case where $A = -\gamma < 0$ and $G = g \neq 0$ and the claim is

$$\pi(\eta) = \eta^2 \gamma / g.$$

For $y \in \Gamma(\eta)$ there is finite $T > 0$ and u with $|u| = \eta$ such that $y_T - \tilde{y}_T = u$. Obviously,

$$S(y) \geq \frac{1}{2g} \int_0^T (\dot{y}_t + \gamma y_t)^2 dt \geq \frac{1}{2g} \widehat{J}_T$$

where

$$\widehat{J}_T = \inf \int_0^T (\dot{v}_t + \gamma v_t)^2 dt,$$

the infimum is taken over all v with $v_0 = y^\circ$ and $v_T = u$. It is attained on the function $v := (v)_{t \leq T}$ with $v_0 := 0$ and

$$v_t := u \frac{e^{\gamma t} - e^{-\gamma t}}{e^{\gamma T} - e^{-\gamma T}}, \quad t > 0,$$

which is the only solution of the Euler equation $v'' = \gamma^2 v$ satisfying the boundary conditions. So,

$$\widehat{J}_T = 2\eta^2 \gamma \frac{e^{2\gamma T} - 1}{(e^{\gamma T} - e^{-\gamma T})^2} = 2\eta^2 \gamma \left(1 + \frac{1 - e^{-2\gamma T}}{(e^{\gamma T} - e^{-\gamma T})^2}\right) \geq 2\eta^2 \gamma$$

and $S(y) \geq \eta^2 \gamma / g$.

On the other hand, $S(y^N) \to \eta^2 \gamma / g$ as $N \to \infty$ for the sequence of functions

$$y_t^N := (\eta + 1/N) \frac{e^{\gamma t} - e^{-\gamma t}}{e^{\gamma N} - e^{-\gamma N}} I_{[0,N]}(t) + (\eta + 1/N) e^{-\gamma(t-N)} I_{]N,\infty[}(t).$$

The proof in the vector case follows the same line and is left to the reader. □

3.2 Deviations in the Metric of $L^2[0, T]$

In this section we shall study, using in $C[0,T]$ the norm $|||.|||_T$ of the space $L^2[0,T]$, the probability of deviation of the process y^ε defined by the equation

(3.0.1) from the constant function identically equal to $\tilde{y}_\infty$, the rest point of the associated equation.

Let us introduce the following hypothesis.

H.3.2.1 There exists a function $Q : \mathbf{R}^n \to] - \infty, 0]$ with the bounded continuous second derivatives such that $B^{-1}(x)F(x) = Q'(x)$.

In other words, we assume that the vector field $B^{-1}(x)F(x)$ is the gradient of a certain scalar potential Q.

Notice that in the scalar case it is always possible to take as Q the function

$$Q(x) = \int_{\tilde{y}_\infty}^x B^{-1}(v)F(v)dv.$$

It is clear from this formula that the second derivative of Q is bounded if F, F', and B' are bounded or B is a constant and F' is bounded.

Let us consider on $C[0, T]$ the functional

$$\Sigma_T(y) := \frac{1}{2}|||B^{-1/2}(y)F(y)|||_T^2 = \frac{1}{2}\int_0^T |B^{-1/2}(y_s)F(y_s)|^2 ds. \qquad (3.2.1)$$

Notice that in contrast with the functional

$$S_T(y) = (1/2)|||B^{-1/2}(y)(\dot{y} - F(y))|||_T^2$$

the above definition does not involve the derivative.

Define the set

$$\Upsilon(\eta) := \{y \in C[0, T] : \ |||y - \tilde{y}_\infty|||_T^2 > \eta\},$$

which is the exterior of the L^2-ball of the radius $\eta^{1/2}$ with center at $\tilde{y}_\infty$ (of course, the last symbol here denotes a constant function).

Theorem 3.2.1 *Assume that* **H.3.1.1**–**H.3.1.3**, **H.3.1.5**, *and* **H.3.2.1** *hold. Then for any* $\eta > 0$

$$\lim_{\varepsilon \to 0} \beta^2 \varepsilon \ln P(|||y^\varepsilon - \tilde{y}_\infty|||_T^2 > \eta) = -\pi_1(\eta) \qquad (3.2.2)$$

where

$$\pi_1(\eta) := \inf_{y \in \Upsilon(\eta)} \Sigma_T(y). \qquad (3.2.3)$$

Example. Let us consider the linear stochastic equation with constant coefficients

$$\varepsilon dy_t^\varepsilon = Ay_t^\varepsilon dt + \beta \varepsilon^{1/2}Gdw_t, \quad y_0^\varepsilon = y^o, \qquad (3.2.4)$$

where A is symmetric with $\lambda(A) < 0$, the matrix $B := GG^\star$ is nondegenerated and commuting with A, and $0 < \beta = o(|\ln \varepsilon|^{-1/2})$. The conditions of the above theorem are fulfilled; in particular, one can take $Q(y) = (1/2)y^\star B^{-1}Ay$

(notice that $(GG^\star)^{-1}A$ is symmetric negative definite). In this model $\tilde{y}_\infty = 0$ and

$$\pi_1(\eta) = \frac{1}{2}\eta\lambda_{\min}(A^\star B^{-1}A). \tag{3.2.5}$$

The latter formula is obvious since

$$\Sigma_T(y) = \frac{1}{2}\int_0^T y_s^\star A^\star B^{-1}Ay_s ds \geq \frac{1}{2}\lambda_{\min}(A^\star B^{-1}A)\int_0^T |y_s|^2 ds$$

with the equality if y is the constant function equal to an eigenvector corresponding to the minimal eigenvalue of $A^\star B^{-1}A$.

Thus, Theorem 3.2.1 claims that

$$\lim_{\varepsilon\to 0}\beta^2\varepsilon\ln P(|||y^\varepsilon|||_T > \eta) = -\frac{1}{2}\eta\lambda_{\min}(A^\star B^{-1}A). \tag{3.2.6}$$

We split the proof of Theorem 3.2.1 into two steps.

Proposition 3.2.2 *Assume that H.3.1.1–H.3.1.3 and H.3.1.5 hold. Let y be a function from $C_0[0, T]$. Then for all ν, $\eta > 0$ there exists $\varepsilon_0 > 0$ such that*

$$P(||y^\varepsilon - y||_T < \eta) \geq \exp\{-\beta^{-2}\varepsilon^{-1}[\Sigma_T(y) + \nu]\} \tag{3.2.7}$$

for all positive $\varepsilon \leq \varepsilon_0$.

Proof. First, we prove (3.2.7) for y from $C_0^a[0, T]$, the space of absolute continuous functions. Put

$$Z_t^\varepsilon := y_t^\varepsilon - y_t, \quad \psi_t^\varepsilon := -y_t + \varepsilon^{-1}\int_0^t F(y_s)ds.$$

It follows from (3.0.1) that

$$dZ_t^\varepsilon = \varepsilon^{-1}f(Z_t^\varepsilon, t)dt + \dot{\psi}_t^\varepsilon dt + \varepsilon^{-1/2}\beta g(Z_t^\varepsilon, t)dW_t, \quad Z_0^\varepsilon = 0, \tag{3.2.8}$$

where $f(z, t) := F(z + y_t) - F(y_t)$, $g(z, t) := G(z + y_t)$.

Put $u_t^\varepsilon = y_{t\varepsilon}$. Making use of Proposition 3.1.$\dot{2}$ we get that

$$P(||y^\varepsilon - y||_T < \eta) = P(||\tilde{y}^\varepsilon - u^\varepsilon||_{T/\varepsilon} < \eta)$$

$$\geq \frac{1}{2}\exp\{-\beta^{-2}S_T^\varepsilon(u^\varepsilon)(1 + \nu)\} \tag{3.2.9}$$

for sufficiently small ε. Here

$$S_T^\varepsilon(u^\varepsilon) = \frac{1}{2}\int_0^{T/\varepsilon} |B^{-1/2}(u_t^\varepsilon)(\dot{u}_t^\varepsilon - F(u_t^\varepsilon))|^2 dt$$

$$= \frac{1}{2\varepsilon}\int_0^T |B^{-1/2}(y_t)(\varepsilon\dot{y}_t - F(y_t))|^2 dt$$

$$= \frac{1}{\varepsilon}\Sigma_T(y) - \int_0^T \dot{y}_t^T B^{-1}(y_t)F(y_t)dt + \frac{\varepsilon}{2}\int_0^T \dot{y}_t^\star B^{-1}(y_t)\dot{y}_t dt. \tag{3.2.10}$$

For absolute continuous functions the inequality (3.2.7) is implied by (3.2.9) and (3.2.10). To extend it to all functions $y \in C_0[0, T]$ we proceed as follows. Let us choose $x^\eta \in C_0^a[0, T]$ such that

$$||y - x^\eta||_T < \min\{\eta/2, \nu/(3L^3 T)\}. \qquad (3.2.11)$$

Then

$$P(||y^\varepsilon - y||_T < \eta) \geq P(||y^\varepsilon - x^\eta||_T < \eta/2)$$
$$\geq \exp\{-\beta^{-2}\varepsilon^{-1}[\Sigma_T(x^\eta) + \nu/2]\}. \qquad (3.2.12)$$

It is easy to check using the "telescope" that if f_i are (matrix-valued) functions bounded by a constant C and satisfying the Lipschitz condition with a constant L then the product $f_1 \ldots f_n$ satisfies the Lipschitz condition with the constant nLC^{n-1}. Thus,

$$|F(y)^\star B^{-1}(y)F(y) - F(z)^\star B^{-1}(z)F(z)| \leq 3L^3|y - z|$$

and we have

$$|\Sigma_T(x^\eta) - \Sigma_T(y)| \leq \frac{3}{2}L^3 \int_0^T |y_t - x_t^\eta| dt \leq \nu/2. \qquad (3.2.13)$$

The needed assertion follows from (3.2.11)–(3.2.13). □

Corollary 3.2.3 *Assume that the conditions of Proposition 3.2.2 are fulfilled. Let y be a function from $C[0, T]$. Then for any ν, $\eta > 0$ there exists $\varepsilon_0 > 0$ such that*

$$P\left(\int_0^T |y_t^\varepsilon - y_t|^2 dt < \eta\right) \geq \exp\{-\beta^{-2}\varepsilon^{-1}[\Sigma_T(y) + \nu]\}$$

for all positive $\varepsilon \leq \varepsilon_0$.

Proof. Let us choose a function $x^\eta \in C_0[0, T]$ such that $\int_0^T |y_t - x_t^\eta|^2 dt < \eta/4$. Applying Proposition 3.2.2 we have:

$$P\left(\int_0^T |y_t^\varepsilon - y_t|^2 dt < \eta\right) \geq P\left(\int_0^T |y_t^\varepsilon - x_t^\eta|^2 dt < \eta/4\right)$$
$$\geq P(||y^\varepsilon - x^\eta||_T < (\eta T/4)^{1/2})$$
$$\geq \exp\{-\beta^{-2}\varepsilon^{-1}[\Sigma_T(x^\eta) + \nu]\}$$

for sufficiently small $\varepsilon > 0$. □

Proposition 3.2.4 *Assume that* **H.3.1.1**–**H.3.1.3**, **H.3.1.5**, *and* **H.3.2.1** *hold. Then for all h, $\nu > 0$ there exists $\varepsilon_0 > 0$ such that*

$$P(\Sigma_T(y^\varepsilon) > h) \leq \exp\{-\beta^{-2}\varepsilon^{-1}(h - \nu)\} \qquad (3.2.14)$$

for all $\varepsilon \leq \varepsilon_0$.

Proof. Let

$$\xi_t^\varepsilon := \int_0^t F(y_s^\varepsilon)^\star B^{-1}(y_s^\varepsilon)F(y_s^\varepsilon)ds - 2\varepsilon \int_0^t F(y_s^\varepsilon)^\star B^{-1}(y_s^\varepsilon)dy_s^\varepsilon$$

$$= -\int_0^t F(y_s^\varepsilon)^\star B^{-1}(y_s^\varepsilon)F(y_s^\varepsilon)ds + 2\beta\varepsilon^{1/2}\int_0^t F(y_s^\varepsilon)^\star B^{-1}(y_s^\varepsilon)G(y_s^\varepsilon)dw_s.$$

It is clear that

$$E \exp\{(1/2)\beta^{-2}\varepsilon^{-1}\xi_T^\varepsilon\} = 1. \tag{3.2.15}$$

It follows from the Ito formula and the gradient condition **H.3.2.1** (requiring that $F(x)^\star B^{-1}(x) = Q'(x)$) that

$$\int_0^T F(y_s^\varepsilon)^\star B^{-1}(y_s^\varepsilon)dy_s^\varepsilon = Q(y_T^\varepsilon) - Q(0) - \frac{\beta^2}{2\varepsilon}\int_0^T \text{tr}\, Q''(y_s^\varepsilon)B(y_s^\varepsilon)ds. \tag{3.2.16}$$

Notice that the absolute value of the last integral is bounded by some constant. According to **H.3.2.1** the function Q is negative. Thus, the left hand side of the above identity is less or equal to $-Q(0) + C\beta^2/\varepsilon$ and we have that

$$P(\Sigma_T(y^\varepsilon) > h) = P\left(\int_0^T F(y_s^\varepsilon)^\star B^{-1}(y_s^\varepsilon)F(y_s^\varepsilon)ds > 2h\right)$$

$$= P\left(\xi_T^\varepsilon > 2h - 2\varepsilon\int_0^T F(y_s^\varepsilon)^\star B^{-1}(y_s^\varepsilon)dy_s^\varepsilon\right)$$

$$\leq P(\xi_T^\varepsilon > 2(h + \varepsilon Q(0) - \beta^2 C))$$

$$\leq \exp\{-\beta^{-2}\varepsilon^{-1}(h + \varepsilon Q(0) - \beta^2 C)\}E \exp\{(1/2)\beta^{-2}\varepsilon^{-1}\xi_T^\varepsilon\}$$

$$\leq \exp\{-\beta^{-2}\varepsilon^{-1}(h - \nu)\}$$

for all sufficiently small $\varepsilon > 0$. $\square$

The assertion of Theorem 3.2.1 follows from Propositions 3.2.2 and 3.2.4 by usual way. These results imply also

Corollary 3.2.5 *Assume that* **H.3.1.1**–**H.3.1.3**, **H.3.1.5**, *and also* **H.3.2.1** *hold. Then*

(a) for any measurable set A in $C[0,T]$ which is open in topology $L^2[0,T]$

$$\liminf_{\varepsilon\to 0} \beta^2\varepsilon \ln P(y^\varepsilon \in A) \geq -\inf_{y\in A} \Sigma_T(y);$$

(b) for any measurable A in $C[0,T]$ such that $\inf_{y\in A} \Sigma_T(y) < \infty$

$$\limsup_{\varepsilon\to 0} \beta^2\varepsilon \ln P(y^\varepsilon \in A) \leq -\inf_{y\in A} \Sigma_T(y);$$

(c) if $\inf_{y\in A} \Sigma_T(y) < \infty$ then

$$\limsup_{\varepsilon\to 0} \beta^2\varepsilon \ln P(y^\varepsilon \in A) = -\infty.$$

4 Uniform Expansions for Two-Scale Systems

We continue here the study of asymptotic expansions started in Chapter 2. Our results are inspired by the Vasil'eva theorem providing, for the Tikhonov system

$$dx_t^\varepsilon = f(t, x_t^\varepsilon, y_t^\varepsilon)dt, \quad x_0^\varepsilon = x^o, \qquad (4.0.1)$$
$$\varepsilon dy_t^\varepsilon = F(t, x_t^\varepsilon, y_t^\varepsilon)dt, \quad y_0^\varepsilon = y^o, \qquad (4.0.2)$$

an asymptotic expansion for both variables, uniform on the whole interval $[0, T]$. This expansion has the form

$$x_t^\varepsilon = \sum_{k \geq 0} \varepsilon^k (x_t^k + \tilde{x}_t^{k,\varepsilon}), \qquad (4.0.3)$$

$$y_t^\varepsilon = \sum_{k \geq 0} \varepsilon^k (y_t^k + \tilde{y}_t^{k,\varepsilon}), \qquad (4.0.4)$$

where $\tilde{x}_t^{k,\varepsilon} = \tilde{x}_{t/\varepsilon}^k$, $\tilde{y}_t^{k,\varepsilon} = \tilde{y}_{t/\varepsilon}^k$. The essential property of the "boundary layer functions" $\tilde{x}^k$ and $\tilde{x}^k$ is that they are exponentially decreasing at infinity and this requirement allows us to define them uniquely, using a rather simple algorithm (but calculations are tedious).

Our main goal is to find a uniform asymptotic approximation on the whole interval $[0, T]$ for the solution of the two-scale system given by the Ito equations

$$dx_t^\varepsilon = f(t, x_t^\varepsilon, y_t^\varepsilon)dt + g(t, x_t^\varepsilon, y_t^\varepsilon)dw_t^x, \quad x_0^\varepsilon = x^o, \qquad (4.0.5)$$
$$\varepsilon dy_t^\varepsilon = F(t, x_t^\varepsilon, y_t^\varepsilon)dt + \beta\varepsilon^{1/2}G(t, x_t^\varepsilon, y_t^\varepsilon)dw_t^y, \quad y_0^\varepsilon = y^o, \qquad (4.0.6)$$

generalizing the Vasil'eva theorem to the stochastic case. Notice that the stochastic model is much more complicated because it contains two small parameters: ε and β (it is, in fact, three-scale!). We begin in Section 4.1 with a particular but important case where $\beta = 0$ and construct an asymptotic expansion, uniform on the whole interval $[0, T]$. It is worth noting its specific features. First of all, it is an expansion in a power series in $\sqrt{\varepsilon}$ and not in ε as in the deterministic case. Moreover, our boundary layer correction functions are stochastic processes which do not vanish at infinity, except the $\tilde{y}^{0,\varepsilon}$ ($\tilde{v}^{0,\varepsilon}$ in notations of Section 4.1), the boundary layer correction term of zero order for the fast variable, removing the discrepancy in the initial conditions, which is a process, exponentially decreasing at infinity. As for all other correction terms, we can guarantee only that they are not of the "fast" growth.

To explain the idea of Section 4.2, recall that in Sections 2.2 and 2.3 we studied the model with only fast variables (and time-invariant coefficients). In stretched time the resulting SDE depends only on the parameter β and we obtained easily for this case the asymptotic expansions of its solution of in a power series in β. The result suggested in Section 4.2 is, actually, an expansion of the solution of (4.0.5), (4.0.6) in power series in β as ε would be frozen. The initial approximation is the solution of (4.0.5), (4.0.6) with $\beta = 0$.

We shall assume that the basic conditions of the Tikhonov theory are fulfilled:

H.4.0.1 The functions f, F and g, G are continuous in all variables and satisfy the linear growth and local Lipschitz conditions in (x, y).

H.4.0.2 There is a function $\varphi : [0, T] \times \mathbf{R}^k \to \mathbf{R}^n$ satisfying the linear growth and local Lipschitz conditions in x such that $F(t, x, \varphi(t, x)) = 0$ for all (t, x).

H.4.0.3 The solution of the problem

$$d\tilde{y}_s = F(0, x^o, \tilde{y}_s)ds, \quad \tilde{y}_0 = y^o, \qquad (4.0.7)$$

tends to $\varphi(0, x^o)$ as $s \to \infty$:

$$\lim_{s \to \infty} \tilde{y}_s = \varphi(0, x^o). \qquad (4.0.8)$$

H.4.0.4 The derivative F_y exists and is a continuous function on the set $[0, T] \times \mathbf{R}^k \times \mathbf{R}^n$ such that for any $N > 0$ there is a constant $\kappa_N > 0$ such that

$$\operatorname{Re} \lambda(F_y(t, x, \varphi(t, x))) < -\kappa_N$$

for all $(t, x) \in [0, T] \times \mathbf{R}^k$ with $|x| \le N$.

Further smoothness conditions, needed to get asymptotic expansions, will be given later. To conclude, in Section 4.3 we consider an example of the Liénard oscillator under slightly different hypotheses to cover some important cases arising in applications.

4.1 No Diffusion at the Fast Variable

4.1.1 Formal Calculations

We consider a particular case of the model (4.0.5), (4.0.6) where $\beta = 0$, i.e., the system of the form

$$du_t^\varepsilon = f(t, u_t^\varepsilon, v_t^\varepsilon)dt + g(t, u_t^\varepsilon, v_t^\varepsilon)dw_t, \quad u_0^\varepsilon = x^o, \qquad (4.1.1)$$

$$\varepsilon dv_t^\varepsilon = F(t, u_t^\varepsilon, v_t^\varepsilon)dt, \quad v_0^\varepsilon = y^o. \qquad (4.1.2)$$

The model of a such type arises, as we mentioned in Chapter 0, in the study of influence of random perturbations upon the Liénard oscillator:

$$\varepsilon \frac{d^2 v}{dt^2} + k\frac{dv}{dt} + h(v) = \frac{dw}{dt}$$

where dw/dt is a formal expression for the "white noise". Putting

$$u = \varepsilon dv/dt + kv,$$

we get an equivalent system

$$\frac{du}{dt} = -h(v) + \frac{dw}{dt}, \qquad \varepsilon\frac{dv}{dt} = u - kv,$$

which can be rewritten in a more familiar (and rigorous) way as

$$du_t = -h(v_t)dt + dw_t, \qquad (4.1.3)$$

$$\varepsilon dv_t = (u_t - kv_t)dt. \qquad (4.1.4)$$

Obviously, the latter system is a particular case of (4.1.1), (4.1.2); an asymptotic expansion for the solution of (4.1.3), (4.1.4) will be analyzed in Section 4.3.

Now we do some formal calculations explaining the structure of the coefficients in the asymptotic expansion for the solution $(u_t^\varepsilon, v_t^\varepsilon)$ of (4.1.1), (4.1.2) taken in the form

$$u_t^\varepsilon = \sum_{k\geq 0} \varepsilon^{k/2}u_t^k + \sum_{k\geq 0} \varepsilon^{k/2}\tilde{u}_t^{k,\varepsilon}, \qquad (4.1.5)$$

$$v_t^\varepsilon = \sum_{k\geq 0} \varepsilon^{k/2}v_t^k + \sum_{k\geq 0} \varepsilon^{k/2}\tilde{v}_t^{k,\varepsilon}. \qquad (4.1.6)$$

Here $\tilde{u}_t^{k,\varepsilon}$ and $\tilde{v}_t^{k,\varepsilon}$ are the boundary layer correction terms of order k.

It is convenient to introduce special notations for "long" vectors by setting

$$z_t^\varepsilon := (u_t^\varepsilon, v_t^\varepsilon), \qquad Z_t^k := (u_t^k, v_t^k), \qquad \tilde{Z}_t^{k,\varepsilon} := (\tilde{u}_t^{k,\varepsilon}, \tilde{v}_t^{k,\varepsilon}). \qquad (4.1.7)$$

Suppose that the coefficients in (4.1.5), (4.1.6) are Ito processes, i.e., that Z_t^k and $\tilde{Z}_t^{k,\varepsilon}$ admit the stochastic differentials

$$dZ_t^k = Z_t^{1,k}dt + Z_t^{2,k}dw_t, \qquad (4.1.8)$$

$$d\tilde{Z}_t^{k,\varepsilon} = \tilde{Z}_t^{1,k,\varepsilon}dt + \tilde{Z}_t^{2,k,\varepsilon}dw_t. \qquad (4.1.9)$$

Put also

$$Z_t^\varepsilon := \sum_{k \geq 0} \varepsilon^{k/2} Z_t^k, \qquad \tilde{Z}_t^\varepsilon := \sum_{k \geq 0} \varepsilon^{k/2} \tilde{Z}_t^{k,\varepsilon}. \qquad (4.1.10)$$

Let us write the formal asymptotic expansions for F, f, and g in power series of $\varepsilon^{1/2}$. As in Section 2.3 we obtain that

$$F_t^\varepsilon := F(t, Z_t^\varepsilon) = \sum_{k \geq 0} \varepsilon^{k/2} F_t^k.$$

In accordance with (2.3.3)–(2.3.5) the coefficients here have the form

$$F_t^0 = F(t, Z_t^0), \qquad F_t^k = F_u^0 u_t^k + F_v^0 v_t^k + R_t^{F,k}, \quad k \geq 1, \qquad (4.1.11)$$

where

$$F_u^0 := \frac{\partial F}{\partial u}(t, Z_t^0),$$

$$F_v^0 := \frac{\partial F}{\partial v}(t, Z_t^0),$$

$R_t^{F,1} = 0$, and the vector function $R_t^{F,k}$ for $k \geq 2$ is of the form

$$R_t^{F,k} := R^{F,k}(t, Z_t^0, \ldots, Z_t^{k-1})$$

for a certain function

$$R^{F,k}(t, c_0, c_1, \ldots, c_{k-1})$$

which is a polynomial of $c_1, \ldots, c_{k-1}$ with the coefficients which are (up to constants) derivatives of F at the point c_0; see the discussion at the beginning of Section 2.3.

In the same way we define the coefficients f_t^k and g_t^k in the expansions in $\varepsilon^{1/2}$ of the functions $f_t^\varepsilon := f(t, Z_t^\varepsilon)$ and $g_t^\varepsilon := g(t, Z_t^\varepsilon)$.

Put

$$\tilde{F}_t^\varepsilon := F(t, Z_t^\varepsilon + \tilde{Z}_t^\varepsilon) - F(t, Z_t^\varepsilon).$$

In the asymptotic expansion

$$\tilde{F}_t^\varepsilon = \sum_{k \geq 0} \varepsilon^{k/2} \tilde{F}_t^{k,\varepsilon}$$

the coefficients have the form

$$\tilde{F}_t^{0,\varepsilon} = F(t, Z_t^{0,\varepsilon} + \tilde{Z}_t^{0,\varepsilon}) - F(t, Z_t^{0,\varepsilon}),$$

$$\tilde{F}_t^{k,\varepsilon} = F_u^{0,\varepsilon} \tilde{u}_t^{k,\varepsilon} + F_v^{0,\varepsilon} \tilde{v}_t^{k,\varepsilon} + R_t^{F,k,\varepsilon}, \quad k \geq 1, \qquad (4.1.12)$$

where

$$F_u^{0,\varepsilon} := \frac{\partial F}{\partial u}(t, Z_t^0 + \tilde{Z}_t^{0,\varepsilon}), \qquad F_v^{0,\varepsilon} := \frac{\partial F}{\partial v}(t, Z_t^0 + \tilde{Z}_t^{0,\varepsilon}),$$

and

$$R_t^{F,k,\varepsilon} := R^{F,k}(t, Z_t^0 + \tilde{Z}_t^{0,\varepsilon}, Z_t^1 + \tilde{Z}_t^{1,\varepsilon}, \ldots, Z_t^k, \ldots, Z_t^{k-1} + \tilde{Z}_t^{k-1,\varepsilon}) - F_t^k.$$

In the same way we define the coefficients $\tilde{f}_t^{k,\varepsilon}$ and $\tilde{g}_t^{k,\varepsilon}$ in the asymptotic expansions in the power series in $\varepsilon^{1/2}$ for the functions

$$\tilde{f}_t^\varepsilon = f(t, Z_t^\varepsilon + \tilde{Z}_t^\varepsilon) - f(t, Z_t^\varepsilon)$$

and

$$\tilde{g}_t^\varepsilon = g(t, Z_t^\varepsilon + \tilde{Z}_t^\varepsilon) - g(t, Z_t^\varepsilon).$$

According to our definitions,

$$F(t, u_t^\varepsilon, v_t^\varepsilon) = F_t^\varepsilon + \tilde{F}_t^\varepsilon = \sum_{k \geq 0} \varepsilon^{k/2} F_t^k + \sum_{k \geq 0} \varepsilon^{k/2} \tilde{F}_t^{k,\varepsilon} \tag{4.1.13}$$

and similar formulae can be written for $f(t, u_t^\varepsilon, v_t^\varepsilon)$ and $F(t, u_t^\varepsilon, v_t^\varepsilon)$.

Substitute formally the series (4.1.6) and (4.1.13) into the ordinary differential equation (4.1.2) describing the behavior of the fast variable. Integrating from zero to t, we get:

$$\varepsilon \sum_{k \geq 0} \varepsilon^{k/2} v_t^k + \varepsilon \sum_{k \geq 0} \varepsilon^{k/2} \tilde{v}_t^{k,\varepsilon} = \varepsilon y^o + \int_0^t F_s^\varepsilon ds + \int_0^t \tilde{F}_s^\varepsilon ds$$

$$= \varepsilon y^o + \sum_{k \geq 0} \varepsilon^{k/2} \int_0^t F_s^k ds + \sum_{k \geq 0} \varepsilon^{k/2} \int_0^t \tilde{F}_s^{k,\varepsilon} ds.$$

In particular, for $t = 0$,

$$\sum_{k \geq 0} \varepsilon^{k/2} (v_0^k + \tilde{v}_0^{k,\varepsilon}) = y^o. \tag{4.1.14}$$

Taking into account (4.1.14) as well as the assumption (4.1.8) we come to the identity

$$\sum_{k \geq 0} \varepsilon^{k/2+1} \int_0^t v_s^{1,k} ds + \sum_{k \geq 0} \varepsilon^{k/2+1} \int_0^t v_s^{2,k} dw_s$$

$$+ \sum_{k \geq 0} \varepsilon^{k/2+1} \int_0^t \tilde{v}_s^{1,k,\varepsilon} ds + \sum_{k \geq 0} \varepsilon^{k/2+1} \int_0^t \tilde{v}_s^{1,k,\varepsilon} dw_s$$

$$= \sum_{k \geq 0} \varepsilon^{k/2} \int_0^t F_s^k ds + \sum_{k \geq 0} \varepsilon^{k/2} \int_0^t \tilde{F}_s^{k,\varepsilon} ds. \tag{4.1.15}$$

The above formula involves three types of processes of a different nature: processes of bounded variation corresponding to outer expansions, processes of bounded variation originated from boundary layer correction terms, and martingales. It seems reasonable to expect that (4.1.15) holds due to the following three identities corresponding to the processes of each type, i.e.,

$$\sum_{k\geq 0} \varepsilon^{(k+2)/2} \int_0^t v_s^{1,k} ds = \sum_{k\geq 0} \varepsilon^{k/2} \int_0^t F_s^k ds, \qquad (4.1.16)$$

$$\sum_{k\geq 0} \varepsilon^{(k+2)/2} \int_0^t \tilde{v}_s^{1,k,\varepsilon} ds = \sum_{k\geq 0} \varepsilon^{k/2} \int_0^t \tilde{F}_s^{k,\varepsilon} ds, \qquad (4.1.17)$$

$$\sum_{k\geq 0} \varepsilon^{(k+2)/2} \int_0^t v_s^{2,k} dw_s + \sum_{k\geq 0} \varepsilon^{(k+2)/2} \int_0^t \tilde{v}_s^{2,k,\varepsilon} dw_s = 0. \qquad (4.1.18)$$

In the deterministic theory boundary layer functions can be identified by a condition that they are exponentially decreasing to zero at infinity. In our stochastic version, *we retain the exponential bound only for* $\tilde{v}_t^{0,\varepsilon}$ assuming that it does not contain a martingale component, that is,

$$\tilde{v}_t^{2,0,\varepsilon} = 0. \qquad (4.1.19)$$

Equalizing the coefficients in (4.1.16) at the same power of $\varepsilon^{1/2}$ we obtain the following equations:

$$F_t^0 = 0, \quad F_t^1 = 0, \quad F_t^k = v_t^{1,k-2}, \quad k \geq 2. \qquad (4.1.20)$$

If we put

$$\varepsilon \tilde{v}_t^{1,k,\varepsilon} = \tilde{F}_t^{k,\varepsilon}, \quad k \geq 0, \qquad (4.1.21)$$

the identity (4.1.17) will be fulfilled. Taking into account (4.1.19), it is easy to see that if

$$v_t^{2,k-1} + \varepsilon^{1/2} \tilde{v}_t^{2,k,\varepsilon} = 0, \quad k \geq 1, \qquad (4.1.22)$$

then (4.1.18) will be also fulfilled.

In a similar way, working with the equation (4.1.1) we come to the initial condition

$$\sum_{k\geq 0} \varepsilon^{k/2} (u_0^k + \tilde{u}_0^{k,\varepsilon}) = x^o \qquad (4.1.23)$$

and the following equations for $u_t^{i,k}$ and $\tilde{u}_t^{i,k,\varepsilon}$:

$$\sum_{k\geq 0} \varepsilon^{k/2} u_t^{1,k} = \sum_{k\geq 0} \varepsilon^{k/2} f_t^k, \qquad (4.1.24)$$

$$\sum_{k\geq 0} \varepsilon^{k/2} u_t^{2,k} = \sum_{k\geq 0} \varepsilon^{k/2} g_t^k, \qquad (4.1.25)$$

$$\sum_{k\geq 0} \varepsilon^{k/2} \tilde{u}_t^{1,k,\varepsilon} = \sum_{k\geq 0} \varepsilon^{k/2} \tilde{f}_t^{k,\varepsilon}, \qquad (4.1.26)$$

$$\sum_{k\geq 0} \varepsilon^{k/2} \tilde{u}_t^{2,k,\varepsilon} = \sum_{k\geq 0} \varepsilon^{k/2} \tilde{g}_t^{k,\varepsilon}. \qquad (4.1.27)$$

We take the function $\tilde{u}_t^{0,\varepsilon}$ identically equal to zero, i.e.,

$$\tilde{u}_t^{1,0,\varepsilon} = \tilde{u}_t^{2,0,\varepsilon} = 0. \tag{4.1.28}$$

The equations (4.1.24)–(4.1.27) are satisfied if

$$u_t^{1,k} = f_t^k, \quad u_t^{2,k} = g_t^k, \quad k \geq 0; \tag{4.1.29}$$

$$\tilde{u}_t^{1,1,\varepsilon} = \tilde{f}_t^{1,\varepsilon}, \quad \tilde{u}_t^{1,2,\varepsilon} = \varepsilon^{-1}\tilde{f}_t^{0,\varepsilon} + \tilde{f}_t^{2,\varepsilon}; \tag{4.1.30}$$

$$\tilde{u}_t^{1,k,\varepsilon} = \tilde{f}_t^{k,\varepsilon}, \quad k \geq 2; \tag{4.1.31}$$

$$\tilde{u}_t^{2,1,\varepsilon} = \varepsilon^{-1/2}\tilde{g}_t^{0,\varepsilon} + \tilde{g}_t^{1,\varepsilon}; \tag{4.1.32}$$

$$\tilde{u}_t^{2,k,\varepsilon} = \tilde{g}_t^{k,\varepsilon}, \quad k \geq 2. \tag{4.1.33}$$

Let us substitute (4.1.11) and (4.1.12) into (4.1.20)–(4.1.21) and the corresponding expressions for f_t^k, $\tilde{f}_t^{k,\varepsilon}$ and g_t^k, $\tilde{g}_t^{k,\varepsilon}$ into (4.1.29)–(4.1.33). According to (4.1.8)

$$du_t^k = u_t^{1,k}dt + u_t^{2,k}dw_t, \quad k \geq 0,$$

and we get from (4.1.29) with $k = 0$ and the first equation in (4.1.20) the system

$$du_t^0 = f(t, u_t^0, v_t^0)dt + g(t, u_t^0, v_t^0)dw_t, \tag{4.1.34}$$

$$v_t^0 = \varphi(t, u_t^0); \tag{4.1.35}$$

from (4.1.29) with $k = 1$ and the second equation in (4.1.20) the system

$$du_t^1 = (f_u^0 u_t^1 + f_v^0 v_t^1)dt + (g_u^0 u_t^1 + g_v^0 v_t^1)dw_t, \tag{4.1.36}$$

$$F_u^0 u_t^1 + F_v^0 v_t^1 = 0; \tag{4.1.37}$$

and, in a similar way, for $k \geq 2$ the system

$$du_t^k = (f_u^0 u_t^k + f_v^0 v_t^k + R_t^{f,k})dt + (g_u^0 u_t^k + g_v^0 v_t^k + R_t^{g,k})dw_t, \tag{4.1.38}$$

$$F_u^0 u_t^k + F_v^0 v_t^k + R_t^{F,k} = v_t^{1,k-2}. \tag{4.1.39}$$

Now we write the equations for $\tilde{Z}_t^{k,\varepsilon}$. We have (from (4.1.28) and (4.1.21) with $k = 0$) for the zero-order boundary layer functions the system

$$d\tilde{u}_t^{0,\varepsilon} = 0, \tag{4.1.40}$$

$$\varepsilon d\tilde{v}_t^{0,\varepsilon} = F(t, Z_t^0 + \tilde{Z}_t^{0,\varepsilon})dt; \tag{4.1.41}$$

for the first-order boundary layer functions the system

$$d\tilde{u}_t^{1,\varepsilon} = (f_u^\varepsilon \tilde{u}_t^{1,\varepsilon} + f_v^\varepsilon \tilde{v}_t^{1,\varepsilon} + R_t^{f,1,\varepsilon})dt + (\varepsilon^{-1/2}\tilde{g}_t^{0,\varepsilon} + g_u^\varepsilon \tilde{u}_t^{1,\varepsilon} + g_v^\varepsilon \tilde{v}_t^{1,\varepsilon} + R_t^{g,1,\varepsilon})dw_t, \tag{4.1.42}$$

$$\varepsilon d\tilde{v}_t^{1,\varepsilon} = (F_u^{0,\varepsilon}\tilde{u}_t^{1,\varepsilon} + F_v^{0,\varepsilon}\tilde{v}_t^{1,\varepsilon} + R_t^{F,1,\varepsilon})dt - \varepsilon^{1/2}v_t^{2,0}dw_t; \qquad (4.1.43)$$

for the second-order boundary layer functions the system

$$
\begin{aligned}
d\tilde{u}_t^{2,\varepsilon} &= (\varepsilon^{-1}\tilde{f}_t^{0,\varepsilon} + f_u^{0,\varepsilon}\tilde{u}_t^{2,\varepsilon} + f_v^{0,\varepsilon}\tilde{v}_t^{2,\varepsilon} + R_t^{f,2,\varepsilon})dt \\
&\quad + (g_u^{0,\varepsilon}\tilde{u}_t^{2,\varepsilon} + g_v^{0,\varepsilon}\tilde{v}_t^{2,\varepsilon} + R_t^{g,2,\varepsilon})dw_t, \qquad (4.1.44)
\end{aligned}
$$

$$\varepsilon d\tilde{v}_t^{2,\varepsilon} = (F_u^{0,\varepsilon}\tilde{u}_t^{2,\varepsilon} + F_v^{0,\varepsilon}\tilde{v}_t^{2,\varepsilon} + R_t^{F,2,\varepsilon})dt - \varepsilon^{1/2}v_t^{2,1}dw_t; \qquad (4.1.45)$$

and for $k > 2$ the system

$$
\begin{aligned}
d\tilde{u}_t^{k,\varepsilon} &= (f_u^{0,\varepsilon}\tilde{u}_t^{k,\varepsilon} + f_v^{0,\varepsilon}\tilde{v}_t^{k,\varepsilon} + R_t^{f,k,\varepsilon})dt \\
&\quad + (g_u^{0,\varepsilon}\tilde{u}_t^{k,\varepsilon} + g_v^{0,\varepsilon}\tilde{v}_t^{k,\varepsilon} + R_t^{g,k,\varepsilon})dw_t, \qquad (4.1.46)
\end{aligned}
$$

$$\varepsilon d\tilde{v}_t^{k,\varepsilon} = (F_u^{0,\varepsilon}\tilde{u}_t^{k,\varepsilon} + F_v^{0,\varepsilon}\tilde{v}_t^{k,\varepsilon} + R_t^{F,k,\varepsilon})dt - \varepsilon^{1/2}v_t^{2,k-1}dw_t. \qquad (4.1.47)$$

The initial conditions for the systems (4.1.34)–(4.1.39) and (4.1.40)–(4.1.47) should be taken to satisfy the asymptotic expansions (4.1.14) and (4.1.23). We set

$$u_0^0 = x^o, \quad \tilde{u}_0^{0,\varepsilon} = 0, \quad v_0^0 = \varphi(0, x^o), \quad \tilde{v}_0^{0,\varepsilon} = y^o - \varphi(0, x^o), \qquad (4.1.48)$$

and

$$u_0^k = \tilde{u}_0^k = 0, \quad v_0^k = -\tilde{v}_0^k, \qquad (4.1.49)$$

implying that

$$u_0^k + \tilde{u}_0^k = 0, \quad v_0^k + \tilde{v}_0^k = 0.$$

Remark. There is a freedom in the choice of initial values. In the deterministic Vasil'eva theorem they are chosen to provide an exponential decay of the boundary layer functions. In order to retain the regular part of the expansions (4.1.5)–(4.1.6) in accordance with the deterministic case we could set the following initial conditions:

$$u_0^{2i} = -\tilde{u}_0^i, \quad v_0^{2i} = -\tilde{v}_0^i, \quad \tilde{u}_0^{2i,\varepsilon} = \tilde{u}_0^i, \quad \tilde{v}_0^{2i,\varepsilon} = \tilde{v}_0^i, \quad i \geq 1, \qquad (4.1.50)$$

and for odd $k \geq 1$

$$u_0^k = v_0^k = \tilde{u}_0^{k,\varepsilon} = \tilde{v}_0^{k,\varepsilon} = 0 \qquad (4.1.51)$$

where $\tilde{u}_0^i$ and $\tilde{v}_0^i$ are the initial conditions in the Vasil'eva theorem.

The above considerations show that it is natural to expect that the asymptotic expansions have the form

$$u_t^\varepsilon = \sum_{i=0}^n \varepsilon^{i/2}(u_t^i + \tilde{u}_t^{i,\varepsilon}) + \varepsilon^{n/2}\Delta_t^{u,n,\varepsilon}, \qquad (4.1.52)$$

$$v_t^\varepsilon = \sum_{i=0}^n \varepsilon^{i/2}(v_t^i + \tilde{v}_t^{i,\varepsilon}) + \varepsilon^{n/2}\Delta_t^{v,n,\varepsilon}, \qquad (4.1.53)$$

where $\Delta_t^{u,n,\varepsilon}$ and $\Delta_t^{v,n,\varepsilon}$ are uniformly small in probability (the above relations are nothing but the definitions of these quantities). This is indeed the case under the following natural assumption on the smoothness of the coefficients:

H.4.1.1 The functions F, f, g and φ have continuous derivatives of a polynomial growth up to the order $n+2$ and the $(n+2)$th derivatives satisfy the local Lipschitz condition.

To simplify the proof we introduce also the following hypothesis:

H.4.1.2 The first derivatives of functions F, f, g, and $F_v^{-1}(t,u,\varphi(t,u))$ are bounded ("-1" denote the inverse of the matrix).

Put

$$\Delta_t^{n,\varepsilon} := (\Delta_t^{u,n,\varepsilon}, \Delta_t^{v,n,\varepsilon}). \tag{4.1.54}$$

Theorem 4.1.1 *Assume the system (4.1.1)–(4.1.2) be such that the conditions* **H.4.0.1**–**H.4.0.4**, **H.4.1.2**, *and* **H.4.1.1** *with some* $n \geq 2$ *hold. Then for the solution of (4.1.1)–(4.1.2) we have the asymptotic expansions (4.1.52)–(4.1.53) with*

$$P\text{-}\lim_{\varepsilon \to 0} \|\Delta^{n,\varepsilon}\|_T = 0. \tag{4.1.55}$$

4.1.2 Integrability of Coefficients

The systems (4.1.34)–(4.1.39) and (4.1.40)–(4.1.47) which is used to determine the coefficients of the asymptotic expansions (4.1.5) and (4.1.6) have a nice recurrent structure and can be solved successively. We summarize some properties of their solutions in the following lemmas.

Lemma 4.1.2 *Assume the conditions* **H.4.0.1**, **H.4.0.2**, **H.4.1.1**, *and also* **H.4.1.2** *be fulfilled. Then the system (4.1.34)–(4.1.39) where $k \leq n+2$ with the initial conditions given by (4.1.48), (4.1.49) has the unique strong solution such that*

$$Z^{2i+1} = 0, \qquad 0 \leq i \leq (n+1)/2, \tag{4.1.56}$$

and, for any $m \geq 1$,

$$E\|Z^{2i}\|_T^m < \infty, \qquad 0 \leq i \leq (n+2)/2, \tag{4.1.57}$$

and

$$E\|Z^{1,2i}\|_T^m + E\|Z^{2,2i}\|_T^m < \infty, \qquad 0 \leq i \leq n/2. \tag{4.1.58}$$

Proof. As we just mentioned, the equations (4.1.34)–(4.1.39) have a recursive structure. The assumption **H.4.1.2** that F_v^0 has the inverse allows us to solve algebraic equations with respect to v^k and reduce the problem to the system of SDEs for $u^0, \ldots, u^{n+2}$ which evidently has the unique strong solution (see Theorem A.1.1). To prove (4.1.56) we observe that the solution of the linear system (4.1.36), (4.1.37) with zero initial conditions (4.1.51) has only the trivial solution. Assume that the equalities (4.1.56) hold for all $i \leq l$ where $l \leq (n-1)/2$ and consider the process Z^{2l+3} defined by the linear system (4.1.38), (4.1.39) with zero initial conditions. Using the property that

$$R^{F,2i-1}(c_0, 0, c_1, \ldots, c_{2i-2}) = 0$$

(see Section 2.3) we get by the induction hypothesis that

$$R_t^{F,2l+3} = R_t^{f,2l+3} = R_t^{g,2l+3} = 0.$$

It follows that the process $Z^{2l+1} = 0$ is the solution of (4.1.38), (4.1.39). Thus, (4.1.56) holds for all $i \leq (n+1)/2$.

First, we prove (4.1.57) for $i = 0$. Substituting the expression (4.1.35) for v^0 into (4.1.34) we get an equation for u^0 with coefficients satisfying the Lipschitz and linear growth conditions. Hence, $||u^0||_T$ has finite moments of any order, and, by the linear growth condition for φ, the same holds for $||v^0||_T$, i.e. (4.1.57) is true for $i = 0$. The similar reasoning can be applied successively to the equations (4.1.38), (4.1.39).

To prove (4.1.58) we notice that by (4.1.34)

$$u_t^{1,0} = f(t, u_t^0, v_t^0), \qquad u_t^{2,0} = g(t, u_t^0, v_t^0).$$

It follows from the linear growth conditions and (4.1.57) that $E||u^{j,0}||_T < \infty$, $j = 1, 2$. To get the expressions for $v^{1,0}$ and $v^{2,0}$ we apply the Ito formula to $\varphi(t, u_t^0)$. By the assumption **H.4.1.1** the first and the second derivatives of φ have a polynomial growth and (4.1.57) implies that $E||v^{j,0}||_T < \infty$, $j = 1, 2$. For $i \geq 1$ the inequalities for $u^{j,2i}$ follow directly from (4.1.38) and (4.1.57). To get the corresponding inequalities for $v^{j,2i}$ we can solve the equation (4.1.39) with respect to v^{2i}. The assumptions **H.4.1.1**, **H.4.1.2** guarantee that the involved processes can be represented as the Ito processes with the needed properties and application of the Ito formula together with (4.1.57) yields the result. $\square$

Lemma 4.1.3 *Let the conditions* **H.4.0.1**, **H.4.0.2**, **H.4.1.1**, *and* **H.4.1.2** *be fulfilled. Then the system (4.1.40)–(4.1.47), where $k \leq n + 2$, with the initial conditions given by (4.1.48), (4.1.49) has the unique strong solution such that for any $\varepsilon > 0$*

$$E||\tilde{Z}^{k,\varepsilon}||_T^m < \infty, \quad k \leq n + 2, \tag{4.1.59}$$

for any $m \geq 1$.

The result follows from Lemma 4.1.2 and Theorem A.1.1.

4.1.3 The Boundary Layer Function of Zero Order

In the asymptotic expansions (4.1.5), (4.1.6) the boundary layer function $\tilde{v}^{0,\varepsilon}$ plays an important role since it compensates the difference between the initial condition y^o for the fast variable v^ε and the initial condition $\varphi(0, x^o)$ for the limiting process v. This function tends to zero at infinity with an exponential rate. To formulate the precise statement put

$$\pi_t^\varepsilon(\gamma) := e^{\gamma t/(2\varepsilon)} |\tilde{v}_t^{0,\varepsilon}|. \tag{4.1.60}$$

Lemma 4.1.4 *Let the conditions* **H.4.0.1**–**H.4.0.2**, **H.4.1.1**, *and* **H.4.1.2** *be fulfilled. Then for any* $N \geq |x^o| + |\varphi(0, x^o)| + 1$ *there are constants* C_N *and* γ_N *such that*

$$\lim_{\varepsilon \to 0} ||\pi^\varepsilon(\gamma_N)||_{\tau_N^0} \leq C_N \tag{4.1.61}$$

where

$$\tau_N^0 := \inf\{t \geq 0 : |Z_t^0| \geq N\} \wedge T. \tag{4.1.62}$$

Proof. We begin with a remark that due to **H.4.0.2** the convergence of the solution $\tilde{y}_s$ of the associated equation (4.0.7) to the rest point $\varphi(0, x^o)$, assumed in (4.0.8), is exponentially fast, i.e., for the function $V_s := \tilde{y}_s - \varphi(0, x^o)$ for all $s \geq 0$ we have the bound

$$|V_s| \leq Ce^{-\alpha s} \tag{4.1.63}$$

with some constants C and α. Indeed, V is the solution of the equation

$$dV_s = F(0, x^o, \varphi(0, x^o) + V_s)ds, \quad V_0 = y^o - \varphi(0, x^o). \tag{4.1.64}$$

By the finite increments formula we can write that

$$dV_s = \widehat{A}_s V_s ds$$

where $\widehat{A}_s$ tends to the constant matrix $F_y(0, x^o, \varphi(0, x^o))$ when $s \to \infty$. By **H.4.0.4** all eigenvalues of the latter have negative real parts and (4.1.63) follows from Lemma A.2.2.

Let us introduce the process $\widehat{v}_s^{0,\varepsilon} := \tilde{v}_{s\varepsilon}^{0,\varepsilon}$. We show that for any s_0

$$\lim_{\varepsilon \to 0} ||\widehat{v}^\varepsilon - V||_{s_0} = 0. \tag{4.1.65}$$

To this aim we notice that according to (4.1.41), (4.1.28), and (4.1.48) the process $\widehat{v}^{0,\varepsilon}$ is the solution of the equation

$$d\widehat{v}_s^{0,\varepsilon} = F(s\varepsilon, u_{s\varepsilon}^0, v_{s\varepsilon}^0 + \widehat{v}_s^{0,\varepsilon})ds, \quad \widehat{v}_0^{0,\varepsilon} = y^o - \varphi(0, x^o). \tag{4.1.66}$$

Put $\Delta^\varepsilon := |\widehat{v}^{0,\varepsilon} - V|$. It follows from (4.1.64), (4.1.66) and the local Lipschitz condition for F (see **H.4.0.1** or **H.4.1.2**) that on $[0, \tau_N^0/\varepsilon]$ we have

$$\Delta_r^\varepsilon \leq \int_0^r |F(s\varepsilon, u_{s\varepsilon}^0, v_{s\varepsilon}^0 + \widehat{v}_s^{0,\varepsilon})) - F(0, x^o, \varphi(0, x^o) + V_s)|ds$$

$$\leq C\left(\int_0^r \Delta_s^\varepsilon ds + \int_0^r |Z_{s\varepsilon}^0 - Z_0^0|ds + r^2\varepsilon\right)$$

where C is a constant depending on N. The relation (4.1.65) follows from here by the Gronwall–Bellman lemma.

The trajectories of $\tilde{v}^{0,\varepsilon}$ are integral curves of the following equation with random coefficients (see (4.1.41) and (4.1.28)):

$$\varepsilon \dot{Y}_t^\varepsilon = A_t Y_t^\varepsilon + K(t, Y_t^\varepsilon) \tag{4.1.67}$$

where $A_t := F_v(t, u_t^0, v_t^0)$ and

$$K(t, y) := F(t, u_t^0, v_t^0 + y) - F_v(t, u_t^0, v_t^0)y. \tag{4.1.68}$$

Certainly, A and K depend also on ω.

Let $\Phi^\varepsilon(t, s)$ be a fundamental matrix corresponding to $\varepsilon^{-1} A$. It follows from **H.4.0.4** and Proposition A.2.5 that there are constants $c = c_N$ and $\gamma = \gamma_N > 0$ such that

$$\limsup_{\varepsilon \to 0} \ \sup_{0 \le s \le t \le \tau_N^0} e^{\gamma(t-s)/\varepsilon} |\Phi^\varepsilon(t, s)| \le c. \tag{4.1.69}$$

Notice that $K(\omega, t, 0) = 0$ and for any $\rho > 0$ there exists a strictly positive number $\eta = \eta(\rho)$ (depending also on N but not on ω) such that

$$|K(\omega, t, y_1) - K(\omega, t, y_2)| \le \rho |y_1 - y_2| \tag{4.1.70}$$

for every y_1 and y_2 such that $|y_1|, |y_2| < \eta$ and all $t \le \tau_N^0(\omega)$. The last property of K can be easily deduced from the finite increments formula and the assumption that F is continuously differentiable (see **H.4.1.2**).

Choose sufficiently small $\rho > 0$ such that $q := 2\rho c \gamma^{-1} < 1$ and take corresponding $\eta = \eta(\rho)$. Fix $\delta > 0$ such that

$$c\delta/(1 - q) \le \eta.$$

For this δ we prove the following property of the integral curves of (4.1.67).

Let Y^ε be a solution of (4.1.67) such that $|Y_{t_0}^\varepsilon| < \delta$. Then there is a constant C_N such that

$$\limsup_{\varepsilon \to 0} \ \sup_{t_0 \le t \le \tau_N^0} e^{\gamma t/(2\varepsilon)} |Y_t^\varepsilon| \le C_N. \tag{4.1.71}$$

By the Cauchy formula the function Y^ε on $[t_0, \tau_N^0]$ admits the representation

$$Y_t^\varepsilon = \Phi^\varepsilon(t, t_0) Y_{t_0}^\varepsilon + \varepsilon^{-1} \int_{t_0}^t \Phi^\varepsilon(t, s) K(s, Y_s^\varepsilon) ds. \tag{4.1.72}$$

We define the successive approximations of Y^ε on the interval $[t_0, \tau_N^0]$ as follows:

$$Y_t^{\varepsilon,0} := \Phi^\varepsilon(t,t_0)Y_{t_0}^\varepsilon, \tag{4.1.73}$$

$$Y_t^{\varepsilon,k} := \Phi^\varepsilon(t,t_0)Y_{t_0}^\varepsilon + \varepsilon^{-1}\int_{t_0}^t \Phi^\varepsilon(t,s)K(s,Y_s^{\varepsilon,k-1})ds, \quad k \geq 1. \tag{4.1.74}$$

From the definitions we easily get that on $[t_0,\tau_N^0]$ for all sufficiently small ε

$$|Y_t^{\varepsilon,0}| \leq c\delta e^{-\gamma(t-t_0)/\varepsilon} \leq c\delta e^{-\gamma(t-t_0)/(2\varepsilon)} \leq c\delta \leq \eta, \tag{4.1.75}$$

and, hence,

$$\begin{aligned}
|Y_t^{\varepsilon,1} - Y_t^{\varepsilon,0}| &\leq \varepsilon^{-1}\int_{t_0}^t |\Phi^\varepsilon(t,s)||K(s,Y_s^{\varepsilon,0}) - K(s,0)|ds \\
&\leq \varepsilon^{-1}\int_{t_0}^t ce^{-\gamma(t-s)/\varepsilon}\rho c\delta e^{-\gamma(s-t_0)/(2\varepsilon)}ds \\
&= c^2\delta\rho e^{-\gamma(t-t_0)/(2\varepsilon)}\varepsilon^{-1}\int_{t_0}^t e^{-\gamma(t-s)/(2\varepsilon)}ds \\
&\leq 2c^2\delta\rho\gamma^{-1}e^{-\gamma(t-t_0)/(2\varepsilon)} \leq c\delta q e^{-\gamma(t-t_0)/(2\varepsilon)}.
\end{aligned} \tag{4.1.76}$$

It follows that

$$|Y_t^{\varepsilon,1}| \leq c\delta(1+q)e^{-\gamma(t-t_0)/(2\varepsilon)} \leq \frac{c\delta}{1-q}e^{-\gamma(t-t_0)/(2\varepsilon)} \leq \eta. \tag{4.1.77}$$

Let us show that on $[t_0,\tau_N^0]$ for all sufficiently small ε we have

$$|Y_t^{\varepsilon,k}| \leq c\delta(1+q+\ldots+q^k)e^{-\gamma(t-t_0)/(2\varepsilon)}, \tag{4.1.78}$$
$$|Y_t^{\varepsilon,k} - Y_t^{\varepsilon,k-1}| \leq c\delta q^k e^{-\gamma(t-t_0)/(2\varepsilon)}. \tag{4.1.79}$$

Indeed, we proved already that the above inequalities hold for $k = 1$. Let us suppose that they are true up to some k. It follows from (4.1.78) that

$$|Y_t^{\varepsilon,k}| \leq \frac{c\delta}{1-q}e^{-\gamma(t-t_0)/(2\varepsilon)} \leq \eta,$$

and, similarly, $|Y_t^{\varepsilon,k-1}| \leq \eta$. Using the property (4.1.70) of the function K and the bound (4.1.79) we get, as in (4.1.76), that

$$\begin{aligned}
|Y_t^{\varepsilon,k+1} - Y_t^{\varepsilon,k}| &\leq \varepsilon^{-1}\int_{t_0}^t |\Phi^\varepsilon(t,s)||K(s,Y_s^{\varepsilon,k}) - K(s,Y_s^{\varepsilon,k-1})|ds \\
&\leq \varepsilon^{-1}\int_{t_0}^t ce^{-\gamma(t-s)/\varepsilon}\rho|Y_s^{\varepsilon,k} - Y_s^{\varepsilon,k-1}|ds \\
&\leq 2c^2\delta\rho\gamma^{-1}q^k e^{-\gamma(t-t_0)/(2\varepsilon)} \leq c\delta q^{k+1}e^{-\gamma(t-t_0)/(2\varepsilon)}.
\end{aligned} \tag{4.1.80}$$

Thus, (4.1.79) holds for $k + 1$. The bounds (4.1.80) and (4.1.78) imply that (4.1.78) also holds for $k + 1$.

It follows from the identity

$$Y_t^{\varepsilon,k} = \sum_{i=1}^{k}(Y_t^{\varepsilon,i} - Y_t^{\varepsilon,i-1}) \tag{4.1.81}$$

and the inequality (4.1.79) that the successive approximations converge uniformly on $[t_0, \tau_N^0]$ to some function which must be a solution of (4.1.72), i.e. to Y^ε. The bound (4.1.78) implies that for sufficiently small ε on $[t_0, \tau_N^0]$ we have

$$|Y_t^\varepsilon| \le \frac{c\delta}{1-q}e^{-\gamma(t-t_0)/(2\varepsilon)}, \tag{4.1.82}$$

implying (4.1.71).

The asymptotic bound (4.1.71) holds whenever t_0 (may be depending on ε) is such that $|Y_{t_0}^\varepsilon| \le \delta$ for sufficiently small ε. We take $t_0 = s_0\varepsilon$ where s_0 is chosen to guarantee the inequality $|V_{s_0}| \le \delta/2$ (which is possible due to (4.1.63)). The relation (4.1.65) shows that for sufficiently small ε we have $|Y_{t_0}^\varepsilon| = |\widehat{v}_{s_0}^\varepsilon| \le \delta$. Moreover, taking into account (4.1.63) and (4.1.65) we infer that

$$\sup_{t\in[0,t_0\wedge\tau_0^N]} e^{\gamma t/(2\varepsilon)}|\tilde{v}_t^{0,\varepsilon}| \le e^{\gamma s_0/2}||\tilde{v}^{0,\varepsilon}||_{s_0} \le e^{\gamma s_0/2}(C+1)$$

for all sufficiently small ε.

Enlarging the constant C_N we get from (4.1.71) and the above bound the desired relation (4.1.61). □

4.1.4 Boundary Layer Functions of Higher Order

The next lemma asserts that the uniform norm of a boundary layer function of a higher order can increase (as $\varepsilon \to 0$) only very slowly.

Lemma 4.1.5 *Assume the conditions* **H.4.0.1 – H.4.0.4, H.4.1.1,** *and* **H.4.1.2** *be fulfilled. Then for any $r > 0$*

$$\lim_{\varepsilon\to 0} P(||\tilde{Z}^{i,\varepsilon}||_T > \varepsilon^{-r}) = 0, \tag{4.1.83}$$

where $1 \le i \le n$.

Proof. Again we start with $\tilde{Z}^{1,\varepsilon}$. Since $Z_t^1 = 0$ for all $t \le T$ (Lemma 4.1.2) we get the following equations for $\tilde{Z}_t^{1,\varepsilon}$:

$$d\tilde{u}_t^{1,\varepsilon} = (f_u^\varepsilon\tilde{u}_t^{1,\varepsilon} + f_v^\varepsilon\tilde{v}_t^{1,\varepsilon})dt + (\varepsilon^{-1/2}\tilde{g}_t^{0,\varepsilon} + g_u^\varepsilon\tilde{u}_t^{1,\varepsilon} + g_v^\varepsilon\tilde{v}_t^{1,\varepsilon})dw_t, \tag{4.1.84}$$
$$\varepsilon d\tilde{v}_t^{1,\varepsilon} = (F_u^{0,\varepsilon}\tilde{u}_t^{1,\varepsilon} + F_v^{0,\varepsilon}\tilde{v}_t^{1,\varepsilon})dt - \varepsilon^{1/2}v_t^{2,0}dw_t, \tag{4.1.85}$$

with the initial conditions

$$\tilde{u}_0^{1,\varepsilon} = 0, \qquad \tilde{v}_0^{1,\varepsilon} = -v_0^1 = 0.$$

Let us define the stopping time

$$\sigma_N^0 := \sigma_N^{0,\varepsilon} := \inf\{t \geq 0 : |\pi_t^\varepsilon(\gamma_N)| \geq C_N\} \wedge \tau_N^0, \qquad (4.1.86)$$

where the process $\pi_t^\varepsilon(\gamma_N)$ and the stopping time τ_N^0 are given by (4.1.60) and (4.1.62), respectively, and the constant C_N is as in Lemma 4.1.4.

We obtain from (4.1.85) that

$$\tilde{v}_t^{1,\varepsilon} = \varepsilon^{-1} \int_0^t \Phi^\varepsilon(t,s) F_u^{0,\varepsilon}(s) \tilde{u}_s^{1,\varepsilon} ds - \xi_t^\varepsilon, \qquad (4.1.87)$$

where $\Phi^\varepsilon(t,s)$ is the fundamental matrix given by

$$\varepsilon \frac{d\Phi^\varepsilon(t,s)}{dt} = F_v^{0,\varepsilon}(t)\Phi^\varepsilon(t,s), \quad \Phi^\varepsilon(s,s) = I_n, \qquad (4.1.88)$$

and the process ξ^ε satisfies the linear SDE

$$\varepsilon d\xi_t^\varepsilon = F_v^{0,\varepsilon}(t)\xi_t^\varepsilon dt + \varepsilon^{1/2} v_t^{2,0} dw, \quad \xi_0^\varepsilon = 0. \qquad (4.1.89)$$

By the hypothesis, we have for $t \leq \sigma_N^0$ that

$$|F_v^{0,\varepsilon}(t) - F_v^0(t)| = |F_v(t, u_t^0, v_t^0 + \tilde{v}_t^{0,\varepsilon}) - F_v(t, u_t^0, v_t^0)|$$
$$\leq L_N |\tilde{v}_t^{0,\varepsilon}| \leq L_N C_N e^{-\gamma_N t/(2\varepsilon)}, \qquad (4.1.90)$$

where L_N is the Lipschitz constant for the matrix-valued function $\partial F/\partial v$.

We shall denote here by L (with indices as appropriate) various constants whose specific values have no importance.

The inequality (4.1.90) and properties of F on the time interval $[0, \sigma_N^0]$ allow us to apply to the equation (4.1.88) Proposition A.2.5 asserting that the following bound holds:

$$\limsup_{\varepsilon \to 0} \sup_{0 \leq s \leq t \leq \sigma_N^0} e^{\kappa_N(t-s)/\varepsilon} |\Phi^\varepsilon(t,s)| < c_N \qquad (4.1.91)$$

with some positive constants κ_N, c_N. Put

$$\varphi_t^{\varepsilon,N} = \sup_{s \leq t} e^{\kappa_N(t-s)/\varepsilon} |\Phi^\varepsilon(t,s)| \qquad (4.1.92)$$

and introduce the stopping time

$$\theta^0 := \theta_N^{0,\varepsilon} := \inf\{t \geq 0 : \varphi_t^{\varepsilon,N} \geq c_N\} \wedge \sigma_N^{0,\varepsilon}. \qquad (4.1.93)$$

It follows from (4.1.91) that

$$\lim_{\varepsilon \to 0} P(\theta_N^{0,\varepsilon} < \sigma_N^{0,\varepsilon}) = 0 \qquad (4.1.94)$$

for any (sufficiently large) $N > 0$. By virtue of (4.1.87) and (4.1.93) for $t \leq \theta^0$ we have

$$|\tilde{v}_t^{1,\varepsilon}| \le L_N(|\xi_t^{\varepsilon,N}| + ||u^{1,\varepsilon}||_t) \tag{4.1.95}$$

with some constant L_N where the process $\xi^{\varepsilon,N}$ is the solution of SDE (4.1.89) with $v_t^{2,0}$ substituted by $v_t^{2,0,N} := v_{t \wedge \tau_N^0}^{2,0}$.

We apply now to (4.1.85) Lemma A.1.3. Taking into account (4.1.95) we get that

$$E||\tilde{Z}^{1,\varepsilon}||_{\theta^0}^2 \le L_N \left(E||\xi^{\varepsilon,N}||_{\theta^0}^2 + \varepsilon^{-1} E \int_0^{\theta^0} |\tilde{g}_t^{0,\varepsilon}|^2 dt \right). \tag{4.1.96}$$

For $t \le \theta^0$ we have

$$|\tilde{g}_t^{0,\varepsilon}| = |g(Z_t^0 + \tilde{Z}_t^{0,\varepsilon}, t) - g(Z_t^0, t)| \le L_N|\tilde{v}_t^{0,\varepsilon}| \le L_N C_N e^{-\gamma_N t/(2\varepsilon)}.$$

Thus, we obtain from (4.1.96) that

$$E||\tilde{Z}^{1,\varepsilon}||_{\theta^0}^2 \le L_N(E||\xi^{\varepsilon,N}||_{\theta^0}^2 + 1) \tag{4.1.97}$$

with some constant L_N.

Notice that

$$||\xi^{\varepsilon,N}||_{\theta^0} \le ||\tilde{\xi}^{\varepsilon,N}||_T$$

where the process $\tilde{\xi}^{\varepsilon,N}$ is given by the linear SDE

$$\varepsilon d\tilde{\xi}_t^{\varepsilon,N} = A_t^{\varepsilon,N} \tilde{\xi}_t^{\varepsilon,N} dt + \varepsilon^{1/2} v_t^{2,0,N} dw_t, \quad \tilde{\xi}_0^{\varepsilon,N} = 0,$$

with

$$A^{\varepsilon,N} := F_v^{0,\varepsilon} I_{[0,\theta_N^0]} - \kappa_N I_{[\theta_N^0,T]}.$$

Applying Proposition 1.2.7 to $\tilde{\xi}^{\varepsilon,N}$ we conclude that

$$E||\xi^{\varepsilon,N}||_{\theta^0}^2 \le E||\tilde{\xi}^{\varepsilon,N}||_T^2 \le L(1 + \ln|\varepsilon|) \tag{4.1.98}$$

with some constant L_N. The inequalities (4.1.97) and (4.1.98) imply that there is a constant L_N such that

$$E||\tilde{Z}^{1,\varepsilon}||_{\theta^0}^2 \le L_N \ln|\varepsilon| \tag{4.1.99}$$

for sufficiently small $\varepsilon > 0$. We get by the Chebyshev inequality that

$$\lim_{\varepsilon \to 0} P(||\tilde{Z}^{1,\varepsilon}||_{\theta^0} > \varepsilon^{-r}) = 0 \tag{4.1.100}$$

for any $r > 0$.

The definition (4.1.62) and Lemma 4.1.2 yield that

$$\lim_{N \to \infty} P(\tau_N^0 < T) = 0. \tag{4.1.101}$$

We infer from (4.1.86) and Lemma 4.1.4 that

$$\lim_{\varepsilon \to 0} P(\sigma_N^{0,\varepsilon} < \tau_N^0) = 0 \qquad (4.1.102)$$

for any $N > 0$. Then

$$P(\|\tilde{Z}^{1,\varepsilon}\|_T > \varepsilon^{-r}) \leq P(\tau_N^0 < T) + P(\sigma_N^{0,\varepsilon} < \tau_N^0)$$
$$+ P(\theta_N^{0,\varepsilon} < \sigma_N^{0,\varepsilon}) + P(\|\tilde{Z}^{1,\varepsilon}\|_{\theta^0} > \varepsilon^{-r}). \qquad (4.1.103)$$

Taking here successively the limits as $\varepsilon \to 0$ and $N \to \infty$ and taking into account (4.1.94) and (4.1.100)–(4.1.102) we get the corresponding inequality in (4.1.83) with $i = 1$.

Now we assume that the seeking limit relation holds for all $i \leq k-1$ and prove it for $i = k$. We set for $k \geq 2$

$$\tau_N^k := \inf\left\{ t \geq 0 : \sum_{i=0}^{k} |Z_t^i| \geq N \right\} \wedge T,$$

$$\sigma_\nu^{k,\varepsilon} := \inf\left\{ t \geq 0 : \sum_{i=1}^{k-1} |\tilde{Z}_t^{i,\varepsilon}| \geq \varepsilon^{-\nu} \right\} \wedge T. \qquad (4.1.104)$$

By Lemma 4.1.2 and the induction hypothesis we have

$$\lim_{N \to \infty} P(\tau_N^k < T) = 0, \qquad \lim_{\varepsilon \to 0} P(\sigma_\nu^{k,\varepsilon} < T) = 0 \qquad (4.1.105)$$

for any $\nu > 0$.

Put

$$\theta^k := \theta_N^{k,\varepsilon} = \theta_N^{0,\varepsilon} \wedge \sigma_\nu^{k,\varepsilon} \wedge \tau_N^k. \qquad (4.1.106)$$

It follows from (4.1.94), (4.1.101), (4.1.102), and (4.1.105) that

$$\lim_{N \to \infty} \limsup_{\varepsilon \to 0} P(\theta_N^{k,\varepsilon} < T) = 0. \qquad (4.1.107)$$

Solving the linear SDE for the function $\tilde{v}_t^{k,\varepsilon}$ we get the representation

$$\tilde{v}_t^{k,\varepsilon} = \Phi^\varepsilon(t,0)\tilde{v}_0^{k,\varepsilon} + \varepsilon^{-1} \int_0^t \Phi^\varepsilon(t,s) F_u^{0,\varepsilon}(s) \tilde{u}_s^{k,\varepsilon} ds$$

$$+ \varepsilon^{-1} \int_0^t \Phi^\varepsilon(t,s) R_s^{F,k,\varepsilon} ds - \xi_t^\varepsilon, \qquad (4.1.108)$$

where $\Phi^\varepsilon(t,s)$ is the fundamental matrix for the system (4.1.88) and the process ξ_t^ε is given by the linear SDE (4.1.89) where $v^{2,0}$ is substituted by $v^{2,k-1}$.

Furthermore, from the definitions of the functions $R^{F,k,\varepsilon}$, $R^{f,k,\varepsilon}$, and $R^{g,k,\varepsilon}$ given by (4.1.12) and (4.3.5) it follows that there is a constant L_N such that

$$||R^{F,k,\varepsilon}||_{\theta^k} + ||R^{f,k,\varepsilon}||_{\theta^k} + ||R^{g,k,\varepsilon}||_{\theta^k} \leq L_N \varepsilon^{-k\nu}. \tag{4.1.109}$$

We get from (4.1.106), (4.1.108), and (4.1.109) that for $t \leq \theta^k$

$$|\tilde{v}_t^{k,\varepsilon}| \leq C_N ||F_u^{0,\varepsilon}||_{\theta^k} ||\tilde{u}^{1,\varepsilon}||_t \varepsilon^{-1} \int_0^t e^{-\kappa_N(t-s)/\varepsilon} ds$$

$$+ L_N C_N \varepsilon^{-1-k\nu} \int_0^t e^{-\kappa_N(t-s)/\varepsilon} ds + |\xi_t^\varepsilon|.$$

It follows that for some constant L_N and for all sufficiently small $\varepsilon > 0$ for $t \leq \theta_N^{k,\varepsilon}$

$$|\tilde{v}_t^{k,\varepsilon}| \leq L_N(\varepsilon^{-k\nu} + |\bar{\xi}_t^{\varepsilon,N}| + ||\tilde{u}^{1,\varepsilon}||_t) \tag{4.1.110}$$

where $\bar{\xi}^{\varepsilon,N}$ is given by the linear SDE

$$\varepsilon d\bar{\xi}_t^{\varepsilon,N} = A_t^{\varepsilon,N} \bar{\xi}_t^{\varepsilon,N} dt + \varepsilon^{1/2} v_t^{2,k-1,N} dw_t, \quad \bar{\xi}_0^{\varepsilon,N} = 0$$

with

$$A^{\varepsilon,N} := F_v^{0,\varepsilon} I_{[0,\theta_N^0]} - \kappa_N I_{[\theta_N^0,T]}.$$

Since $v_t^{2,k-1}$ depends continuously on $Z_t^0, \ldots, Z_t^{k-1}$ we have the bound

$$||v^{2,k-1,N}||_T \leq L_N \tag{4.1.111}$$

for some constant L_N.

Let us apply Lemma A.1.3 to the equations (4.1.44) (when $k = 2$) and (4.1.46) for the process $\tilde{u}^{k,\varepsilon}$ taking into account the inequality (4.1.100). According to this lemma for sufficiently small $\varepsilon > 0$ we have

$$E||\tilde{Z}^{k,\varepsilon}||_{\theta^k}^2 \leq L_N \varepsilon^{-2k\nu} + L_N E||\xi^{\varepsilon,N}||_{\theta^k}^2$$

$$+ L_N E \left(\delta_{k2} \varepsilon^{-1} \int_0^{\theta^k} |\tilde{f}_t^{0,\varepsilon}| dt + \int_0^{\theta^k} |R_t^{f,k,\varepsilon}| dt \right)^2$$

$$+ L_N E \int_0^{\theta^k} |R_t^{g,k,\varepsilon}|^2 dt \tag{4.1.112}$$

where $\delta_{k2} = 1$ if $k = 2$ and zero otherwise.

Using the local Lipschitz condition for f we get for $t \leq \theta_N^{k,\varepsilon}$ that

$$|\tilde{f}_t^{0,\varepsilon}| = |f(Z_t^0 + \tilde{Z}_t^{0,\varepsilon}, t) - f(Z_t^0, t)| \leq L_N |\tilde{v}_t^{0,\varepsilon}| \leq L_N C_N e^{-\gamma_N t/(2\varepsilon)}. \tag{4.1.113}$$

Applying the bounds (4.1.109) and (4.1.113) we infer from (4.1.102) that there is a constant L_N such that

$$E||\tilde{Z}^{k,\varepsilon}||_{\theta^k}^2 \leq L_N(\varepsilon^{-2k\nu} + E||\bar{\xi}^{\varepsilon,N}||_T^2) \tag{4.1.114}$$

for sufficiently small $\varepsilon > 0$. Taking into account the definition of the stopping time θ^k and the bound (4.1.111), apply Proposition 1.2.4 to the process $\bar{\xi}_t^{\varepsilon,N}$. The inequality (4.1.114) yields that there is a constant L_N such that

$$E\|\tilde{Z}^{k,\varepsilon}\|_{\theta^k}^2 \le L_{N,m}(\varepsilon^{-2k\nu} + |\ln\varepsilon|) \tag{4.1.115}$$

for sufficiently small $\varepsilon > 0$. It follows from the Chebyshev inequality and (4.1.115) that

$$P(\|\tilde{Z}^{k,\varepsilon}\|_{\theta^k} > \varepsilon^{-r}) \le L_N(\varepsilon^{2r-2k\nu} + \varepsilon^{2r}|\ln\varepsilon|).$$

This bound implies (with a suitable choice of the parameters $\nu > 0$) that

$$\lim_{\varepsilon\to 0} P(\|\tilde{Z}^{k,\varepsilon}\|_{\theta^k} > \varepsilon^{-r}) = 0$$

for any $r > 0$. The above relation together with (4.1.107) gives (4.1.83) for $i = k$. $\square$

4.1.5 Proof of Theorem 4.1.1

As usual, we deduce asymptotic properties of the residual from its representation as a solution of a certain linear equation. Put

$$Q_t^{n,\varepsilon} := \sum_{k=0}^n \varepsilon^{k/2} Z_t^k + \sum_{k=0}^n \varepsilon^{k/2} \tilde{Z}_t^k.$$

It follows from (4.1.2), (4.1.20), (4.1.21), and (4.1.19), (4.1.22) that

$$\varepsilon^{n/2} d\Delta_t^{v,n,\varepsilon} := dv_t^\varepsilon - \sum_{k=0}^n \varepsilon^{k/2}(v_t^{1,k}dt + v_t^{2,k}dw_t + \tilde{v}_t^{1,k,\varepsilon}dt + \tilde{v}_t^{2,k,\varepsilon}dw_t)$$

$$= \varepsilon^{-1}F(t, z_t^\varepsilon)dt - \sum_{k=0}^n \varepsilon^{k/2}(F_t^{k+2} + \varepsilon^{-1}\tilde{F}_t^{k,\varepsilon})dt$$

$$+ \sum_{k=0}^n \varepsilon^{k/2}(v_t^{2,k} - \varepsilon^{-1/2}v_t^{2,k-1})dw_t$$

where we put $v_t^{2,-1} = 0$. Taking into account that $F_t^0 = F_t^1 = 0$ we get by simple transformations the formula

$$\varepsilon^{n/2}d\Delta_t^{v,n,\varepsilon} = \varepsilon^{-1}(F(t,z_t^\varepsilon) - F(t,Q_t^{n,\varepsilon}))dt + \varepsilon^{n/2-1}B_t^{F,n,\varepsilon}dt$$
$$-\varepsilon^{(n-1)/2}F_t^{n+1}dt - \varepsilon^{n/2}F_t^{n+2}dt - \varepsilon^{n/2}v_t^{2,n}dw_t \tag{4.1.116}$$

with

$$B_t^{F,n,\varepsilon} := \varepsilon^{-n/2}\left(F(t,Q_t^{n,\varepsilon}) - \sum_{k=0}^n \varepsilon^{k/2}F_t^k - \sum_{k=0}^n \varepsilon^{k/2}\tilde{F}_t^{k,\varepsilon}\right). \tag{4.1.117}$$

Similarly, for the slow variable we can write using the equation (4.1.1) and identities (4.1.29)–(4.1.33) that

$$\varepsilon^{n/2} d\Delta_t^{u,n,\varepsilon} := du_t^\varepsilon - \sum_{k=0}^{n} \varepsilon^{k/2} (u_t^{1,k} dt + u_t^{2,k} dw_t + \tilde{u}_t^{1,k,\varepsilon} dt + \tilde{u}_t^{2,k,\varepsilon} dw_t)$$

$$= f(t, z_t^\varepsilon) dt + g(t, z_t^\varepsilon) dw_t - \sum_{k=0}^{n} \varepsilon^{k/2} f_t^k dt - \sum_{k=0}^{n} \varepsilon^{k/2} g_t^{k,\varepsilon} dw_t$$

$$- \varepsilon^{1/2} \tilde{f}^{1,\varepsilon} dt - \varepsilon(\varepsilon^{-1} \tilde{f}_t^{0,\varepsilon} + \tilde{f}_t^{2,0,\varepsilon}) dt - \sum_{k=3}^{n} \varepsilon^{k/2} \tilde{f}_t^{k,\varepsilon} dt$$

$$- \varepsilon^{1/2} (\varepsilon^{-1/2} \tilde{g}_t^{0,\varepsilon} + \tilde{g}_t^{1,\varepsilon}) dw_t - \sum_{k=2}^{n} \varepsilon^{k/2} \tilde{g}_t^{k,\varepsilon} dw_t$$

and, hence,

$$d\Delta_t^{u,n,\varepsilon} = \varepsilon^{-n/2} (f(t, z_t^\varepsilon) - f(t, Q_t^{n,\varepsilon})) dt + \varepsilon^{-n/2} (g(t, z^\varepsilon) - g(t, Q_t^{n,\varepsilon})) dw_t$$
$$+ B_t^{f,n,\varepsilon} dt + B_t^{g,n,\varepsilon} dw_t \qquad (4.1.118)$$

where $B_t^{f,n,\varepsilon}$ and $B_t^{g,n,\varepsilon}$ are defined as in (4.1.117).

Now we linearize the "leading term" on the right-hand side of (4.1.116). The formula of finite increments yields:

$$\varepsilon^{-n/2} (F(t, z_t^\varepsilon) - F(t, Q_t^{n,\varepsilon})) = \widehat{F}_u^{0,\varepsilon} \Delta_t^{u,n,\varepsilon} + F_v^{0,\varepsilon} \Delta_t^{v,n,\varepsilon} + H_v^\varepsilon \Delta_t^{v,n,\varepsilon}$$

where $H_v^\varepsilon := \widehat{F}_v^{0,\varepsilon} - F_v^{0,\varepsilon}$ and where, e.g., the elements of the ith row of $\widehat{F}_v^{0,\varepsilon}$ are partial derivatives in v of the corresponding component calculated at the point $(t, Q_t^{n,\varepsilon} + \zeta_i \varepsilon^{n/2} \Delta_t^{n,\varepsilon})$, $\zeta_i \in [0,1]$. Applying also the formula of finite increments to the first two terms in the right-hand side of (4.1.118) we get from (4.1.118) and (4.1.116) the system

$$d\Delta_t^{u,n,\varepsilon} = (\widehat{f}_u^{0,\varepsilon} \Delta_t^{u,n,\varepsilon} + \widehat{f}_v^{0,\varepsilon} \Delta_t^{v,n,\varepsilon}) dt + (\widehat{g}_u^{0,\varepsilon} \Delta_t^{u,n,\varepsilon} + \widehat{g}_v^{0,\varepsilon} \Delta_t^{v,n,\varepsilon}) dw_t$$
$$+ B_t^{f,n,\varepsilon} dt + B_t^{g,n,\varepsilon} dw_t, \qquad (4.1.119)$$
$$\varepsilon d\Delta_t^{v,n,\varepsilon} = \widehat{F}_u^{0,\varepsilon} \Delta_t^{u,n,\varepsilon} dt + F_v^{0,\varepsilon} \Delta_t^{v,n,\varepsilon} dt + H_v^\varepsilon \Delta_t^{v,n,\varepsilon} dt$$
$$+ B_t^{F,n,\varepsilon} dt - \varepsilon^{1/2} F_t^{k+1} dt - \varepsilon F_t^{k+2} dt - \varepsilon v_t^{2,n} dw_t. \qquad (4.1.120)$$

Let us introduce the stopping times

$$\eta := \eta_\mu^{n,\varepsilon} := \inf\{t \geq 0 : |\Delta_t^{n,\varepsilon}| \geq \mu\} \wedge T, \qquad (4.1.121)$$

$$\rho := \rho_{N,\mu}^\varepsilon := \tau_N^n \wedge \sigma_N^{0,\varepsilon} \wedge \sigma_v^{n+2,\varepsilon} \wedge \eta_\mu^{n,\varepsilon} \qquad (4.1.122)$$

where $\mu \in [0,1]$ and the stopping times $\sigma_N^{0,\varepsilon}$, τ_N^n, and $\sigma_v^{n+2,\varepsilon}$ are defined in (4.1.86), (4.1.62), and (4.1.104). Assume that $\nu \in [0, 1/(2(n+1))[$. Then for $t \leq \rho$ we have the following bounds:

$$|Q_t^{n,\varepsilon}| \le \sum_{k=0}^{n} \varepsilon^{k/2}|Z_t^k| + |\tilde{Z}_t^0| + \sum_{k=1}^{n} \varepsilon^{k/2}|\tilde{Z}_t^k| \le N + C_N e^{-\gamma_N t/\varepsilon} + \varepsilon^{1/2-\nu}$$
$$\le N + C_N + 1, \tag{4.1.123}$$
$$|z_t^\varepsilon| \le |Q_t^{n,\varepsilon}| + \varepsilon^{n/2}|\Delta_t^{n,\varepsilon}| \le |Q_t^{n,\varepsilon}| + \mu \varepsilon^{n/2} \le N + C_N + 2. \tag{4.1.124}$$

It follows from the local Lipschitz condition for F_v that if $t \le \rho$ then

$$\begin{aligned}
|H_v^\varepsilon(t)| &= |\widehat{F}_v^{0,\varepsilon}(t) - F_v^{0,\varepsilon}(t)| \\
&\le L_N(|Q_t^{n,\varepsilon} - Z_t^0 - \tilde{Z}_t^0| + \varepsilon^{n/2}|\Delta_t^{n,\varepsilon}|) \\
&\le L_N \left(\sum_{k=1}^{n} \varepsilon^{k/2}|Z_t^k| + \sum_{k=1}^{n} \varepsilon^{k/2}|\tilde{Z}_t^k| + \mu \varepsilon^{n/2} \right) \\
&\le L_N(\varepsilon^{1/2-\nu} + N\varepsilon^{1/2} + \varepsilon^{1/2}).
\end{aligned}$$

Thus, for any N there is a constant L_N such that

$$\|H_v^\varepsilon\|_\rho \le L_N \varepsilon^{1/2-\nu}. \tag{4.1.125}$$

It follows from Lemma 2.3.5 that for some L_N

$$\|B^{F,n,\varepsilon}\|_\rho + \|B^{f,n,\varepsilon}\|_\rho + \|B^{g,n,\varepsilon}\|_\rho \le L_N \varepsilon^{1/2-(n+1)\nu}. \tag{4.1.126}$$

One can choose a constant L_N large enough to satisfy the bound

$$\|F^{n+1}\|_\rho + \|F^{n+2}\|_\rho + \|v^{2,n}\|_\rho \le L_N; \tag{4.1.127}$$

this is possible since by **H.4.1.1** the involved functions are continuous.

By the Cauchy formula we get from (4.1.125) the following representation on the interval $[0, \rho]$:

$$\begin{aligned}
\Delta_t^{v,n,\varepsilon} &= \varepsilon^{-1} \int_0^t \Phi^\varepsilon(t,s)\widehat{F}_u^{0,\varepsilon}(s)\Delta_s^{u,n,\varepsilon}ds + \varepsilon^{-1}\int_0^t \Phi^\varepsilon(t,s)H_v^\varepsilon(s)\Delta_s^{v,n,\varepsilon}ds \\
&\quad + \varepsilon^{-1}\int_0^t \Phi^\varepsilon(t,s)B_s^{F,n,\varepsilon}ds - \varepsilon^{-1/2}\int_0^t \Phi^\varepsilon(t,s)F_s^{n+1}ds \\
&\quad - \int_0^t \Phi^\varepsilon(t,s)F_s^{n+2}ds - \xi_t^\varepsilon
\end{aligned} \tag{4.1.128}$$

where $\Phi^\varepsilon(t,s)$ is the fundamental matrix corresponding to $F_v^{0,\varepsilon}$ and

$$d\xi_t^\varepsilon = \varepsilon^{-1}F_v^{0,\varepsilon}(t)\xi_t^\varepsilon dt + \tilde{v}_t^{2,n}dw_t, \quad \xi_0^\varepsilon = 0 \tag{4.1.129}$$

where

$$A_t^{\varepsilon,N} := F_v^{0,\varepsilon}(t)I_{[0,\theta_N^0]} - \kappa_N I_{[\theta_N^0,T]}.$$

and $\tilde{v}_t^{2,n} = \tilde{v}_{t\wedge\rho}^{2,n}$.

Let us introduce the stopping time $\theta := \theta_N^\varepsilon := \theta_N^{0,\varepsilon} \wedge \rho$ where $\theta_N^{0,\varepsilon}$ is defined in (4.1.93). It follows from the definition (4.1.122) and the property (4.1.94) that

$$\lim_{\varepsilon \to 0} P(\theta_N^\varepsilon < \rho_{N,\mu}^\varepsilon) = 0. \tag{4.1.130}$$

Recall that the stopping time $\theta_N^{0,\varepsilon}$ is constructed to provide an exponential bound for $\Phi^\varepsilon(t, s)$:

$$|\Phi^\varepsilon(t, s)| \leq c_N e^{-\kappa_N(t-s)/\varepsilon} \tag{4.1.131}$$

when $0 \leq s \leq t \leq \theta_N^{0,\varepsilon}$.

We obtain from the representation (4.1.128) and the bounds (4.1.125)–(4.1.127) and (4.1.131) that for $t \leq \theta_N^\varepsilon$

$$|\Delta_t^{v,n,\varepsilon}| \leq c_N \|\Delta^{u,n,\varepsilon}\|_t \varepsilon^{-1} \int_0^t e^{-\kappa_N(t-s)/\varepsilon} |\widehat{F}_u^{0,\varepsilon}(s)| ds$$

$$+ \mu c_N L_N \varepsilon^{1/2-\nu} \varepsilon^{-1} \int_0^t e^{-\kappa_N(t-s)/\varepsilon} ds$$

$$+ c_N L_N \varepsilon^{1/2-(n+1)\nu} \varepsilon^{-1} \int_0^t e^{-\kappa_N(t-s)/\varepsilon} ds$$

$$+ c_N L_N \varepsilon^{1/2-(n+1)\nu} (1 + \varepsilon^{-1/2}) \int_0^t e^{-\kappa_N(t-s)/\varepsilon} ds + |\xi_t^\varepsilon|.$$

Since F_u is bounded we have from above that

$$|\Delta_t^{v,n,\varepsilon}| \leq L_N(\|\Delta^{u,n,\varepsilon}\|_t + \varepsilon^{1/2-(n+1)\nu} + |\xi_t^\varepsilon|) \tag{4.1.132}$$

for some constant L_N.

Using Lemma A.1.3, we infer from the equations (4.1.119), (4.1.120) and the bounds (4.1.126), (4.1.132) that

$$\|\Delta^{n,\varepsilon}\|_\theta \leq L_N(\varepsilon^{1/2-2(n+1)\nu} + \|\xi^\varepsilon\|_\theta).$$

Proposition 1.2.4 and the inequality (4.1.127) imply that for some constant L_N and sufficiently small $\varepsilon > 0$ we have

$$E\|\xi^\varepsilon\|_T \leq L_N \sqrt{\varepsilon |\ln \varepsilon|}.$$

Thus, for any N there is a constant L_N such that for all sufficient small $\varepsilon > 0$ we have

$$E\|\Delta^{n,\varepsilon}\|_\theta^2 \leq L_N(\varepsilon^{1/2-2(n+1)\nu} + \varepsilon |\ln \varepsilon|). \tag{4.1.133}$$

It follows from (4.1.102), (4.1.105) that

$$\lim_{N \to \infty} \limsup_{\varepsilon \to 0} P(\rho_{N,\mu}^\varepsilon < \eta_\mu^{n,\varepsilon}) = 0. \tag{4.1.134}$$

Notice that

$$P(\|\Delta^{n,\varepsilon}\|_T > \mu) \le P(\|\Delta^{n,\varepsilon}\|_\eta \ge \mu)$$
$$\le P(\rho^\varepsilon_{N,\mu} < \eta^{n,\varepsilon}_\mu) + P(\theta^\varepsilon_N < \rho^\varepsilon_{N,\mu}) + P(\|\Delta^{n,\varepsilon}\|_\theta \ge \mu)$$
$$\le P(\rho^\varepsilon_{N,\mu} < \eta^{n,\varepsilon}_\mu) + P(\theta^\varepsilon_N < \rho^\varepsilon_{N,\mu}) + \mu^{-2} E \|\Delta^{n,\varepsilon}\|^2_\theta.$$

Taking here the limits as $\varepsilon \to 0$ and then as $N \to \infty$ and using (4.1.134), (4.1.130), and (4.1.133) we get that

$$\lim_\varepsilon P(\|\Delta^{n,\varepsilon}\|_T > \mu) = 0.$$

Theorem 4.1.1 is proved. $\square$

4.2 Expansions for the General Model

4.2.1 Formulations

The aim of this section is to construct an asymptotic expansion for the system (4.0.5), (4.0.6) by taking as the initial approximation the solution $(u^\varepsilon, v^\varepsilon)$ of the system (4.1.1) and (4.1.2) (with $w = w^x$).

We shall impose here the hypothesis

H.4.2.1 There is $\delta > 0$ such that $\beta = O(\varepsilon^\delta)$ as $\varepsilon \to 0$.

Considering the system (4.0.5) and (4.0.6) as "regularly" perturbed with respect to the parameter β we expand the solution of the latter in the asymptotic series in β:

$$x^\varepsilon_t = \sum_{k \ge 0} \beta^k x^{k,\varepsilon}_t, \qquad y^\varepsilon_t = \sum_{k \ge 0} \beta^k y^{k,\varepsilon}_t. \qquad (4.2.1)$$

The coefficients here are determined through formal substitution of the series (4.2.1) into the system (4.1.1) and (4.1.2) by equalizing the coefficients at the same powers of β. This procedure yields the following equations:

$$x^{0,\varepsilon}_t = u^\varepsilon_t, \qquad (4.2.2)$$
$$y^{0,\varepsilon}_t = v^\varepsilon_t, \qquad (4.2.3)$$
$$dx^{k,\varepsilon}_t = f^{k,\varepsilon}_t dt + g^{k,\varepsilon}_t dw^x_t, \qquad x^{k,\varepsilon}_0 = 0, \qquad (4.2.4)$$
$$\varepsilon dy^{k,\varepsilon}_t = F^{k,\varepsilon}_t dt + \varepsilon^{1/2} G^{k-1,\varepsilon}_t dw^y_t, \qquad y^{k,\varepsilon}_0 = 0, \qquad k \ge 1. \qquad (4.2.5)$$

The functions $f^{k,\varepsilon}_t$, $g^{k,\varepsilon}_t$, etc. are the coefficients at powers of β in the asymptotic series for $f(t, x^\varepsilon_t, y^\varepsilon_t)$, $g(t, x^\varepsilon_t, y^\varepsilon_t)$, etc.

To specify the structure of coefficients we shall use abbreviated "technical" notations $h^\varepsilon := (x^\varepsilon, y^\varepsilon)$, $h^{k,\varepsilon} := (x^{k,\varepsilon}, y^{k,\varepsilon})$. Then, for example,

$$F^{k,\varepsilon}_t = F_k(h^{0,\varepsilon}_t, \ldots, h^{k,\varepsilon}_t) \qquad (4.2.6)$$

with the function $F_k(C_0, \ldots, C_k)$ given by (2.3.4). More specifically,

$$F_t^{k,\varepsilon} = F_x(t, \varepsilon)x_t^{k,\varepsilon} + F_y(t, \varepsilon)y_t^{k,\varepsilon} + R_t(F, k, \varepsilon), \qquad (4.2.7)$$

where $F_x(t, \varepsilon)$ and $F_y(t, \varepsilon)$ are matrices of partial derivatives calculated at the point $h_t^{0,\varepsilon} = (u_t^\varepsilon, v_t^\varepsilon)$, and $R_t(F, k, \varepsilon)$ is a polynomial of $h^{0,\varepsilon}, \ldots, h^{k-1,\varepsilon}$, $R_t(F, 1, \varepsilon) = 0$.

Let us rewrite the equations (4.2.4) and (4.2.5) as follows:

$$dx_t^{k,\varepsilon} = \Big(f_x(t, \varepsilon)x_t^{k,\varepsilon} + f_y(t, \varepsilon)y_t^{k,\varepsilon} + R_t(f, k, \varepsilon)\Big)dt$$
$$+ (g_x(t, \varepsilon)x_t^{k,\varepsilon} + g_y(t, \varepsilon)y_t^{k,\varepsilon} + R_t(g, k, \varepsilon))dw_t^x, \qquad (4.2.8)$$
$$\varepsilon dy_t^{k,\varepsilon} = \Big(F_x(t, \varepsilon)x_t^{k,\varepsilon} + F_y(t, \varepsilon)y_t^{k,\varepsilon} + R_t(F, k, \varepsilon)\Big)dt + \varepsilon^{1/2}G_t^{k-1,\varepsilon}du_t^y \quad (4.2.9)$$

Define the processes $\delta_t^{x,n+2,\varepsilon}$ and $\delta_t^{y,n+2,\varepsilon}$ by the equalities

$$x_t^\varepsilon = u_t^\varepsilon + \sum_{k=1}^{n+2} \beta^k x_t^{k,\varepsilon} + \beta^{n+2}\delta_t^{x,n+2,\varepsilon},$$

$$y_t^\varepsilon = v_t^\varepsilon + \sum_{k=1}^{n+2} \beta^k y_t^{k,\varepsilon} + \beta^{n+2}\delta_t^{y,n+2,\varepsilon}. \qquad (4.2.10)$$

We set $\delta_t^{n+2,\varepsilon} := (\delta_t^{x,n+2,\varepsilon}, \delta_t^{y,n+2,\varepsilon})$.

Theorem 4.2.1 *Assume that for the system (4.0.5), (4.0.6) the conditions* **H.4.0.1 – H.4.0.4, H.4.1.1** *with some* $n \geq 0$, **H.4.1.2**, *and* **H.4.2.1** *are fulfilled. Then*

$$P\text{-}\lim_{\varepsilon \to 0} \|\delta^{n+2,\varepsilon}\|_T = 0. \qquad (4.2.11)$$

Theorems 4.1.1 and 4.2.1 imply the following result.

Theorem 4.2.2 *Assume that for the system (4.0.5), (4.0.6) the conditions* **H.4.0.1 – H.4.0.4, H.4.1.1** *with some* $n \geq 2$, **H.4.1.2**, *and* **H.4.2.1** *are fulfilled. Then*

$$x_t^\varepsilon = \sum_{k=0}^{n} \varepsilon^{k/2}(u_t^{k,\varepsilon} + \tilde{u}_t^{k,\varepsilon}) + \sum_{k=1}^{n+2} \beta^k x_t^{k,\varepsilon} + \varepsilon^{n/2}\Delta_t^{u,n,\varepsilon} + \beta^{n+2}\delta_t^{x,n+2,\varepsilon}, \quad (4.2.12)$$

$$y_t^\varepsilon = \sum_{k=0}^{n} \varepsilon^{k/2}(v_t^{k,\varepsilon} + \tilde{v}_t^{k,\varepsilon}) + \sum_{k=1}^{n+2} \beta^k y_t^{k,\varepsilon} + \varepsilon^{n/2}\Delta_t^{v,n,\varepsilon} + \beta^{n+2}\delta_t^{y,n+2,\varepsilon}, \quad (4.2.13)$$

where $\|\Delta^{n,\varepsilon}\|_T$ *and* $\|\delta^{n+2,\varepsilon}\|_T$ *tends to zero in probability as* $\varepsilon \to 0$, *the coefficients* $u_t^{k,\varepsilon}, \tilde{u}_t^{k,\varepsilon}, v_t^{k,\varepsilon}, \tilde{v}_t^{k,\varepsilon}$, *and* $x_t^{k,\varepsilon}, y_t^{k,\varepsilon}$ *are defined in (4.1.28)–(4.1.47) and (4.2.4), (4.2.5).*

4.2.2 Growth of Coefficients

For the proof of Theorem 4.2.1 we need an estimate ensuring that the coefficients in the expansions (4.2.1) are not "exploding" as $\varepsilon \to 0$.

Lemma 4.2.3 *Assume that the hypothesis of Theorem 4.2.1 are fulfilled. Then for any $r > 0$ we have*

$$\lim_{\varepsilon \to 0} P(\|h^{k,\varepsilon}\|_T > \varepsilon^{-r}) = 0, \quad 1 \le k \le n+2. \tag{4.2.14}$$

Proof. We use the same technique as in the proof of Lemma 4.1.5 and give arguments only for the case $k = 1$.

Solving the equation (4.2.8) with respect to $y_t^{1,\varepsilon}$, we get the representation

$$y_t^{1,\varepsilon} = \varepsilon^{-1} \int_0^t \Phi^\varepsilon(t,s) x_s^{1,\varepsilon} ds + \xi_t^\varepsilon$$

where $\Phi^\varepsilon(t,s)$ is the transition martix corresponding to $\varepsilon^{-1} F_y(t,\varepsilon)$, i.e. it is the solution of the linear equation

$$\varepsilon \frac{d\Phi^\varepsilon(t,s)}{dt} = \varepsilon^{-1} F_y(t,\varepsilon)\Phi^\varepsilon(t,s), \quad \Phi^\varepsilon(s,s) = I,$$

and the process ξ^ε is the solution of the linear SDE

$$\varepsilon d\xi_t^\varepsilon = F_y(t,\varepsilon)\xi_t^\varepsilon dt + \varepsilon^{1/2} G_t^{0,\varepsilon} dw_t^y, \quad \xi_0^\varepsilon = 0.$$

Let us consider the asymptotic expansion of $(u^\varepsilon, v^\varepsilon)$ up to the terms of the first order:

$$u_t^\varepsilon = u_t^0 + \varepsilon^{1/2} \tilde{u}_t^{1,\varepsilon} + \varepsilon^{1/2} \Delta_t^{u,1,\varepsilon}, \tag{4.2.15}$$
$$v_t^\varepsilon = v_t^0 + \tilde{v}_t^{0,\varepsilon} + \varepsilon^{1/2} \tilde{v}_t^{1,\varepsilon} + \varepsilon^{1/2} \Delta_t^{v,1,\varepsilon} \tag{4.2.16}$$

(notice that $\tilde{u}_t^{0,\varepsilon} = 0$ by (4.1.28) and $u_t^{1,\varepsilon} = 0$, $v_t^{1,\varepsilon} = 0$ by Lemma 4.1.2; by Theorem 4.1.1

$$\lim_{\varepsilon \to 0} P(\|\Delta^{1,\varepsilon}\|_T \ge \mu) = 0 \tag{4.2.17}$$

for any $\mu \to 0$.

We consider the stopping times

$$\eta := \eta_\mu^{1,\varepsilon} := \inf\{t \ge 0 : |\Delta_t^{1,\varepsilon}| \ge \mu\} \wedge T, \tag{4.2.18}$$
$$S := S_{N,\mu}^\varepsilon := \sigma_N^{0,\varepsilon} \wedge \sigma_v^{2,\varepsilon} \wedge \eta_\mu^{1,\varepsilon} \tag{4.2.19}$$

where the stopping times $\sigma_N^{0,\varepsilon}$ and $\sigma_v^{2,\varepsilon}$ are defined in (4.1.86) and (4.1.104). Assume that $\nu \in \,]0,1/2[$. Put $F_v^0(t) := (\partial/\partial v) F(t, u_t^0, v_t^0)$. Then for $t \le S$ we have:

$$|F_y(t,\varepsilon) - F_v^0(t)| \le L_N(|u_t^\varepsilon - u_t^0| + |v_t^\varepsilon - v_t^0|)$$
$$\le L_N(|\tilde{v}_t^{0,\varepsilon}| + \varepsilon^{1/2}|\tilde{Z}^{1,\varepsilon}| + \varepsilon^{1/2}|\Delta_t^{1,\varepsilon}|)$$
$$\le L_N(C_N e^{-\gamma_N t/(2\varepsilon)} + \varepsilon^{1/2-\nu} + \mu\varepsilon^{1/2}).$$

It follows that the hypothesis of Proposition A.2.5 is fulfilled and hence

$$\limsup_{\varepsilon\to 0}\ \sup_{0\le s\le S}\ e^{\kappa_N(t-s)/\varepsilon}|\Phi^\varepsilon(t,s)| \le C_N$$

for certain constants C_N and $\kappa_N > 0$. With this property we proceed further exactly as in the proof of Lemma 4.1.5. $\square$

Remark. By **H.4.2.1** we have the relation $\beta \le L\varepsilon^\delta$ for some $\delta > 0$ and, hence, (4.2.14) implies that for any $r > 0$

$$\lim_{\varepsilon\to 0} P(\|h^{k,\varepsilon}\|_T > \beta^{-r}) = 0, \quad 1 \le k \le n+2.$$

4.2.3 Proof of Theorem 4.2.1

Put

$$Q_t^{x,n,\varepsilon} := \sum_{k=0}^n \beta^k x_t^{k,\varepsilon}, \quad Q_t^{y,n,\varepsilon} := \sum_{k=0}^n \beta^k y_t^{k,\varepsilon}, \tag{4.2.20}$$

and $Q_t^{h,n,\varepsilon} := (Q_t^{x,n,\varepsilon}, Q_t^{y,n,\varepsilon})$ where the coefficients are given by (4.2.2)–(4.2.5).

It follows from (4.0.5), the first definition in (4.2.20), and (4.2.2), (4.2.4) that

$$\beta^{n+2} d\delta_t^{x,n+2,\varepsilon} := dx_t^\varepsilon - dQ_t^{x,n+2,\varepsilon}$$
$$= \left(f(t,x_t^\varepsilon,y_t^\varepsilon) - \sum_{k=0}^{n+2} \beta^k f_t^{k,\varepsilon} \right) dt + \left(g(t,x_t^\varepsilon,y_t^\varepsilon) - \sum_{k=0}^{n+2} \beta^k g_t^{k,\varepsilon} \right) dw_t^x,$$

and, similarly, from (4.0.6), the second definition in (4.2.20), and (4.2.3), (4.2.5) that

$$\beta^{n+2} d\delta_t^{y,n+2,\varepsilon} := dy_t^\varepsilon - dQ_t^{y,n+2,\varepsilon} = \varepsilon^{-1}\left(F(t,x_t^\varepsilon,y_t^\varepsilon) - \sum_{k=0}^{n+2} \beta^k F_t^{k,\varepsilon} \right) dt$$

$$+ \beta\varepsilon^{-1/2}\left(G(t,x_t^\varepsilon,y_t^\varepsilon) - \sum_{k=1}^{n+2} \beta^{k-1} G_t^{k-1,\varepsilon} \right) dw_t^y$$

(with $f_t^{0,\varepsilon} := f(t,u_t^\varepsilon,v_t^\varepsilon)$, etc.).

Using the formula of finite increments and introducing some abbreviations we get from here the following representations:

$$d\delta_t^{x,n+2,\varepsilon} := \widehat{f}_x(t,\varepsilon)\delta_t^{x,n+2,\varepsilon}dt + \widehat{f}_y(t,\varepsilon)\delta_t^{y,n+2,\varepsilon}dt + D_t^{f,n+2,\varepsilon}dt$$
$$+ (\widehat{g}_x(t,\varepsilon)\delta_t^{x,n+2,\varepsilon} + \widehat{g}_y(t,\varepsilon)\delta_t^{y,n+2,\varepsilon} + D_t^{g,n+2,\varepsilon})dw_t^x, \quad (4.2.21)$$

$$\varepsilon d\delta_t^{y,n+2,\varepsilon} := \widehat{F}_x(t,\varepsilon)\delta_t^{x,n+2,\varepsilon}dt + \widehat{F}_y(t,\varepsilon)\delta_t^{y,n+2,\varepsilon}dt + D_t^{F,n+2,\varepsilon}dt$$
$$+ \varepsilon^{1/2}\bar{D}_t^{G,n+1,\varepsilon}dw_t^y, \quad (4.2.22)$$

where $\widehat{f}_x(t,\varepsilon)$ is the predictable process the components of which are partial derivatives of f calculated at some points $(t, Q_t^{h,n+2,\varepsilon} + \zeta\beta^{n+2}\delta_t^{h,n+2,\varepsilon})$ with $\zeta \in [0,1]$,

$$D_t^{f,n+2,\varepsilon} := \beta^{-(n+2)}\left(f(t, Q_t^{h,n+2,\varepsilon}) - \sum_{k=0}^{n+2}\beta^k f_t^{k,\varepsilon}\right), \quad (4.2.23)$$

etc.,

$$\bar{D}_t^{G,n+1,\varepsilon} := \beta^{-(n+1)}(G(t, h_t^\varepsilon) - G(t, Q_t^{h,n+1,\varepsilon})) + D_t^{G,n+1,\varepsilon}. \quad (4.2.24)$$

Define the stopping times

$$\vartheta_\mu^{n,\varepsilon} := \inf\{t \geq 0 : |\delta_t^{h,n,\varepsilon}| \geq \mu\} \wedge T, \quad (4.2.25)$$

$$r_\nu^{n,\varepsilon} := \inf\left\{t \geq 0 : \sum_{k=1}^n |h_t^{k,\varepsilon}| \geq \varepsilon^{-\nu}\right\} \wedge T, \quad (4.2.26)$$

$$\chi := \chi_N^\varepsilon := S \wedge \vartheta_\mu^{n,\varepsilon} \wedge r_\nu^{n,\varepsilon} \quad (4.2.27)$$

where the stopping time $S := S_{N,\mu}^\varepsilon$ is given by (4.2.19) (as usual, we omit dependence of some parameters).

By (4.2.15), (4.2.16), (4.2.19), (4.1.86), (4.1.104)

$$\|u^\varepsilon\|_S + \|v^\varepsilon\|_S \leq L_N \quad (4.2.28)$$

for some constant L_N. It follows from here and our definitions that

$$\|Q^{h,n+2,\varepsilon}\|_\chi \leq \|u^\varepsilon\|_S + \|v^\varepsilon\|_S + \sum_{k=1}^{n+2}\beta^k\|h^{k,\varepsilon}\|_\chi \leq L_N + \beta\varepsilon^{-\nu},$$

$$\|h^\varepsilon\|_\chi \leq \|Q^{h,n,\varepsilon}\|_\chi + \beta^{n+2}\|\delta^{h,n,\varepsilon}\|_\chi \leq L_N + \beta\varepsilon^{-\nu} + \mu\beta^{n+2}.$$

Since $\beta = O(\varepsilon^\delta)$ we conclude that there is a constant L_N such

$$\|h^\varepsilon\|_S + \|Q^{h,n+2,\varepsilon}\|_S \leq L_N \quad (4.2.29)$$

for any $\varepsilon \in {]}0,1]$ and $\nu \in {]}0, \delta \wedge (1/2)[$.

It follows from the local Lipschitz condition for the first derivative of F and (4.2.15)–(4.2.16) that for $t \leq \chi$

$$\|\widehat{F}_y(t, \varepsilon) - F_v^0(t)\| \le L_N(|Q_t^{h,n+2,\varepsilon} - Z_t^0| + \beta^{n+2}|\delta_t^{y,n+2,\varepsilon}|)$$

$$\le L_N\left(|h_t^{0,\varepsilon} - Z_t^0| + \sum_{k=1}^{n+2} \beta^k |h_t^{k,\varepsilon}| + \mu\beta^{n+2}\delta_t^{y,n+2,\varepsilon}\right)$$

$$\le L_N(|\tilde{v}_t^{0,\varepsilon}| + \varepsilon^{1/2}|\tilde{Z}^{1,\varepsilon}| + \varepsilon^{1/2}|\Delta_t^{1,\varepsilon}| + \beta\varepsilon^{-\nu} + \mu\beta^{n+2})$$

$$\le L_N(C_N e^{-\gamma_N t/(2\varepsilon)} + \varepsilon^{1/2-\nu} + \mu\varepsilon^{1/2} + \beta\varepsilon^{-\nu} + \mu\beta^{n+2}).$$

$$(4.2.30)$$

Let us introduce the fundamental matrix $\widehat{\Phi}^\varepsilon(t, s)$ which is the solution of the linear equation

$$\varepsilon\frac{d\widehat{\Phi}^\varepsilon(t, s)}{dt} = \varepsilon^{-1}\widehat{F}_y(t, \varepsilon)\widehat{\Phi}^\varepsilon(t, s), \quad \widehat{\Phi}^\varepsilon(s, s) = I, \quad (4.2.31)$$

The inequality (4.2.30) allows us to apply Proposition A.2.5 which asserts that there are some constants C_N and $\kappa_N > 0$ such that

$$\limsup_{\varepsilon\to 0} \sup_{0\le s\le t\le\chi} e^{\kappa_N(t-s)/\varepsilon}|\widehat{\Phi}^\varepsilon(t, s)| < C_N. \quad (4.2.32)$$

Put

$$\widehat{\theta}_N^\varepsilon := \inf\{t \ge 0 : \widehat{\phi}_t^{N,\varepsilon} \ge C_N\} \wedge \chi_N^\varepsilon, \quad (4.2.33)$$

where

$$\widehat{\phi}_t^{N,\varepsilon} := \sup_{s\le t} e^{\kappa_N(t-s)/\varepsilon}|\widehat{\Phi}^\varepsilon(t, s)|.$$

Obviously,

$$\lim_{\varepsilon\to 0} P(\widehat{\theta}_N^\varepsilon < \chi_N^\varepsilon) = 0. \quad (4.2.34)$$

It follows from Lemma 2.3.5 that

$$\|D^{f,n+2,\varepsilon}\|_\chi + \|D^{g,n+2,\varepsilon}\|_\chi + \|D^{F,n+2,\varepsilon}\|_\chi + \|D^{G,n+1,\varepsilon}\|_\chi \le L_N\beta\varepsilon^{-(n+3)\nu}.$$

$$(4.2.35)$$

Using the local Lipschitz condition for G we get from the definition (4.2.24) and the above bound that

$$\|\bar{D}^{G,n+1,\varepsilon}\|_\chi \le \beta^{-(n+1)} \sup_{t\le\chi} |G(t, h_t^\varepsilon) - G(t, Q_t^{h,n+1,\varepsilon})| + \|D^{G,n+1,\varepsilon}\|_\chi$$

$$\le L_N\beta^{-(n+1)}\|h^\varepsilon - Q^{h,n+1,\varepsilon}\|_\chi + \|D^{G,n+1,\varepsilon}\|_\chi$$

$$\le L_N\beta^{-(n+1)}(\beta^{n+2}\|h^{n+2,\varepsilon}\|_\chi + \beta^{n+2}\|\delta^{h,n+2,\varepsilon}\|_\chi)$$

$$+ L_N\beta\varepsilon^{-(n+3)\nu}$$

$$\le L_N(\beta\varepsilon^{-\nu} + \mu\beta + \beta\varepsilon^{-(n+3)\nu}) \le L_N\varepsilon^{\delta-(n+3)\nu} \quad (4.2.36)$$

where the last inequality holds when $\nu < \delta/(n+3)$; we assume that such a ν is chosen.

Applying the Cauchy formula to (4.2.22), we obtain for $t \leq \widehat{\theta}_N^\varepsilon$ the representation

$$\delta_t^{y,n+2,\varepsilon} = \frac{1}{\varepsilon} \int_0^t \widehat{\Phi}^\varepsilon(t,s) \widehat{F}_x(s,\varepsilon) \delta_s^{x,n+2,\varepsilon} ds + \frac{1}{\varepsilon} \int_0^t \widehat{\Phi}^\varepsilon(t,s) D_s^{F,n+2,\varepsilon} ds + \xi_t^\varepsilon$$

$$(4.2.37)$$

where the process ξ^ε satisfies the equation

$$\varepsilon d\xi_t^\varepsilon = A_t^\varepsilon \widehat{F}_y(t,\varepsilon) \xi_t^\varepsilon dt + \varepsilon^{1/2} \bar{D}_{t \wedge \chi}^{G,n+1,\varepsilon} dw_t^y, \quad \xi_0^\varepsilon = 0,$$

$$(4.2.38)$$

with

$$A_t^\varepsilon := \widehat{F}_y(t,\varepsilon) I_{[0,\widehat{\theta}_N^\varepsilon]}(t) - I_{]\widehat{\theta}_N^\varepsilon, T]}(t).$$

Using the boundedness of $\widehat{F}_x(t,\varepsilon)$ (see **H.4.1.2**) and the estimate (4.2.36) we deduce from (4.2.37) that for $t \leq \widehat{\theta}_N^\varepsilon$

$$|\delta_t^{y,n+2,\varepsilon}| \leq L_N(\kappa_N^{-1} \|\delta^{x,n+2,\varepsilon}\|_t + \kappa_N^{-1} \varepsilon^{\delta_0} + |\xi_t^\varepsilon|)$$

where $\delta_0 := \delta - (n+3)\nu$. From Lemma A.1.3 and the last bound we have

$$E\|\delta^{h,n+2,\varepsilon}\|_{\widehat{\theta}}^2 \leq L_N(\varepsilon^{2\delta_0} + E\|\xi^\varepsilon\|_T^2).$$

Furthermore, Proposition 1.2.4 implies that there is a constant L (depending on N) such that for all sufficiently small $\varepsilon > 0$

$$E\|\xi^\varepsilon\|_T \leq L\varepsilon^{\delta_0} |\ln \varepsilon|.$$

Thus,

$$E\|\delta^{h,n+2,\varepsilon}\|_{\widehat{\theta}}^2 \leq L_N(\varepsilon^{2\delta_0} + \varepsilon^{\delta_0} |\ln \varepsilon|) \to 0$$

and, by the Chebyshev inequality,

$$\lim_{\varepsilon \to 0} P(\|\delta^{h,n+2,\varepsilon}\|_{\widehat{\theta}} \geq \mu) = 0.$$

$$(4.2.39)$$

It follows from (4.2.14), (4.2.26) that for arbitrary $\nu > 0$

$$\lim_{\varepsilon \to 0} P(r_\nu^{n+2,\varepsilon} < T) = 0.$$

$$(4.2.40)$$

By (4.1.102), (4.1.105), (4.2.17)–(4.2.19)

$$\lim_{N \to \infty} \limsup_{\varepsilon \to 0} P(S_{N,\mu}^\varepsilon < T) = 0,$$

$$(4.2.41)$$

and by (4.2.39), (4.2.40), and (4.2.41)

$$\lim_{N \to \infty} \limsup_{\varepsilon \to 0} P(\chi_N^\varepsilon < \vartheta_\mu^{n+2,\varepsilon}) = 0.$$

$$(4.2.42)$$

At last,

$$P(\|\delta^{h,n+2,\varepsilon}\|_T \geq \mu) \leq P(\|\delta^{h,n+2,\varepsilon}\|_\vartheta \geq \mu)$$
$$\leq P(\chi_N^\varepsilon < \vartheta_\mu^\varepsilon) + P(\widehat{\theta}_N^\varepsilon < \chi_N^\varepsilon)$$
$$+ P(\|\delta^{h,n+2,\varepsilon}\|_{\widehat{\theta}} \geq \mu)$$

$$(4.2.43)$$

and we get from (4.2.34), (4.2.42), and (4.2.43) that $\|\delta^{h,n+2,\varepsilon}\|_T \to 0$ in probability as $\varepsilon \to 0$. $\square$

4.3 Liénard Oscillator Driven by a Random Force

We consider an asymptotic expansion for the two one-dimensional system

$$du_t = -h(v_t)dt + dw_t, \quad u_0 = x^o, \qquad (4.3.1)$$

$$\varepsilon dv_t = (u_t - kv_t)dt, \quad v_0 = y^o, \qquad (4.3.2)$$

describing, in the Liénard coordinates, the Liénard oscillator perturbed by a white noise. Since this this system is a particular case of (4.1.1), (4.1.2), the coefficients in the asymptotic expansions are given by the formulae (4.1.28)–(4.1.47) with corresponding initial conditions. For the regular components we get the following sequence of equations which can be solved recursively.

System 1:

$$\begin{cases} du_t^0 = -h(v_t^0)dt + dw_t, \\ u_t^0 - kv_t^0 = 0, \end{cases} \quad \begin{aligned} u_0^0 &= x^0, \\ v_0^0 &= k^{-1}x^0; \end{aligned}$$

$$\begin{cases} du_t^2 = -h_t^0 v_t^2 dt, \\ u_t^2 - kv_t^2 = -k^{-1}h(v_t^0), \end{cases} \quad \begin{aligned} u_0^2 &= 0, \\ v_0^2 &= k^{-2}h(k^{-1}x^0); \end{aligned}$$

$$\begin{cases} du_t^{2i} = -h^0(t)v_t^{2i}dt - R_t^{h,2i}dt, \\ u_t^{2i} - kv_t^{2i} = v_t^{1,2(i-1)}, \end{cases} \quad \begin{aligned} u_0^{2i} &= 0, \\ v_0^2 &= k^{-1}v_0^{1,2(i-1)}, \quad i \geq 2, \end{aligned}$$

where $h_t^0 := h'(v_t^0)$ and $R_t^{h,2i}$ is the function of $(u_t^0, v_t^t, \ldots, u_t^{2(i-1)}, v_t^{2(i-1)})$ given by (4.1.11); all odd coefficients (u_t^{2i+1}, v_t^{2i+1}) vanish.

Boundary layer correction terms are given by another recurrent structure.

System 2:

$$\begin{cases} d\tilde{u}_t^{0,\varepsilon} = 0, \\ \varepsilon d\tilde{v}_t^{0,\varepsilon} = -k\tilde{v}_t^{0,\varepsilon}dt, \end{cases} \quad \begin{aligned} \tilde{u}_0^{0,\varepsilon} &= 0, \\ \tilde{v}_0^{0,\varepsilon} &= y^0 - k^{-1}x^0; \end{aligned}$$

$$\begin{cases} d\tilde{u}_t^{1,\varepsilon} = -h_t^{0,\varepsilon}\tilde{v}_t^{1,\varepsilon}dt - R_t^{h,1,\varepsilon}dt, \\ \varepsilon d\tilde{v}_t^{1,\varepsilon} = (\tilde{u}_t^{1,\varepsilon} - k\tilde{v}_t^{1,\varepsilon})dt - \varepsilon^{1/2}v_t^{2,0}dw_t, \end{cases} \quad \begin{aligned} \tilde{u}_0^{1,\varepsilon} &= 0, \\ \tilde{v}_0^{1,\varepsilon} &= 0; \end{aligned}$$

$$\begin{cases} d\tilde{u}_t^{2,\varepsilon} = -h_t^{0,\varepsilon}\tilde{v}_t^{2,\varepsilon}dt - (\varepsilon^{-1}\tilde{h}_t^{0,\varepsilon} + R_t^{h,2,\varepsilon})dt, \\ \varepsilon d\tilde{v}_t^{2,\varepsilon} = (\tilde{u}_t^{2,\varepsilon} - k\tilde{v}_t^{2,\varepsilon})dt - \varepsilon^{1/2}v_t^{2,1}dw_t, \end{cases} \quad \begin{aligned} \tilde{u}_0^{2,\varepsilon} &= 0, \\ \tilde{v}_0^{2,\varepsilon} &= -k^{-2}h(x^0/k); \end{aligned}$$

$$\begin{cases} d\tilde{u}_t^{j,\varepsilon} = -h_t^{0,\varepsilon}\tilde{v}_t^{j,\varepsilon}dt - R_t^{h,j,\varepsilon}dt, \\ \varepsilon d\tilde{v}_t^{j,\varepsilon} = (\tilde{u}_t^{j,\varepsilon} - k\tilde{v}_t^{j,\varepsilon})dt - \varepsilon^{1/2}v_t^{2,j-1}dw_t, \end{cases} \quad \begin{aligned} \tilde{u}_0^{j,\varepsilon} &= 0, \\ \tilde{v}_0^{2,\varepsilon} &= k^{-1}v^{1,j-2}, \end{aligned}$$

where $j \geq 2$,

$$h_t^{0,\varepsilon} := h'(v_t^0 + \tilde{v}_t^{0,\varepsilon}),$$
$$\tilde{h}_t^{0,\varepsilon} := h(v_t^0 + \tilde{v}_t^{0,\varepsilon}) - h(v_t^0),$$

and where $R_t^{h,j,\varepsilon}$ is the remainder of the jth coefficient in the asymptotic expansion of h defined in (4.1.12).

The hypotheses **H.4.0.1-H.4.0.4** hold automatically and we have as an obvious corollary of Theorem 4.1.1:

Theorem 4.3.1 *Assume that the function h has $n+2$ continuous derivatives such that h' is bounded and all others are of a polynomial growth. Then*

$$u_t^\varepsilon = \sum_{i=0}^n \varepsilon^{i/2}(u_t^i + \tilde{u}_t^{i,\varepsilon}) + \varepsilon^{n/2}\Delta_t^{u,n,\varepsilon}, \tag{4.3.3}$$

$$v_t^\varepsilon = \sum_{i=0}^n \varepsilon^{i/2}(v_t^i + \tilde{v}_t^{i,\varepsilon}) + \varepsilon^{n/2}\Delta_t^{v,n,\varepsilon}, \tag{4.3.4}$$

with

$$P\text{-}\lim_{\varepsilon\to 0}(||\Delta^{u,n,\varepsilon}||_T + ||\Delta^{v,n,\varepsilon}||_T) = 0. \tag{4.3.5}$$

However, the assumption that h' is bounded does not cover cases which are considered in the standard deterministic setting as the most interesting, e.g., when $h(v) = v^3$. Below we give a result with hypotheses allowing us to include this example.

H.4.3.1 The function h is continuously differentiable and for some constant $\alpha > 0$ we have

$$-xh(x) \le \alpha \quad \forall x \in \mathbf{R}.$$

Clearly, **H.4.3.1** is fulfilled for all functions $h(x) = ax^{2k-1}$ where $a \ge 0$, $k \in \mathbf{N}$, and their sums. On the other hand, for the linear function $h(x) = -x$ the condition does not hold.

The first equation of System 1 is nonlinear with a function h which may be not of the linear growth. The following lemma shows that under **H.4.3.1** everything is going well.

Lemma 4.3.2 *Assume that h satisfies **H.4.3.1**. Then the equation*

$$dv_t = -h(v_t)dt + dw_t, \quad v_0 = v^o, \tag{4.3.6}$$

with $v^o \in L^2$ has a unique strong solution $v = (v_t)$ and $||v||_T \in L^2$.

Proof. Let $\mathcal{L}V := (1/2)V'' - hV'$. For $V(x) = x^2/2 + \alpha + 1/2$ we have

$$\mathcal{L}V(x) = -xh(x) + 1/2 \le \alpha + 1/2 \le V(x)$$

and Theorem A.1.2 ensures the existence and uniqueness of the solution.

By the Ito formula

$$v_t^2 = v_0^2 + \int_0^t (1 - 2v_s h(v_s))ds + 2\int_0^t v_s dw_s.$$

Localizing and taking the expectation, we get easily from here, using the Gronwall–Bellman inequality and the Fatou lemma, that $E||v||_T^2 < \infty$. $\square$

We need to check that the parametric family of solutions of System 2 is of a "moderate" growth as $\varepsilon \to 0$.

Lemma 4.3.3 *Assume that h has $n+2$ continuous derivative of a polynomial growth. Then System 2 has a unique strong solution and for all $\nu > 0$*

$$\lim_{\varepsilon \to 0} P(||\tilde{u}^{i,\varepsilon}||_T + ||\tilde{v}^{i,\varepsilon}||_T > \varepsilon^{-\nu}) = 0 \tag{4.3.7}$$

for every $i \leq n$.

Proof. Define the stopping times

$$\tau_N := \left\{ t \geq 0 : \sum_{i=0}^{n+2}(|u_t^i| + |v_t^i|) \geq N \right\}.$$

Taking into account that $v^{2,0} = k^{-1}$, we get the representation

$$\tilde{v}_t^{1,\varepsilon} = \frac{1}{\varepsilon} \int_0^t \tilde{u}_s^{1,\varepsilon} e^{-k(t-s)/\varepsilon} ds - \xi_t^\varepsilon \tag{4.3.8}$$

with

$$\xi_t^\varepsilon := \frac{1}{k\sqrt{\varepsilon}} \int_0^t e^{-k(t-s)/\varepsilon} dw_s,$$

implying, in particular, that

$$E||\tilde{v}^{1,\varepsilon}||_t \leq k^{-1} E||\tilde{u}^{1,\varepsilon}||_t + E||\xi^\varepsilon||_T.$$

By virtue of Theorem 1.1.7 there is a constant L such that

$$E||\xi^\varepsilon||_T \leq L|\ln \varepsilon|^{1/2} \tag{4.3.9}$$

for all $\varepsilon \in [0, 1/2]$. (We shall keep this assumption on ε needed here to ensure that $\ln \varepsilon$ is bounded away from zero). Notice that for some constant L_N we have

$$||\tilde{u}^{1,\varepsilon}||_{t \wedge \tau_N} \leq L_N \left(\int_0^t ||\tilde{u}^{1,\varepsilon}||_{s \wedge \tau_N} ds + ||\xi^\varepsilon||_T \right).$$

By the standard use of the Gronwall–Bellman inequality and the Fatou lemma we infer that

$$||\tilde{u}^{1,\varepsilon}||_{\tau_N} \leq L_N |\ln \varepsilon|^{1/2}$$

This inequality implies, via the above bounds, the similar one for $||\tilde{u}^{1,\varepsilon}||_{\tau_N}$. Now, using the Chebyshev inequality, we observe that

$$P(||\tilde{u}^{1,\varepsilon}||_T + ||\tilde{v}^{1,\varepsilon}||_T > \varepsilon^{-\nu}) \leq P(\tau_N < T) + P(||\tilde{u}^{1,\varepsilon}||_{\tau_N} + ||\tilde{v}^{1,\varepsilon}||_{\tau_N} > \varepsilon^{-\nu})$$

$$\leq P(\tau_N < T) + \varepsilon^\nu |\ln \varepsilon|^{1/2}.$$

Taking here successively the limits as $\varepsilon \to 0$ and $N \to \infty$ we get the claim for $i = 1$.

We proceed further by induction. Assume that the assertion of the lemma holds for all $i \leq l$ and check that it holds also for $i = l + 1$.

For arbitrary $\mu > 0$ we define the stopping time

$$\sigma = \sigma_{N,\mu}^{l,\varepsilon} := \left\{ t \geq 0 : \sum_{i=0}^{l} (|u_t^{i,\varepsilon}| + |v_t^{i,\varepsilon}|) \geq N \right\} \wedge \tau_N.$$

By the induction hypothesis we have

$$\lim_{\varepsilon \to 0} \lim_{N \to \infty} P(\sigma_{N,\mu}^{l,\varepsilon} < T) = 0. \tag{4.3.10}$$

Define also the set $\Gamma_N = \{|x^\circ| + |y^\circ| \leq N\}$. It follows from the equation for $\tilde{v}^{l+1,\varepsilon}$ that on the set $\{t \leq \sigma\} \cap \Gamma_N$

$$\tilde{v}_t^{l+1,\varepsilon} = e^{-kt/\varepsilon} \tilde{v}_0^{l+1,\varepsilon} + \frac{1}{\varepsilon} \int_0^t \tilde{u}_s^{l+1,\varepsilon} e^{-k(t-s)/\varepsilon} ds - \xi_t^{N,l,\varepsilon}, \tag{4.3.11}$$

where

$$\xi_t^{N,l,\varepsilon} := \varepsilon^{-1/2} \int_0^t v_s^{2,N,l} e^{-k(t-s)/\varepsilon} dw_s,$$

$v_t 2, N, l := v_{t \wedge \tau_N} 2, l$ and, therefore,

$$\|\tilde{v}^{l+1,\varepsilon}\|_t \leq L_N (1 + \|\tilde{u}^{l+1,\varepsilon}\|_t + \|\xi^{N,l,\varepsilon}\|_T) \tag{4.3.12}$$

It follows from the definitions of the stopping time τ_N and the function $v^{2,l}$ that $\|v^{2,N,l}\|_T \leq L_N$ for some constant L_N. Applying Theorem 1.1.7 we infer that

$$E\|\xi^{N,l,\varepsilon}\|_T \leq L_N |\ln \varepsilon|^{1/2} \tag{4.3.13}$$

for all $\varepsilon \in \,]0, 1/2]$.

We infer also from the definition of $R_t^{h,l+1,\varepsilon}$ that on the set $\{t \leq \sigma\} \cap \Gamma_N$

$$\|R^{h,l+1,\varepsilon}\|_t \leq L_N \varepsilon^{-(l+1)\mu}$$

and, also,

$$\tilde{h}_t^{0,\varepsilon} = |h(v_t^0 + \tilde{v}_t^{0,\varepsilon}) - h(v_t^0)| \leq L_N |\tilde{v}_t^{0,\varepsilon}| \leq L_N e^{-kt/\varepsilon}.$$

Using the equation for the function $\tilde{u}^{l,\varepsilon}$ and the above bounds we get on the set $\{t \leq \sigma\} \cap \Gamma_N$ that

$$\|\tilde{u}^{l+1,\varepsilon}\|_t \leq L_N \left(\int_0^t \|\tilde{u}^{l+1,\varepsilon}\|_s ds + \delta_{1l} \frac{1}{\varepsilon} \int_0^t |\tilde{h}_s^{0,\varepsilon}| ds \right.$$
$$\left. + 1 + \varepsilon^{-(l+1)\mu} + \|\xi^{N,l,\varepsilon}\|_T \right)$$
$$\leq L_N \left(\int_0^t \|\tilde{u}^{l+1,\varepsilon}\|_s ds + 1 + \varepsilon^{-(l+1)\mu} + \|\xi^{N,l,\varepsilon}\|_T \right),$$

where $\delta_{1l} = 1$ if $l = 1$ and zero otherwise.

Again by Gronwall–Bellman inequality we get that on Γ_N

$$||\tilde{u}^{l+1,\varepsilon}||_\sigma + ||\tilde{v}^{l+1,\varepsilon}||_\sigma \leq L_N(1 + \varepsilon^{-(l+1)\mu} + ||\xi^{N,l,\varepsilon}||_T).$$

Using the similar arguments as in the proof of the first step of induction, we have:

$$
\begin{aligned}
P(||\tilde{u}^{i,\varepsilon}||_T + ||\tilde{v}^{i,\varepsilon}||_T > \varepsilon^{-\nu}) &\leq P(\sigma < T) + P(||\tilde{u}^{i,\varepsilon}||_\sigma + ||\tilde{v}^{i,\varepsilon}||_\sigma > \varepsilon^{-\nu}) \\
&\leq P(\sigma < T) + P(\Gamma_N^c) \\
&\quad + P(L_N(1 + \varepsilon^{-(l+1)\mu} + ||\xi^{N,l,\varepsilon}||_T) > \varepsilon^{-\nu}) \\
&\leq P(\sigma < T) + P(\Gamma_N^c) \\
&\quad + L_N\left(\varepsilon^\nu + \varepsilon^{\nu-(l+1)\mu} + \varepsilon^\nu|\ln e|^{1/2}\right).
\end{aligned}
$$

If we choose the free parameter μ to be in the interval $]0, \nu/(l+1)[$ we get the result by taking the limits as $\varepsilon \to 0$ and $N \to \infty$. $\square$

Theorem 4.3.4 *Assume that the function h has continuous derivatives of a polynomial growth up to order $n + 2$, the hypothesis* **H.3.1** *is fulfilled, and*

$$\lim_{|y|\to\infty} \int_0^y h(u)du = +\infty. \tag{4.3.14}$$

Then the relations (4.3.3), (4.3.4) hold with the remainders satisfying (4.3.5).

Proof. First of all, notice that Systems 1 and 2 have a unique strong solution. This follows from Theorem A.1.2. Indeed, by virtue of (4.3.14) we can choose $\mu = \mu_\varepsilon$ such that

$$\varepsilon \int_0^y h(u)du + \varepsilon\mu \geq k\alpha + 1/2.$$

Then for

$$V_\varepsilon(x, y) := x^2/2 + \varepsilon \int_0^y h(u)du + \varepsilon\mu$$

we have that

$$\mathcal{L}V_\varepsilon(x, y) = -xh(y) + (x - ky)h(y) + \frac{1}{2} = -kyh(y) + 1/2 \leq k\alpha + \frac{1}{2} \leq V_\varepsilon(x, y).$$

The condition (4.3.14) ensures that $V_\varepsilon(x, y) \to \infty$ as $|x| + |y| \to \infty$. Thus, $V_\varepsilon(x, y)$ is the Lyapunov function.

The proof of (4.3.5) is the same as in Theorem 4.1.1. $\square$

5 Two-Scale Optimal Control Problems

In this chapter we study the limiting behavior of the optimal value of a cost functional for controlled two-scale stochastic systems with a small parameter tending to zero. In Section 5.1 we consider the Bolza problem where the cost functional contains an integral part ("running cost") and a part depending only on terminal values of the phase variables of both types: slow and fast. The result is proved for the case where the model is "semilinear" in the sense that the coefficients depend linearly on the fast variable. This structure is essential: even in the deterministic setting the general nonlinear models are hardly tractable. It is assumed also that the diffusion coefficient of the fast variable is $\beta\varepsilon^{1/2}$ with $\beta = o(|\ln \varepsilon|^{-1/2})$ which is our usual hypothesis. The admissible controls are open loop, i.e., adapted to the driving Wiener process. Such a setting seems to be the most developed: it allows to consider SDEs in the strong sense and use techniques similar to that of the classical theory of optimal control of ordinary differential equations. In the next sections we discuss more delicate subjects, namely, the behavior of the attainability sets for SDEs, aiming to prove a stochastic version of the Dontchev–Veliov theorem. In Section 5.2 we consider rather general models with open loop and closed loop (feedback) controls and compare the structure of attainability sets in these two settings. The main theorem of Section 5.2 clarifies the difference between these two concepts and shows why the model with closed loop controls (where solutions of SDEs are understood in the weak sense) suits more to the question we address later.

In Sections 5.3 and 5.4 we prove convergence results for the attainability sets for linear two-scale systems when the diffusion coefficient of the fast variable is of order $\varepsilon^{1/2+\delta}$ and $\varepsilon^{1/2}$, respectively, and apply the results to the Mayer problem, i.e., to the problem of terminal cost minimization.

Our techniques is based on direct probabilistic methods. However, the approach does not rely on the theory of weak convergence in functional spaces. It is worth noting that we use simultaneously both the basic concepts of stochastic optimal control which is rather rare in the existing literature.

5.1 Semilinear Controlled System

5.1.1 The Model and Main Result

We are given a fixed stochastic basis $(\Omega, \mathcal{F}, \mathbf{F} = (\mathcal{F}_t)_{t \geq 0}, P)$ with a standard Wiener process $W = (w^x, w^y)$ taking values in $\mathbf{R}^{k+n}$. Assume that the filtration $\mathbf{F}$ is generated by W and the P-null sets. In the system (5.1.1), (5.1.2) below the slow component x^ε takes values in $\mathbf{R}^k$ while the fast component y^ε evolves in $\mathbf{R}^n$. The set $\mathcal{U}$ of admissible controls consists of all predictable functions $u = (u_t)$ with values in a compact subset $U \subseteq \mathbf{R}^d$. As usual, $\varepsilon \in]0, 1]$.

The dynamic of the system is described by the following SDEs:

$$dx_t^\varepsilon = A_1(t, x_t^\varepsilon, u_t)dt + A_2(t)y_t^\varepsilon dt + \sigma(t, x_t^\varepsilon, u_t)dw_t^x, \quad x_0^\varepsilon = x^o, \quad (5.1.1)$$

$$\varepsilon dy_t^\varepsilon = A_3(t)x_t^\varepsilon dt + A_4(t)y_t^\varepsilon dt + B(t)u_t dt + \beta\sqrt{\varepsilon}dw_t^y, \quad y_0^\varepsilon = 0. \quad (5.1.2)$$

Our aim is to investigate the limit behavior, as ε tends to zero, of the minimal value

$$J_*^\varepsilon := \inf_{u \in \mathcal{U}} J^\varepsilon(u)$$

of the cost functional

$$J^\varepsilon(u) := E \int_0^T [f(t, x_t^\varepsilon, u_t) + b(t)y_t^\varepsilon] \, dt + Eg(x_T^\varepsilon, y_T^\varepsilon). \quad (5.1.3)$$

We show that, under some natural assumptions, this minimal value converges to the minimal value

$$\bar{J}_* := \inf_{u \in \mathcal{U}} \bar{J}(u)$$

of the cost functional of the "reduced" stochastic control problem:

$$\bar{J}(u) := E \int_0^T f_0(t, \bar{x}_t, u_t) \, dt + Eg_0(\bar{x}_T) \to \min, \quad (5.1.4)$$

$$d\bar{x}_t = A_0(t, \bar{x}_t, u_t)dt + \sigma(t, \bar{x}_t, u_t)dw^x, \quad \bar{x}_0 = x_0, \quad (5.1.5)$$

where

$$A_0(t, x, u) := A_1(t, x, u) - A_2(t)A_4^{-1}(t)[A_3(t)x + B(t)u], \quad (5.1.6)$$

$$f_0(t, x, u) := f(t, x, u) - b(t)A_4^{-1}(t)[A_3(t)x + B(t)u], \quad (5.1.7)$$

$$g_0(x) := \inf_{y \in Y} g(x, -A_4^{-1}(T)A_3(T)x + y) \quad (5.1.8)$$

with

$$Y := \left\{ y : y = \int_0^\infty e^{A_4(T)s}B(T)v_s \, ds, \quad v \in \mathcal{B}_U \right\}$$

and $\mathcal{B}_U$ denoting the set of all Borel functions $v : \mathbf{R}_+ \to U$. Furthermore, the proof indicates that an "almost" optimal control for the limit problem can be modified to get an "almost" optimal control for the prelimit problem for sufficiently small ε.

Assumptions on the coefficients:

H.5.1.1 The functions A_2, A_3, A_4, B, b are continuous on $[0, T]$ (A_2 is $k \times n$ matrix, B is $k \times d$, b is $1 \times k$, etc.).

H.5.1.2 The functions A_1, σ and f are continuous in all their arguments, Lipschitz in x (uniformly in t, u), and of linear growth. More precisely: there exists a constant L such that

$$|A_1(t, x_1, u) - A_1(t, x_2, u)| + |f(t, x_1, u) - f(t, x_2, u)| \le L|x_1 - x_2|, \quad (5.1.9)$$

$$|\sigma(t, x_1, u) - \sigma(t, x_2, u)| \le L|x_1 - x_2|, \quad (5.1.10)$$

$$|A_1(t, x, u)| + |f(t, x, u)| \le L(1 + |x|), \quad (5.1.11)$$

$$|\sigma(t, x, u)| \le L(1 + |x|) \quad (5.1.12)$$

for all t, x, x_1, x_2, u.

H.5.1.3 There exists a positive constant γ such that all real parts of the eigenvalues of $A_4(t)$ are strictly less then -2γ: for all $t \in [0, T]$

$$\mathrm{Re}\,\lambda(A_4(t)) < -2\gamma. \quad (5.1.13)$$

H.5.1.4 The function $\beta = \beta_\varepsilon$ is bounded on $]0, 1]$ and $\beta = o(1/|\ln \varepsilon|^{1/2})$ as $\varepsilon \to 0$.

H.5.1.5 The positive continuous function g on $\mathbf{R}^{k+n}$ is of a polynomial growth:

$$g(x, y) \le L(1 + |x|^l + |y|^l). \quad (5.1.14)$$

Theorem 5.1.1 *Under the assumptions* **H.5.1.1**–**H.5.1.5** *we have*

$$\lim_{\varepsilon \to 0} J_*^\varepsilon = \bar{J}_*. \quad (5.1.15)$$

The proof of Theorem 5.1.1 is based on the following two statements.

Proposition 5.1.2 *Let u^ε be any admissible control, let $(x^\varepsilon, y^\varepsilon)$ be the solution of (5.1.1), (5.1.2) corresponding to u^ε, and let $\bar{x}^\varepsilon$ be the solution of (5.1.5) corresponding to the same u^ε. Define*

$$\bar{y}_t^\varepsilon := -A_4^{-1}(t)[A_3(t)\bar{x}_t^\varepsilon + B(t)u_t^\varepsilon], \quad t < T, \quad (5.1.16)$$

$$\bar{y}_T^\varepsilon := -A_4^{-1}(T)A_3(T)\bar{x}_T^\varepsilon + \int_0^\infty e^{A_4(T)s}B(T)v_s^\varepsilon\,ds, \quad (5.1.17)$$

where

$$v_s^\varepsilon := u_{T-\varepsilon s}^\varepsilon I_{[0, T/\varepsilon^{1/2}]}(s). \quad (5.1.18)$$

Then in every L^p, $p < \infty$, we have that

(a) *the following sets are bounded:*

$$M^x := \{||x^\varepsilon||_T : \ u^\varepsilon \in \mathcal{U}, \ \varepsilon \in \,]0,1]\},$$

$$M^y := \{||y^\varepsilon||_T : \ u^\varepsilon \in \mathcal{U}, \ \varepsilon \in \,]0,1]\},$$

$$M_h^y := \left\{ \left\| \int_0^\cdot h_s y^\varepsilon \, ds \right\|_T : \ u^\varepsilon \in \mathcal{U}, \ \varepsilon \in \,]0,1] \right\},$$

where h is a fixed bounded and measurable process;

(b) $\lim\limits_{\varepsilon \to 0} \sup\limits_{\mathcal{U}} ||x^\varepsilon - \bar{x}^\varepsilon||_T = 0;$

(c) $\lim\limits_{\varepsilon \to 0} \sup\limits_{\mathcal{U}} |y_T^\varepsilon - \bar{y}_T^\varepsilon| = 0;$

(d) $\lim\limits_{\varepsilon \to 0} \sup\limits_{\mathcal{U}} || \int_0^\cdot V_s(y_s^\varepsilon - \bar{y}_s^\varepsilon) \, ds ||_T = 0$ *for any bounded measurable function*

V.

Proposition 5.1.3 *Let $\bar{x}$ be the solution of (5.1.5), corresponding to some $u \in \mathcal{U}$. Put*

$$\bar{y}_t := -A_4^{-1}(t)[A_3(t)\bar{x}_t + B(t)u_t], \qquad t < T, \tag{5.1.19}$$

$$\bar{y}_T := -A_4^{-1}(T)A_3(T)\bar{x}_T + \int_0^\infty e^{A_4(T)s} B(T)v_s \, ds \tag{5.1.20}$$

where v is a $\mathcal{F} \otimes \mathcal{B}_+$-measurable process with values in U. Then there exists a family $u^\varepsilon \in \mathcal{U}$ such that for any $S < T$

$$\lim_{\varepsilon \to 0} ||u^\varepsilon - u||_S = 0; \tag{5.1.21}$$

for the corresponding solutions of (5.1.1), (5.1.2) we have in any L^p, $p < \infty$,

$$\lim_{\varepsilon \to 0} ||x^\varepsilon - \bar{x}||_T = 0, \tag{5.1.22}$$

$$\lim_{\varepsilon \to 0} |y_T^\varepsilon - \bar{y}_T| = 0, \tag{5.1.23}$$

$$\lim_{\varepsilon \to 0} \left\| \int_0^\cdot V_s(y_s^\varepsilon - \bar{y}_s) \, ds \right\|_T = 0 \tag{5.1.24}$$

for any bounded measurable function V.

5.1.2 Proof of Proposition 5.1.2

We consider the fundamental matrix $\Psi^\varepsilon(t,s)$ given by the equation

$$\varepsilon \frac{\partial \Psi^\varepsilon(t,s)}{\partial t} = A_4(t)\Psi^\varepsilon(t,s), \quad \Psi^\varepsilon(s,s) = I, \tag{5.1.25}$$

where I is the identity matrix of order $n \times n$.

The assumption **H.5.1.3** implies (see Proposition A.2.3) that there exists a constant γ_0 such that

$$|\Psi^\varepsilon(t,s)| \leq \gamma_0 e^{-\gamma(t-s)/\varepsilon} \qquad (5.1.26)$$

for all $s \leq t \leq T$, $\varepsilon \in]0,1]$.

It follows that for all $\varepsilon \in]0,1]$

$$\sup_{t \leq T} \varepsilon^{-1} \int_0^t |\Psi^\varepsilon(t,s)|\, ds \leq \gamma_0/\gamma. \qquad (5.1.27)$$

We introduce the process

$$\eta_t^\varepsilon := \varepsilon^{-1/2} \int_0^t \Psi^\varepsilon(t,s)\, dw_s^y \qquad (5.1.28)$$

which is the solution of the linear SDE

$$d\eta_t^\varepsilon = \varepsilon^{-1} A_4(t)\eta_t^\varepsilon dt + \varepsilon^{-1/2} dw_t^y, \quad \eta_0^\varepsilon = 0. \qquad (5.1.29)$$

It follows from (5.1.26) and Lemma 1.2.6 that there is a constant C_p such that

$$\sup_{t \geq 0} E|\eta_t^\varepsilon|^p \leq C_p \qquad (5.1.30)$$

for any $p \in [1,\infty[$. By virtue of Proposition 1.2.7

$$E||\eta^\varepsilon||_T^p \leq C_p \varepsilon^{-1} \qquad (5.1.31)$$

for any $p \in [4,\infty[$.

Let u^ε be an admissible control. We have the following representation for the process y^ε satisfying (5.1.2) with $u = u^\varepsilon$:

$$y_t^\varepsilon = \varepsilon^{-1} \int_0^t \Psi^\varepsilon(t,s)[A_3(s)x_s^\varepsilon + B(s)u_s^\varepsilon]\, ds + \beta\eta_t^\varepsilon. \qquad (5.1.32)$$

Plugging (5.1.32) into (5.1.1) we get the integral stochastic equation for x^ε:

$$x_t^\varepsilon = x^o + \int_0^t A_1(s,x_s^\varepsilon,u_s^\varepsilon)\, ds + \int_0^t A_2(s)\left[\varepsilon^{-1}\int_0^s \Psi^\varepsilon(s,r)[A_3(r)x_r^\varepsilon\right.$$

$$\left. + B(r)u_r^\varepsilon]dr\right]ds + \beta \int_0^t A_2(s)\eta_s^\varepsilon ds + \int_0^t \sigma(s,x_s^\varepsilon,u_s^\varepsilon)\, dw_s^x. \qquad (5.1.33)$$

Lemma 5.1.4 *Let A be a bounded matrix-valued function on $[0,T]$ and let*

$$\zeta_t^\varepsilon := \int_0^t A(s)\eta_s^\varepsilon\, ds.$$

Then for any $p \in [1,\infty[$ there exists a constant c_p such that for all $\varepsilon \in]0,1]$

$$E||\zeta^\varepsilon||_T^p \leq c_p. \qquad (5.1.34)$$

If, moreover, A is continuous, then

$$\lim_{\varepsilon \to 0} E||\zeta^\varepsilon||_T^p = 0. \qquad (5.1.35)$$

Proof. Since A is bounded, (5.1.34) follows immediately from the Jensen inequality and (5.1.30). To prove (5.1.35) we consider the approximation of $D := AA_4^{-1}$ by the step functions

$$D^N := \sum_{i=1}^{N} D_{t_i} I_{]t_{i-1}, t_i]}$$

where $t_i := iT/N$. Using (5.1.29) we have:

$$\zeta_t^{\varepsilon} = \int_0^t D_s^N A_4(s) \eta_s^{\varepsilon} \, ds + \int_0^t (D_s - D_s^N) A_4(s) \eta_s^{\varepsilon} \, ds$$

$$= \varepsilon \sum_{i=1}^{N} D_{t_i} [\eta_{t_i \wedge t}^{\varepsilon} - \eta_{t_{i-1} \wedge t}^{\varepsilon} - \varepsilon^{-1/2} (w_{t_i \wedge t}^y - w_{t_{i-1} \wedge t}^y)]$$

$$+ \int_0^t (D_s - D_s^N) A_4(s) \eta_s^{\varepsilon} \, ds.$$

This implies the bound

$$||\zeta^{\varepsilon}||_T \le 2CN\varepsilon^{1/2}(\varepsilon^{1/2}||\eta^{\varepsilon}||_T + ||w^y||_T) + C\delta_N \int_0^T |\eta_s^{\varepsilon}| \, ds$$

where $\delta_N := ||D - D^N||_T \to 0$ as $N \to \infty$ due to continuity of A.

Notice that (5.1.31) ensures that the family of random variables

$$\{\varepsilon^{1/2}||\eta^{\varepsilon}||_T, \ \varepsilon \in \,]0, 1]\}$$

is bounded in L^p (for any finite p). It follows from (5.1.30) that the family of the integrals in the right-hand side of last inequality is also bounded in L^p. Thus,

$$\limsup_{\varepsilon \to 0} E||\zeta^{\varepsilon}||_T^p \le C\delta_N^p$$

and (5.1.35) holds. □

Using the linear growth conditions (5.1.11) and (5.1.12), the bounds (5.1.27) and (5.1.34), and the Burkholder–Gundy inequality we easily obtain from (5.1.33) that

$$E||x^{\varepsilon}||_t^{2m} \le c \left(1 + \int_0^t E||x^{\varepsilon}||_s^{2m} \, ds\right)$$

for some constant c independent on u. By the Gronwall–Bellman lemma we can conclude that there exists a constant C (obviously depending on m) such that

$$\sup_{\varepsilon} \sup_{u \in \mathcal{U}} E||x^{\varepsilon}||_T^{2m} \le C. \tag{5.1.36}$$

Thus, M^x is bounded in L^p for any finite p.

The representation (5.1.32), the bounds (5.1.26), (5.1.34), and the hypothesis **H.5.1.4** imply that for any h with $||h||_T \le c_1$

$$\sup_{\varepsilon} \sup_{u \in \mathcal{U}} E \left\| \int_0^{\cdot} h_s y_s^{\varepsilon} \, ds \right\|_T^{2m} \le C. \tag{5.1.37}$$

(the constant C depends on c_1). Hence, M_h^y is bounded in L^p for any finite p.

It follows from the representation (5.1.32), and bounds (5.1.26), (5.1.30), and (5.1.36) that the set M^y is also bounded in L^p.

The assertion (a) of Proposition 5.1.2 is established.

Lemma 5.1.5 *Let $A(t)$ be a continuous matrix function on $[0, T]$ and let $h(t)$ be a bounded measurable vector function. Then for every $\varepsilon \in \,]0, 1]$ and every $\eta > 0$ the following inequality holds:*

$$\left\| \int_0^{\cdot} A(s) \left[\varepsilon^{-1} \int_0^s \Psi^{\varepsilon}(s, r) h_r dr \right] ds + \int_0^{\cdot} A(s) A_4^{-1}(s) h_s \, ds \right\|_T$$
$$\le ||h||_T T [C_1 \eta + \varepsilon C_2(\eta) T. \tag{5.1.38}$$

where C_1 and $C_2(\eta)$ depend on A and A_4.

Proof. Put $\alpha_t := A(t) A_4^{-1}(t)$. Let α^{η} be a continuously differentiable matrix function such that $||\alpha - \alpha^{\eta}||_T \le \eta$. Let $C(\eta) := ||\dot{\alpha}^{\eta}||_T$. Using the definition (5.1.25) and integrating by parts we obtain that

$$\int_0^t A(s) \left[\varepsilon^{-1} \int_0^s \Psi^{\varepsilon}(s, r) h_r dr \right] ds = \int_0^t \left[\varepsilon^{-1} \int_r^t A(s) \Psi^{\varepsilon}(s, r) ds \right] h_r \, dr$$

$$= \int_0^t \left[\int_r^t \alpha_s \frac{\partial \Psi^{\varepsilon}}{\partial s}(s, r) ds \right] h_r dr$$

$$= \int_0^t \left[\int_r^t \alpha_s^{\eta} \frac{\partial \Psi^{\varepsilon}}{\partial s}(s, r) ds \right] h_r \, dr + \int_0^t \left[\int_r^t (\alpha_s - \alpha_s^{\eta}) \frac{\partial \Psi^{\varepsilon}}{\partial s}(s, r) ds \right] h_r \, dr$$

$$= \int_0^t [\alpha_t^{\eta} \Psi^{\varepsilon}(t, r) - \alpha_r^{\eta}] h_r \, dr - \int_0^t \left[\int_r^t \dot{\alpha}_s^{\eta} \Psi^{\varepsilon}(s, r) ds \right] h_r \, dr$$

$$+ \int_0^t \left[\int_r^t (\alpha_s - \alpha_s^{\eta}) \frac{\partial \Psi^{\varepsilon}}{\partial s}(s, r) ds \right] h_r \, dr.$$

Thus, the sum of the integrals on the left-hand side of (5.1.38) can be transformed into the following expression:

$$\int_0^t \alpha_t^{\eta} \Psi^{\varepsilon}(t, r) h_r \, dr + \int_0^t (\alpha_s - \alpha_s^{\eta}) h_s \, ds - \int_0^t \left[\int_r^t \dot{\alpha}_s^{\eta} \Psi^{\varepsilon}(s, r) \, ds \right] h_r \, dr$$

$$+ \int_0^t \left[\int_r^t (\alpha_s - \alpha_s^\eta) \varepsilon^{-1} A_4(s) \Psi^\varepsilon(s, r) \, ds \right] h_r \, dr$$
$$= I_t^1 + I_t^2 - I_t^3 + I_t^4.$$

Obviously, $||I^2||_T \le \eta ||h||_T T$. The inequalities (5.1.26) and (5.1.27) imply that

$$||I^1||_T \le ||h||_T (||\alpha||_T + \eta) \varepsilon \gamma_0 / \gamma,$$
$$||I^3||_T \le ||h||_T T \varepsilon C(\eta) \gamma_0 / \gamma,$$
$$||I^4||_T \le \eta ||h||_T T ||A_4||_T \gamma_0 / \gamma.$$

The desired inequality (5.1.38) follows from these bounds. $\square$

Corollary 5.1.6 *Let $A(t)$ be an integrable matrix function and let $\mathcal{H}$ be a set of measurable processes such that the set of random variables $\{||h||_T : h \in \mathcal{H}\}$ is bounded in L^p. Then*

$$\lim_{\varepsilon \to 0} \sup_{h \in \mathcal{H}} \left\| \int_0^\cdot A(s) \left[\varepsilon^{-1} \int_0^s \Psi^\varepsilon(s, r) h_r \, dr \right] ds + \int_0^\cdot A(s) A_4^{-1}(s) h_s \, ds \right\|_T = 0 \tag{5.1.39}$$

in L^p.

Proof. For continuous A the relation (5.1.39) follows directly from (5.1.38). In the general case we can approximate A in $L^1([0, T])$ by continuous functions. $\square$

Returning to Proposition 5.1.2(b), we define the process

$$R_t^\varepsilon(u^\varepsilon) := x_t^\varepsilon - x^o - \int_0^t A_0(s, x_s^\varepsilon, u_s^\varepsilon) ds - \int_0^t \sigma(s, x_s^\varepsilon, u_s^\varepsilon) \, dw_s^x. \tag{5.1.40}$$

Lemma 5.1.7 *We have $\sup_{\mathcal{U}} ||R^\varepsilon(u^\varepsilon)||_T \to 0$ in any L^p, $p < \infty$, as $\varepsilon \to 0$.*

Proof. Plugging the expression (5.1.33) into the definition (5.1.40) we get that

$$R_t^\varepsilon(u^\varepsilon) := \int_0^t A_2(s) \left[\varepsilon^{-1} \int_0^s \Psi^\varepsilon(s, r) A_3(r) x_r^\varepsilon \, dr + A_4^{-1}(s) A_3(s) x_s^\varepsilon \right] ds$$
$$+ \int_0^t A_2(s) \left[\varepsilon^{-1} \int_0^s \Psi^\varepsilon(s, r) B(r) u_r^\varepsilon \, dr + A_4^{-1}(s) B(s) u_s^\varepsilon \right] ds$$
$$+ \beta \int_0^t A_2(s) \eta_s^\varepsilon \, ds. \tag{5.1.41}$$

Since β is bounded on $]0, 1]$, it follows from Lemma 5.1.4 that the uniform norm of the third term in the right-hand side of (5.1.41) tends to zero in L^p

as $\varepsilon \to 0$. The same holds for the first and the second terms (uniformly in $u^\varepsilon \in \mathcal{U}$) by virtue of Corollary 5.1.6 and the assertion (a) proven above. $\square$

From the definition (5.1.40) we have

$$x_t^\varepsilon = x^o + \int_0^t A_0(s, x_s^\varepsilon, u_s^\varepsilon)\, ds + \int_0^t \sigma(s, x_s^\varepsilon, u_s^\varepsilon)\, dw_s^x + R_t^\varepsilon(u^\varepsilon). \qquad (5.1.42)$$

Comparing this formula with the expression for the solution $\bar{x}^\varepsilon$ of (5.1.5) (with $u = u^\varepsilon$) we get that

$$\Delta x_t^\varepsilon := x_t^\varepsilon - \bar{x}_t^\varepsilon = \int_0^t [A_0(s, x_s^\varepsilon, u_s^\varepsilon) - A_0(s, \bar{x}_s^\varepsilon, u_s^\varepsilon)]\, ds$$

$$+ \int_0^t [\sigma(s, x_s^\varepsilon, u_s^\varepsilon) - \sigma(s, \bar{x}_s^\varepsilon, u_s^\varepsilon)]\, dw_s^x + R_t^\varepsilon(u^\varepsilon). \qquad (5.1.43)$$

Making use of the Lipschitz conditions (5.1.9), (5.1.10) and the Doob inequality we obtain that

$$E\|\Delta x^\varepsilon\|_t^p \leq C \left(\int_0^t E\|\Delta x^\varepsilon\|_s^p\, ds + E\|R^\varepsilon(u^\varepsilon)\|_T^p \right).$$

Thus, by the Gronwall–Bellman lemma

$$E\|\Delta x^\varepsilon\|_T^p \leq E\|R^\varepsilon(u^\varepsilon)\|_T^p e^{CT}$$

and the assertion (b) follows from Lemma 5.1.7.

To prove (d), we notice that (5.1.16) implies that

$$B(t)u_t^\varepsilon = -A_4(t)\bar{y}_t^\varepsilon - A_3(t)\bar{x}_t^\varepsilon, \quad t < T,$$

and we obtain from (5.1.32) for the process $Y^\varepsilon := y^\varepsilon - \bar{y}^\varepsilon$ the following representation on $[0, T[$:

$$Y_t^\varepsilon = \varepsilon^{-1} \int_0^t \Psi^\varepsilon(t, s) A_3(s)(x_s^\varepsilon - \bar{x}_s^\varepsilon)\, ds + \beta \eta_t^\varepsilon + r_t^\varepsilon \qquad (5.1.44)$$

where

$$r_t^\varepsilon := -\varepsilon^{-1} \int_0^t \Psi^\varepsilon(t, s) A_4(s) \bar{y}_s^\varepsilon\, ds - \bar{y}_t^\varepsilon. \qquad (5.1.45)$$

It follows from (b) and (5.1.27) that the first term in the right-hand side of (5.1.44) admits a majorant which tends to zero in L^p. By virtue of (5.1.30) and **H.5.1.4** the same holds also for the second term. But

$$-\int_0^t V_s r_s^\varepsilon\, ds = \int_0^t V_s \left[\frac{1}{\varepsilon} \int_0^s \Psi^\varepsilon(s, r) A_4(r) \bar{y}_r^\varepsilon\, dr \right] ds + \int_0^t V_s A_4^{-1}(s) A_4(s) \bar{y}_s^\varepsilon\, ds$$

and we can apply Corollary 5.1.6 with $A = V$ and the set

$$\mathcal{H} := \{h : h = A_4 \tilde{y}^\varepsilon, \ u^\varepsilon \in \mathcal{U}, \ \varepsilon \in]0,1]\}.$$

The last set is bounded in probability since $A_4(t)\bar{y}_t^\varepsilon = -A_3(t)\bar{x}_t^\varepsilon - B(t)\bar{u}_t^\varepsilon$ for $t < T$.

Note that the values of $\bar{y}^\varepsilon$ at the final time T are not involved in our considerations.

Thus, (d) is also proved.

To prove the remaining assertion (c) we proceed by considering the following linear equation with constant coefficients:

$$\varepsilon d\tilde{y}_t^\varepsilon = (A_4(T)\tilde{y}_t^\varepsilon + A_3(T)x_T^\varepsilon + B(T)u_t^\varepsilon)dt + \beta\varepsilon^{1/2}dw_t^y, \quad \tilde{y}_0^\varepsilon = 0. \quad (5.1.46)$$

Let us define
$$\Delta_t^\varepsilon := y_t^\varepsilon - \tilde{y}_t^\varepsilon, \quad \hat{x}_t^\varepsilon := x_t^\varepsilon - x_T^\varepsilon, \quad (5.1.47)$$

$$\hat{A}_i(t) := A_i(t) - A_i(T), \quad i = 3, \ 4, \quad \hat{B}(t) := B(t) - B(T). \quad (5.1.48)$$

Clearly, the process Δ^ε satisfies the ordinary differential equation

$$\varepsilon d\Delta_t^\varepsilon = (A_4(T)\Delta_t^\varepsilon + \hat{A}_4(t)y_t^\varepsilon + \hat{A}_3(t)x_t^\varepsilon + A_3(T)\hat{x}_t^\varepsilon + \hat{B}(t)u_t^\varepsilon)dt, \quad \Delta_0^\varepsilon = 0.$$

We represent Δ^ε by the Cauchy formula as

$$\Delta_T^\varepsilon = \varepsilon^{-1} \int_0^T e^{A_4(T)(T-t)/\varepsilon} \varphi_t^\varepsilon \, dt \quad (5.1.49)$$

where
$$\varphi_t^\varepsilon := \hat{A}_4(t)y_t^\varepsilon + \hat{A}_3(t)x_t^\varepsilon + A_3(T)\hat{x}_t^\varepsilon + \hat{B}(t)u_t^\varepsilon. \quad (5.1.50)$$

It follows from (a) that the set $\{||\varphi^\varepsilon||_T : \ u^\varepsilon \in \mathcal{U}, \ \varepsilon \in]0,1]\}$ is bounded in L^p, i.e.

$$\sup_\varepsilon \sup_\mathcal{U} E||\varphi^\varepsilon||_T^p \le C < \infty. \quad (5.1.51)$$

Let $T_\varepsilon := (1 - \varepsilon^{1/2})T$ and

$$\rho_\varepsilon := \max_{t \in [T_\varepsilon, T]} (|\hat{A}_3(t)| + |\hat{A}_3(t)| + |\hat{B}(t)|).$$

On the interval $[0, T_\varepsilon[$ we estimate the L^p-norm of φ_t^ε by a constant from (5.1.51), while on $[T_\varepsilon, T]$ we use the bound

$$||\varphi_t^\varepsilon||_p \le C(\rho_\varepsilon + ||\bar{x}_T^\varepsilon - \bar{x}_t^\varepsilon||_p + ||||x^\varepsilon - \bar{x}^\varepsilon||_T||_p). \quad (5.1.52)$$

It follows from the linear growth of coefficients and the Burkholder inequality that

$$||\bar{x}_T^\varepsilon - \bar{x}_t^\varepsilon||_p \le C(T - t)^{1/2}. \quad (5.1.53)$$

Summarizing, we obtain from (5.1.49) that

$$||\Delta_T^\varepsilon||_p \leq C \left(e^{-\gamma T/\sqrt{\varepsilon}} + \rho_\varepsilon + \sqrt{\varepsilon} + |||x^\varepsilon - \bar{x}^\varepsilon||_T||_p \right).$$

Since $\rho_\varepsilon \to 0$ by virtue of **H.5.1.1**, we conclude from here, using also (b), that

$$\limsup_{\varepsilon \atop u} E|\Delta_T^\varepsilon|^p = 0. \tag{5.1.54}$$

It follows from definitions and (5.1.46) that

$$
\begin{aligned}
y_T^\varepsilon &= \Delta_T^\varepsilon + \tilde{y}_T^\varepsilon = \Delta_T^\varepsilon + \frac{1}{\varepsilon} \int_0^T e^{A_4(T)(T-t)/\varepsilon} \, dt A_3(T) x_T^\varepsilon \\
&\quad + \frac{1}{\varepsilon} \int_0^T e^{A_4(T)(T-t)/\varepsilon} B(T) u_t^\varepsilon \, dt + \beta \frac{1}{\varepsilon^{1/2}} \int_0^T e^{A_4(T)(T-t)/\varepsilon} \, dw_t^y \\
&= \Delta_T^\varepsilon - A_4^{-1}(T)(1 - e^{A_4(T)T/\varepsilon}) A_3(T) x_T^\varepsilon + \frac{1}{\varepsilon} \int_0^{T_\varepsilon} e^{A_4(T)(T-t)/\varepsilon} B(T) u_t^\varepsilon \, dt \\
&\quad + \frac{1}{\varepsilon} \int_{T_\varepsilon}^T e^{A_4(T)(T-t)/\varepsilon} B(T) u_t^\varepsilon dt + \beta \frac{1}{\varepsilon^{1/2}} \int_0^T e^{A_4(T)(T-t)/\varepsilon} \, dw_t^y \\
&= \bar{y}_T^\varepsilon + q_T^\varepsilon
\end{aligned}
$$

where

$$
\begin{aligned}
q_T^\varepsilon &:= \Delta_T^\varepsilon - A_4^{-1}(T) A_3(T)(x_T^\varepsilon - \bar{x}_T^\varepsilon) + A_4^{-1}(T) e^{A_4(T)T/\varepsilon} A_3(T) x_T^\varepsilon \\
&\quad + \frac{1}{\varepsilon} \int_0^{T_\varepsilon} e^{A_4(T)(T-t)/\varepsilon} B(T) u_t^\varepsilon \, dt + \beta \frac{1}{\varepsilon^{1/2}} \int_0^T e^{A_4(T)(T-t)/\varepsilon} \, dw_t^y \\
&= \Delta_T^\varepsilon - A_4^{-1}(T) A_3(T)(x_T^\varepsilon - \bar{x}_T^\varepsilon) + A_4^{-1}(T) e^{A_4(T)T/\varepsilon} A_3(T) x_T^\varepsilon \\
&\quad + \int_{T/\varepsilon^{1/2}}^{T/\varepsilon} e^{A_4(T)t} B(T) u_{T-t\varepsilon}^\varepsilon \, dt + \beta \frac{1}{\varepsilon^{1/2}} \int_0^T e^{A_4(T)(T-t)/\varepsilon} \, dw_t^y.
\end{aligned}
$$

The relations (5.1.54), (a) and (b), (5.1.30), **H.5.1.3**, and the boundedness of the coefficients and controls imply that for every finite p

$$\limsup_{\varepsilon \atop u} E|q_T^\varepsilon|^p = 0. \tag{5.1.55}$$

Proposition 5.1.2 is proved. □

Remark. Analysis of the above proof shows that the claims (b) and (a), except the assertion on the boundedness of M^y, hold also for the model with constant β which will be studied in Section 5.4.

5.1.3 Proof of Proposition 5.1.3

Let us define the control u^ε by the formula

$$u_t^\varepsilon := u_t I_{[0,T_\varepsilon]}(t) + \tilde{v}_t^\varepsilon I_{]T_\varepsilon,T]}(t) \tag{5.1.56}$$

where $\tilde{v}^\varepsilon$ is the predictable projection of the process $(v_{(T-t)/\varepsilon})_{t\in[0,T]}$ and, as above, $T_\varepsilon := (1 - \varepsilon^{1/2})T$.

The definition of the predictable projection and the "left continuity" of the natural filtration of the Wiener process (remember that it is augmented by the P-null sets) imply that for any $t \in [0, T]$

$$\tilde{v}_t^\varepsilon = E(v_{(T-t)/\varepsilon}|\mathcal{F}_{t-}) = E(v_{(T-t)/\varepsilon} \mid \mathcal{F}_t). \qquad (5.1.57)$$

The relation (5.1.21) is an immediate consequence of (5.1.56). Recall that by virtue of Proposition 5.1.2 we have:

$$\lim_{\varepsilon \to 0} ||x^\varepsilon - \bar{x}^\varepsilon||_T^p = 0, \qquad (5.1.58)$$

$$\lim_{\varepsilon \to 0} |y_T^\varepsilon - \bar{y}_T^\varepsilon| = 0, \qquad (5.1.59)$$

$$\lim_{\varepsilon \to 0} \left|\left| \int_0^\cdot V_s(y_s^\varepsilon - \bar{y}_s^\varepsilon) \, ds \right|\right|_T = 0 \qquad (5.1.60)$$

in every L^p for any bounded measurable function V where $\bar{x}^\varepsilon$ and $\bar{y}^\varepsilon$ are defined by (5.1.5), (5.1.16), and (5.1.17) for the given u^ε.

By the definition (5.1.56) the process $(\bar{x}^\varepsilon, \bar{y}^\varepsilon)$ coincides with $(\bar{x}, \bar{y})$ on $[0, T_\varepsilon]$. Using **H.5.1.2** and the Burkholder inequality it is easy to check that

$$E||\bar{x}^\varepsilon - \bar{x}||_T^p = E||\bar{x}^\varepsilon - \bar{x}||_{T_\varepsilon,T}^p \leq C(1 + ||||\bar{x}^\varepsilon||_T||_p + ||||\bar{x}||_T||_p) \to 0 \quad (5.1.61)$$

as $\varepsilon \to 0$.

The relation (5.1.24) is an immediate consequence of (5.1.60) and (5.1.56).

Taking into account (5.1.22) and (5.1.59), to prove (5.1.23) it suffices to show that

$$\int_0^{T/\sqrt{\varepsilon}} e^{A_4(T)s} B(T)\tilde{v}_{(T-\varepsilon s)}^\varepsilon \, ds \to \int_0^\infty e^{A_4(T)s} B(T) v_s \, ds \qquad (5.1.62)$$

in L^p. Notice that in accordance with (5.1.57)

$$\tilde{v}_{T-\varepsilon s}^\varepsilon = E(v_s \mid \mathcal{F}_{(T-\varepsilon s)-}) = E(v_s|\mathcal{F}_{T-\varepsilon s}).$$

The relation (5.1.62) now follows from the obvious bound (implied by **H.5.1.3** and the Jensen inequality):

$$E \left| \int_0^{T/\sqrt{\varepsilon}} e^{A_4(T)s} B(T)\tilde{v}_{(T-\varepsilon s)}^\varepsilon ds - \int_0^{T/\sqrt{\varepsilon}} e^{A_4(T)s} B(T) v_s \, ds \right|^p$$

$$\leq C_\gamma \int_0^\infty e^{-2\gamma s} E|E(v_s|\mathcal{F}_{T-\varepsilon s}) - v_s|^p \, ds.$$

The right-hand side here tends to zero since Levy's theorem implies the convergence (a.s. and in L^p) of $E(v_s|\mathcal{F}_{T-\varepsilon s})$ to $E(v_s|\mathcal{F}_{T-}) = E(v_s|\mathcal{F}_T) = v_s$.

Proposition 5.1.3 is proved. $\square$

5.1.4 Proof of Theorem 5.1.1

Let u^ε be an admissible control and let $(x^\varepsilon, y^\varepsilon)$ be the solution of (5.1.1), (5.1.2) corresponding to u^ε. Let $\bar{x}^\varepsilon$ be the solution of (5.1.5) corresponding to the same u^ε and let $\bar{y}^\varepsilon$ be given by (5.1.16), (5.1.17).

From the definition (5.1.7) of the limiting running cost and the assumption (5.1.9) (the Lipschitz condition for f) as well as from (5.1.16) we have

$$\left| \int_0^T [f(t, x_t^\varepsilon, u_t) + b(t) y_t^\varepsilon] \, dt - \int_0^T f_0(t, \bar{x}_t^\varepsilon, u_t) \, dt \right|$$

$$\leq L \int_0^T |x_t^\varepsilon - \bar{x}_t^\varepsilon| \, dt + \left| \int_0^T b(t)(y_t^\varepsilon - \bar{y}_t^\varepsilon) \, dt \right|. \tag{5.1.63}$$

The right-hand side of this inequality tends in L^p to zero as $\varepsilon \to 0$ by Proposition 5.1.2.

From the definitions (5.1.8) and (5.1.17) we get that

$$g(x_T^\varepsilon, y_T^\varepsilon) - g_0(\bar{x}_T^\varepsilon) \geq g(x_T^\varepsilon, y_T^\varepsilon) - g(\bar{x}_T^\varepsilon, \bar{y}_T^\varepsilon) \to 0$$

in probability by virtue of Proposition 5.1.2 and the continuity of g.

By **H.5.1.2** and **H.5.1.5** (the linear and polynomial growth conditions) and Proposition 5.1.2(a) (L^p-boundedness) we can conclude that the expectations of the right-hand sides of the above inequalities also tend to zero. Thus,

$$J^\varepsilon(u^\varepsilon) \geq \bar{J}(u^\varepsilon) + \alpha(\varepsilon) \geq \bar{J}_* + \alpha(\varepsilon) \tag{5.1.64}$$

where $\alpha(\varepsilon) \to 0$ as $\varepsilon \to 0$.

It follows from (5.1.64) that

$$\liminf_{\varepsilon \to 0} J_*^\varepsilon \geq \bar{J}_*. \tag{5.1.65}$$

Now let u be an η-optimal control for the limit problem. This means that

$$\bar{J}(u) \leq \bar{J}_* + \eta. \tag{5.1.66}$$

Let $\bar{x}$ be the solution of (5.1.5) corresponding to the given u. Choose a process $v \in \mathcal{F} \otimes \mathcal{B}_+$ in such a way that

$$g_0(\bar{x}_T) + \eta \geq g(\bar{x}_T, \bar{y}_T) \tag{5.1.67}$$

for $\bar{y}_T$ given by (5.1.20). According to Proposition 5.1.3 there exists a family of admissible controls u^ε such that for the corresponding solutions of the prelimit problem (5.1.1)–(5.1.3) the relations (5.1.22)–(5.1.24) hold. As in (5.1.25), we get the convergence in L^p of the running costs

$$\int_0^T [f(t, x_t^\varepsilon, u_t) + b(t) y_t^\varepsilon] \, dt \to \int_0^T f_0(t, \bar{x}_t, u_t) \, dt. \tag{5.1.68}$$

For the terminal costs we have by (5.1.67) the bound

$$Eg(\bar{x}_T, \bar{y}_T) \leq Eg_0(\bar{x}_T) + \eta. \tag{5.1.69}$$

We infer from Proposition 5.1.2, making use also of Proposition 5.1.3, that

$$\lim_{\varepsilon \to 0} Eg(x_T^\varepsilon, y_T^\varepsilon) \leq Eg_0(\bar{x}_T) + \eta.$$

Thus, we found a family u^ε of admissible controls such that

$$\lim_{\varepsilon \to 0} J^\varepsilon(u^\varepsilon) \leq \bar{J}_* + 2\eta. \tag{5.1.70}$$

Hence,

$$\limsup_{\varepsilon \to 0} J_*^\varepsilon \leq \bar{J}_*. \tag{5.1.71}$$

Combining (5.1.65) and (5.1.71) we obtain the assertion of the theorem. □

5.2 Structure of the Attainability Sets

5.2.1 Weak and Strong Solutions of SDEs

Until now we have worked always with the strong solutions of stochastic differential equations. This concept has arisen at early stages of historical development of the theory which followed the same line of ideas as the theory of ordinary differential equations inheriting a lot of techniques of the latter, e.g., Picard's method of successive approximation in the proof of existence of solution. The standard setting is the following.

We are given a stochastic basis $(\Omega, \mathcal{F}, \mathbf{F} = (\mathcal{F}_t), P)$ with a Wiener process W adapted to $\mathbf{F}$ and two functions $f(x, t)$ and $\sigma(x, t)$ on $\mathbf{R}^n \times [0, T]$ with values in $\mathbf{R}^n$ and the set of $n \times n$ matrices, respectively. The stochastic differential equation with respect to an unknown process X is the formal expression

$$dX_t = f(X_t, t)dt + \sigma(X_t, t)dW_t, \quad X_0 = 0. \tag{5.2.1}$$

The problem is to find on the same stochastic basis an $\mathbf{F}$-adapted continuous process X satisfying (5.2.1) in the sense that on $[0, T]$ there is the identity

$$X_t = \int_0^t f(X_s, s)ds + \int_0^t \sigma(X_s, s)dW_s. \tag{5.2.2}$$

It is said that such an X (adapted to the filtration generated by W, at least, if $\sigma(t, x)$, is nondegenerate) is a strong solution of (5.2.1). The first results of the Ito theory of SDEs were based on suitably modified recipes from ordinary differential equations: existence and uniqueness theorems for strong solutions under linear growth and local Lipschitz conditions. This approach can be readily generalized to a more general situation where the coefficients functionally depend on trajectories in a causal (non-anticipating) way, i.e. for the equation

$$dX_t = f(X,t)dt + \sigma(X,t)dW_t, \quad X_0 = 0. \tag{5.2.3}$$

However, it happens that, in fact, a stochastic equation differs essentially from an ordinary differential equation: e.g., in the early 1970s Zvonkin discovered that the one-dimensional equation

$$dX_t = f(X_t,t)dt + dW_t, \quad X_0 = 0, \tag{5.2.4}$$

has a unique strong solution for any bounded measurable function f. One may guess that the same holds for the equation

$$dX_t = f(X,t)dt + dW_t, \quad X_0 = 0, \tag{5.2.5}$$

where $f : C[0,T] \times [0,T] \to \mathbf{R}^1$ is a bounded measurable non-anticipating function. Enormous efforts were made to prove this conjecture, which would be of a great importance for the stochastic control theory. But in 1975 Tsyrel'son constructed his famous counterexample of the function f for which (5.2.5) does not have a solution in a strong sense.

Remarkably, the formal expression (5.2.1) (as well as (5.2.3)) can be interpreted as problem to find a "weak" solution, existing under rather mild assumptions. The only difference in the definition is that now we are given only the coefficients while the stochastic basis and the Wiener process are not fixed in advance. To solve the problem, one should find a stochastic basis $(\Omega, \mathcal{F}, \mathbf{F}, P)$ with a Wiener processes W and a continuous process X, both adapted with respect to $\mathbf{F}$, such that the integral identity (5.2.2) is fulfilled. In this case X may not be adapted to the filtration generated by W.

For the equation (5.2.5) the weak solution can be constructed in a rather explicit (and easy) way using the Girsanov transform. Indeed, let $(C[0,T], \mathcal{C}, \mathbf{C} = (\mathcal{C}_t), P)$ be the space of continuous functions $W = (W_t)$ with the natural measurable structure and let P be the Wiener measure, this means that the coordinate mappings $W \mapsto W_t$ defines a Wiener process under P. Let $f : C[0,T] \times [0,T] \to \mathbf{R}^1$ be a bounded measurable function such that $f(W,t)$ is $\mathcal{C}_t$-measurable for all $t \in [0,T]$. Define a positive random variable

$$\rho_T := \exp\left\{\int_0^T f(W,s)dW_s - \frac{1}{2}\int_0^T f^2(W,s)ds\right\}. \tag{5.2.6}$$

It is an easy exercise to check that $E\rho_T = 1$ and hence $\tilde{P} := \rho_T P$ is a probability measure. By the Girsanov theorem the process

$$\tilde{W}_t := W_t - \int_0^t f(W,s)ds \tag{5.2.7}$$

is the Wiener process with respect to $\tilde{P}$. It remains to take $(C[0, T], \mathcal{C}, \mathbf{C}), \tilde{P}$ as the stochastic basis we are looking for and to change the notations W for X and $\tilde{W}$ for W.

The relation (5.2.7) can be written in the differential form as

$$dW_t = f(W, t)dt + d\tilde{W}_t, \quad W_0 = 0; \tag{5.2.8}$$

this is, up to notations, exactly the same expression as (5.2.5). Notice that in the above construction only the measure $\tilde{P}$ is a new element. Sometimes this measure is referred to as a weak solution to the problem (5.2.5).

5.2.2 Closed Loop Controls Versus Open Loop

Now we recall some basic concepts from the control theory.

In the deterministic setting the dynamics of a control object is usually given by the ordinary differential equation

$$dX_t = f(X_t, t, u_t)dt, \quad X_0 = 0, \tag{5.2.9}$$

We are given a function $f : \mathbf{R}^n \times [0, t] \times U \rightarrow \mathbf{R}^n$ describing the dynamics, and a certain set $\mathcal{U}$ of admissible controls u. The equation (5.2.9) is understood as an integral one. It would be desirable that the decision u_t at time t be a point in the action phase space U depending on a current position X_t, i.e., $u_t = u_t(X_t)$ for a suitable function $u_t(x)$ and one may admit only such controls. However, for such a definition of admissibility ("closed loop control") we ought to impose some smoothness assumptions on $u_t(x)$ to guarantee the existence and uniqueness for the equation

$$dX_t = f(X_t, t, u_t(X_t))dt, \quad X_0 = 0. \tag{5.2.10}$$

Certainly, this is not appropriate: very often the optimal control is a switching from one regime to another which does not depend continuously on the state of the system. To avoid any difficulty arising, in the deterministic theory one takes as admissible controls the "open loop" controls or "programs", i.e. the measurable functions $u : [0, T] \rightarrow U$ (of course, further constraints can be imposed). The problem, whether the optimal control can be represented as a "closed loop control" ("synthesis"), is a separate question.

The straightforward generalization of this standard approach leads, in the context of SDEs, to the model with the "stochastic open loop" controls where the dynamics of the system is given by the equation

$$dX_t = f(X_t, t, u_t)dt + \sigma(X_t, t, u_t)dW_t, \quad X_0 = 0, \tag{5.2.11}$$

understood in the strong sense and the class of admissible controls contains all adapted processes $u = u_t(\omega)$ with values in U (eventually, with constraints).

If the filtration $\mathbf{F}$ is generated by W, the controls are non-anticipating functionals of the "driving noise" W.

However, in contrast with the deterministic setting there is an alternative approach to consider as admissible controls the class of "closed loop stochastic controls" since we may understand the equation (5.2.11) in the weak sense. In this case our control action changes the underlying probability measure rather than a sample trajectory.

Though the first approach seems to dominate in the literature, the second one has its own advantages. The aim of this section is to introduce the concept of "tubes" and attainability sets in both the basic stochastic control models and to compare their geometric and topological properties, having in mind further applications to the problem of approximation of the optimal cost.

Notations. Let $(X, \mathcal{X})$ be a measurable space. We shall denote by $\mathbf{P}(X, \mathcal{X})$ (or simply $\mathbf{P}(X)$ the set of all probability measures on it. If $(X, \mathcal{X}, P)$ is a probability space, $(Y, \mathcal{Y})$ is a measurable space, and $f : X \to Y$ is a measurable mapping, then Pf^{-1} is the measure on $(Y, \mathcal{Y})$ defined by

$$Pf^{-1}(\Gamma) = P(x : f(x) \in \Gamma),$$

i.e. the distribution of the random variable f. In the sequel we shall use also the notation $\mathcal{L}(f)$ ("law of f") when there is no ambiguity for the underlying probability measure. When X is a Polish space, we consider $\mathbf{P}(X)$ with the topology of the weak convergence equipped with the Prohorov metric. Under this metric $\mathbf{P}(X)$ is also a Polish space. Further information concerning properties of $\mathbf{P}(X)$ is given in Appendix A.6.

Let $C[0, T]$ be the space of continuous functions $W : [0, T] \to \mathbf{R}^n$ with the uniform norm $||W||_T$, $\mathcal{C}_t^o := \sigma\{W_s : s \le t\}$, $\mathcal{C}_t := \bigcap_{\varepsilon > 0} \mathcal{C}_{t+\varepsilon}^o$ if $t < T$, and $\mathcal{C}_T = \mathcal{C}_T^o$, $\mathbf{C} := (\mathcal{C}_t)$. Let $\mathcal{P} = \mathcal{P}(\mathbf{C})$ be the predictable σ-algebra in $C[0, T] \times [0, T]$ (generated by all left-continuous $\mathbf{C}$-adapted processes).

Let U be a compact set in $\mathbf{R}^d$ (the phase space of controls).

The set of all admissible controls $\mathcal{U}$ is the set of all predictable processes on $C[0, T]$ with values in U.

Model with feedback controls. We are given a function

$$f : C[0, T] \times [0, T] \times U \to \mathbf{R}^n$$

which satisfies the following hypotheses:

H.5.2.1 The function $u \mapsto f(W, t, u)$ is continuous on U for all (W, t); the function $(W, t) \mapsto f(W, t, u)$ is predictable for all u.

H.5.2.2 There is a constant k such that

$$|f(W, t, u)|^2 \le k(1 + ||W||_t^2) \quad \forall (W, t, u).$$

We associate with every admissible control $u = (u_t(W))_{t \le T}$ the positive process

$$\rho_t^u = \rho_t^u(f) := \exp\left\{\int_0^t f(W, s, u_s)^* W_s - \frac{1}{2}\int_0^t |f(W, s, u_s)|^2 ds\right\}. \quad (5.2.12)$$

Lemma 5.2.1 *The process $(\rho_t^u)_{t\leq T}$ is a martingale with $E\rho_t^u = 1$.*

Proof. By the Ito formula we have that

$$d\rho_t^u = \rho_t^u f(W, s, u_s)^* dW_s, \quad \rho_0^u = 1. \quad (5.2.13)$$

It follows that $(\rho_t^u)_{t\leq T}$ is a local martingale. Being positive, it is a super-martingale by the Fatou lemma. We need only to check that **H.5.2.2** ensures its uniform integrability. To this aim recall the Novikov condition:

Let X is a predictable process such that

$$E\exp\left\{\frac{1}{2}\int_0^T |X_r|^2 dr\right\} < \infty$$

and let Z is the solution on $[0, T]$ of the linear SDE

$$dZ_t = Z_t X_t^* dW_t, \quad Z_0 = 1. \quad (5.2.14)$$

Then Z is a martingale on $[0, T]$ with $EZ_t = 1$ for all $t \leq T$; see Subsection A.1.4.

The Novikov conditions implies another sufficient condition which suits our problem well: if

$$E\exp\left\{\frac{1}{2}\int_{t_j}^{t_{j+1}} |X_r|^2 dr\right\} < \infty, \quad j = 0, \ldots, n-1, \quad (5.2.15)$$

where $0 = t_0 < t_1 < \ldots < t_n = T$, then Z is a martingale on $[0, T]$.

Indeed, put $X^j := I_{]t_j, t_{j+1}]}X$ and define the processes Z^j with

$$dZ_t^j = Z_t^j X_t^{j*} dW_t, \quad Z_0^j = 1.$$

Obviously, Z_t^j are martingales by the Novikov condition. Since

$$E(Z_T^{n-1}|\mathcal{C}_{t_{n-1}}) = 1,$$

we have:

$$EZ_T = EZ_T^1 \ldots Z_T^{n-2} Z_T^{n-1} = EZ_T^1 \ldots Z_T^{n-2} E(Z_T^{n-1}|\mathcal{C}_{t_{n-1}}) = EZ_T^1 \ldots Z_T^{n-2}$$

and the induction shows that $EZ_T = 1$.

For $t_j := j\Delta$ we have by **H.5.2.2** that

$$E\exp\left\{\frac{1}{2}\int_{t_j}^{t_{j+1}} |f(W, r, u_r)|^2 dr\right\} \leq CEe^{(1/2)k\Delta||W||_T^2} < \infty$$

for sufficiently small Δ by virtue of Fernique's Lemma 1.3.6. Thus, (5.2.15) holds and for any $u \in \mathcal{U}$ we can define a probability measure $P^u := \rho_T^u P$.

By the Girsanov theorem the process

$$\tilde{W}_t := W_t - \int_0^t f(W, s, u_s)ds$$

is a standard Wiener process under P^u and we can rewrite this relation as

$$dW_t = f(W, t, u_t(W))dt + d\tilde{W}_t, \quad W_0 = 0. \tag{5.2.16}$$

As we discussed above, (5.2.16) is a stochastic differential equation considered in the weak sense and P^u provides its solution.

Define the sets

$$\mathcal{K} := \{P^u : u \in \mathcal{U}\},$$
$$\mathcal{K}(T) := \{P^u W_T^{-1} : u \in \mathcal{U}\},$$

where W_T is the projection mapping $W \mapsto W_T$. The set $\mathcal{K} \subseteq \mathbf{P}(C[0,T])$ is an analog of the tube (or funnel) of trajectories of the deterministic theory, while the set $\mathcal{K}(T) \subseteq \mathbf{P}(\mathbf{R}^n)$ is an analog of the attainability set. We shall say that $\mathcal{K}(T)$ is the attainability set in the model with feedback controls.

Model with open loop controls. In this model we are also given a function $f : C[0,T] \times [0,T] \times U \to \mathbf{R}^n$ satisfying **H.5.2.1–H.5.2.2**. The definition of admissible controls remains exactly the same as in the previous model but the resulting probability measure corresponding to $u \in \mathcal{U}$ is constructed in a different way.

Let us consider on the probability space $(C[0,T], \mathcal{C}, \mathbf{C}, P)$ the stochastic equation

$$dx_t = f(x, t, u_t)dt + dW_t, \quad x_0 = 0, \tag{5.2.17}$$

where $u \in \mathcal{U}$ and W_t is the coordinate mapping in $C[0,T]$. By definition of the measure P the process $W = (W_t)_{t \leq T}$ is standard Wiener.

In addition to the hypotheses **H.5.2.1–H.5.2.2** we assume that for every $u \in \mathcal{U}$ the equation (5.2.17) has the unique strong solution $x = x^u$; this assumption is fulfilled if $f(., t, u)$ is Lipschitz for each (t, u).

Let P_x^u denote the distribution of the solution x^u in the space $(C[0,T], \mathcal{C})$, i.e. the probability measure $P_x^u = P(x^u)^{-1}$ where $x^u : C[0,T] \to C[0,T]$ is the function which maps a point W to $x^u(W)$ (the trajectory of the solution of (5.2.17)).

Define the sets

$$\tilde{\mathcal{K}} := \{P_x^u : u \in \mathcal{U}\},$$
$$\tilde{\mathcal{K}}(T) := \{P(x_T^u)^{-1} : u \in \mathcal{U}\}.$$

We shall say that $\tilde{\mathcal{K}}(T)$ is the attainability set in the model with open loop controls.

For the open loop control model we can consider also the set of attainable random variables. However, it will play in our presentation only an auxiliary role.

5.2.3 "Tubes" and Attainability Sets for Feedback Controls

Now we show that the "tubes" and the attainability sets for the model with feedback controls have nice topological and geometric properties.

Define the set of attainable densities $\mathcal{D}_T := \{\rho_T^u : \ u \in \mathcal{U}\} \subseteq L^1(P)$.

Lemma 5.2.2 *The set $\mathcal{D}_T$ is uniformly integrable, relatively weakly sequentially compact and relatively weakly compact (in the topology $\sigma(L^1, L^\infty)$).*

Proof. Since the properties in the formulation are equivalent by the Dunford–Pettis theorem, we need to check only a sufficient condition for the uniform integrability. Namely, we show that for some $\alpha > 1$

$$\sup_{u \in \mathcal{U}} E(\rho_T^u)^\alpha < \infty. \tag{5.2.18}$$

Denoting by $\rho_T^u(\alpha f)$ the density corresponding to the function αf which, of course, satisfies **H.5.2.1–H.5.2.2**, we get that

$$(\rho_T^u)^\alpha = \rho_T^u(\alpha f) \exp\left\{\frac{\alpha^2 - \alpha}{2} \int_0^T |f(W, t, u_t)|^2 dt\right\} \leq C\rho_T^u(\alpha f) e^{C(\alpha^2 - \alpha)\|W\|_T^2}$$

and, therefore,

$$E(\rho_T^u)^\alpha \leq CE^{u,\alpha} e^{C(\alpha^2 - \alpha)\|W\|_T^2} \tag{5.2.19}$$

where $E^{u,\alpha}$ is the expectation with respect to the probability

$$P^{u,\alpha} := \rho_T^u(\alpha f)P.$$

Put

$$X_t := W_t - \alpha \int_0^t f(W, s, u_s) ds$$

Then

$$|W_t|^2 \leq 2|X_t|^2 + 2\alpha^2 k \left(1 + \int_0^t |W_s|^2 ds\right)$$

and the Gronwall–Bellman lemma leads to the bound

$$|W_t|^2 \leq 2(\alpha^2 k + \|X\|_T^2) e^{2\alpha^2 kT}.$$

Thus, we obtain from (5.2.18) that

$$E(\rho_T^u)^\alpha \le h(\alpha)E^{u,\alpha}e^{g(\alpha)||X||_T^2} \tag{5.2.20}$$

where $h(\alpha)$ is a function bounded for $\alpha \in [1,2]$ and

$$g(\alpha) := C(\alpha^2 - \alpha)e^{2\alpha^2 kT} \to 0, \quad \alpha \to 1.$$

Under $P^{u,\alpha}$ the process X is Wiener. Hence, by Fernique's Lemma 1.3.6 the right-hand side of (5.2.20) is finite for some $\alpha > 1$ and we get the result. $\square$

Let us introduce the Roxin condition

H.5.2.3. The set $f(W,t,U) := \{f(W,t,u) : u \in U\}$ is convex for all (W,t).

Lemma 5.2.3 *Assume that the Roxin condition is satisfied. Then $\mathcal{K}$ (hence $\mathcal{K}(T)$) is a convex set.*

Proof. Let $\alpha_1, \alpha_2 > 0$, $\alpha_1 + \alpha_2 = 1$. For $u^i \in \mathcal{U}$, $i = 1, 2$, we define the density processes $\rho^i := \rho^{u^i}$ in accordance with (5.2.12).

Put $\rho := \alpha_1\rho^1 + \alpha_2\rho^2$. Using the representation (5.2.13) for ρ^i we get that

$$\begin{aligned}
d\rho_t &= [\alpha_1\rho_t^1 f(W,t,u_t^1)^\star + \alpha_2\rho_t^2 f(W,t,u_t^2)^\star]dW_t \\
&= \rho_t[\gamma_t^1 f(W,t,u_t^1) + \gamma_t^2 f(W,t,u_t^2)]dW_t,
\end{aligned}$$

where

$$\gamma_t^1 := \frac{\alpha_1\rho_t^1}{\alpha_1\rho_t^1 + \alpha_2\rho_t^2}, \qquad \gamma_t^2 := \frac{\alpha_2\rho_t^2}{\alpha_1\rho_t^1 + \alpha_2\rho_t^2}.$$

The convexity of $f(W,t,U)$ yields that for any (W,t) the point

$$\gamma_t^1 f(W,t,u_t^1) + \gamma_t^2 f(W,t,u_t^2) \in f(W,t,U),$$

i.e. of the form $f(W,t,u(W,t))$. The Filippov implicit function lemma guarantees that the representing points $u(W,t)$ can be chosen in such a way that the mapping $(W,t) \to u(W,t)$ will be a predictable process. Thus, we obtain that $\rho = \rho^u$ is given by (5.2.13). Consequently, the set of attainable densities $\mathcal{D}_T := \{\rho_T^u : u \in \mathcal{U}\}$ is convex and so are the sets $\mathcal{K}$ and $\mathcal{K}(T)$. $\square$

Lemma 5.2.4 *The set $\mathcal{D}_T$ is closed in $L^1(P)$.*

Proof. Let u^n be a sequence of admissible controls such that the corresponding densities ρ_T^n converge in $L^1(P)$ to some random variable ρ_T. First, notice that $\rho_T > 0$ a.s. Indeed, it follows from **H.5.2.2** that the ordinary integrals in the definition (5.2.12) are bounded by a finite random variable. As to the stochastic integrals

$$M_T^n := \int_0^T f(W,s,u_s^n)^\star dW_s,$$

we have from **H.5.2.2** that

$$E|M_T^n|^2 = E \int_0^T |f(W, s, u_s^n)|^2 ds \leq C.$$

But they should diverge to $-\infty$ on the set $\{\rho_T = 0\}$ and hence for any $N > 0$

$$P(\rho_T = 0, \ M_T^n > -N) \to 0, \quad n \to \infty.$$

Since

$$P(\rho_T = 0) \leq P(\rho_T = 0, \ M_T^n > -N) + P(-M_T^n \geq N)$$
$$\leq P(\rho_T = 0, \ M_T^n > -N) + C/N^2,$$

this implies that $P(\rho_T = 0) = 0$.

Now we consider the martingale $\rho_t = E(\rho_T|\mathcal{C}_t)$ which is strictly positive. For any stopping time τ with values in $[0, T]$ we have $\rho_\tau^n = E(\rho_T^n|\mathcal{C}_\tau)$ and $\rho_\tau = E(\rho_T|\mathcal{C}_\tau)$. It follows from Lemma 5.2.2 that $\rho_\tau = \lim_n \rho_\tau^n$ a.s.

To apply the theory of square integrable martingales we introduce the stopping times

$$\tau_N := \inf\{s: \ |W_s| \geq N \text{ or } \rho_s \leq 1/N\} \wedge T$$

and consider the processes $\rho_t^{n,N} := \rho_{t \wedge \tau_N}^n$ and $\rho_t^N := \rho_{t \wedge \tau_N}$ stopped at τ_N.

It follows from the bound (5.2.19) (written for τ_N instead of T) that $E(\rho_{\tau_N}^n)^\alpha$ is bounded for any α. Thus, $\rho_{\tau_N}^n \to \rho_{\tau_N}$ in $L^2(P)$ as $n \to \infty$. The Doob inequality

$$E\|\rho^{n,N} - \rho^N\|_T^2 \leq 4E|\rho_T^{n,N} - \rho_T^N|^2 = 4E|\rho_{\tau_N}^n - \rho_{\tau_N}|^2$$

yields that $\|\rho^{n,N} - \rho^N\|_T \to 0$ in $L^2(P)$ as $n \to \infty$ and, taking the subsequence, we may assume without loss of generality that

$$\|\rho^{n,N} - \rho^N\|_T^2 \to 0 \quad \text{a.s.}$$

Put $m(dW, dt) := P(dW)dt$. By the integral representation theorem there is a unique predictable function φ such that $I_{[0,\tau_N]}\varphi \in L^2(C[0,T] \times [0,T], m)$ for all N and

$$\rho_t^N = 1 + \int_0^t I_{[0,\tau_N]}(s)\varphi(W, s)^\star dW_s.$$

Put $\psi = \varphi/\rho$. Then the above formula can be rewritten as

$$\rho_t^N = 1 + \int_0^t I_{[0,\tau_N]}(s)\rho_s\psi(W, s)^\star dW_s. \tag{5.2.21}$$

Making use of the representation (5.2.21) as well as (5.2.13) we have:

$$E|\rho_T^{n,N} - \rho_T^N|^2 = E|\rho_{\tau_N}^n - \rho_{\tau_N}|^2 = E \int_0^{\tau_N} |(\rho_s^{n,N} f(W, s, u_s^n) - \rho_s\psi(W, s))|^2 ds.$$

Thus,

$$I_{[0,\tau_N]}(s)(\rho_s^{n,N} f(W, s, u_s^n) - \rho_s \psi(W, s)) \to 0$$

in $L^2(m)$ as $n \to \infty$ and we may assume, taking a subsequence, that the convergence holds also m-a.s. We get from here that $f(W, s, u_s^n) \to \psi(W, s)$ m-a.s. Moreover, we can modify on the exceptional set our functions to have the convergence everywhere. By the Filippov lemma there exists a function $u \in \mathcal{U}$ such that $\psi(W, s) = f(W, s, u_s)$. We infer from (5.2.21) that ρ satisfies (5.2.12) and hence ρ is an attainable density process. $\square$

Lemma 5.2.5 *Under the Roxin condition the set $\mathcal{D}_T$ is compact in the topology $\sigma(L^1, L^\infty)$.*

Proof. The assertion follows from Lemmas 5.2.2–5.2.4 since a convex and norm-closed set in L^1 is closed in $\sigma(L^1, L^\infty)$. $\square$

Notice that $\rho_T \mapsto \rho_T P$ is a continuous mapping from $\mathcal{D}_T$ equipped with the topology $\sigma(L^1, L^\infty)$ into $\mathbf{P}(C[0,T])$, the space of probability measures over $C[0,T]$ with the topology of weak convergence. It is evident also that $P^u \mapsto P^u W^{-1}$ is a continuous mapping from $\mathbf{P}(C[0,T])$ into $\mathbf{P}(R^m)$. The continuous image of a compact set is also compact. Therefore, we get the following:

Corollary 5.2.6 *Under the Roxin condition the set $\mathcal{K}$ is a compact in the space $\mathbf{P}(C[0,T])$, the set $\mathcal{K}(T)$ is a compact in $\mathbf{P}(R^m)$.*

5.2.4 Extreme Points of the Set of Attainable Densities

Let us denote by $\mathrm{ex}\, A$ the set of all extreme points of a convex set A, i.e. points of A which cannot be represented as a nontrivial convex combination of any two points of A.

Proposition 5.2.7 *The set $\mathrm{ex}\, \mathcal{K} = \{P^u : u \in E(\mathcal{U})\}$ where*

$$E(\mathcal{U}) := \{u \in \mathcal{U} : f(W, t, u(W, t)) \in \mathrm{ex}\, f(W, t, U) \;\; \forall(W, t)\}.$$

Proof. Assume that $u \in E(\mathcal{U})$. Let $u^i \in \mathcal{U}$, $i = 1, 2$, be such that for some $\alpha^i \in [0, 1]$ with $\alpha^1 + \alpha^2 = 1$ we have $\rho_T^u = \alpha^1 \rho_T^{u^1} + \alpha^2 \rho_T^{u^2}$, or, equivalently,

$$\rho_T^u - \alpha^1 \rho_T^{u^1} - \alpha^2 \rho_T^{u^2} = 0.$$

The last identity means that the terminal value of a continuous martingale is equal to zero. Hence,

$$\rho_t^u - \alpha^1 \rho_t^{u^1} - \alpha^2 \rho_t^{u^2} = 0 \quad \forall t \in [0, T],$$

and we can write, using the representation of the density process as the solution of the corresponding linear SDE, that

$$\int_0^t [\alpha^1 \rho_s^{u^1}(f_s^u - f_s^{u^1}) + \alpha^2 \rho_s^{u^2}(f_s^u - f_s^{u^2})]dW_s = 0 \quad \forall t \in [0, T],$$

where we use the abbreviations $f_s^u := f(W, s, u_s)$, etc. It follows that the integrand here must be equal to zero m-almost everywhere for $m(dW, dt) := P(dW)dt$. Thus,

$$f^u = f^{u^1}\frac{\alpha^1\rho^{u^1}}{\alpha^1\rho^{u^1} + \alpha^2\rho^{u^2}} + f^{u^2}\frac{\alpha^2\rho^{u^2}}{\alpha^1\rho^{u^1} + \alpha^2\rho^{u^2}} \quad m\text{-a.e.},$$

implying, by the assumption, that $f^u = f^{u^1} = f^{u^2}$ m-a.e. Consequently,

$$\rho_T^u = \rho_T^{u^1} = \rho_T^{u^2},$$

i.e. $P^u \in \mathrm{ex}\,\mathcal{K}$.

Let $P^u \in \mathrm{ex}\,\mathcal{K}$. For any (W, t) the set $f(W, t, U)$ is a convex compact subset in $\mathbf{R}^d$ and by the Carathéodory theorem the point $f(W, t, u_t(W)) \in f(W, t, U)$ can be represented as a convex linear combination of $d + 1$ points in $\mathrm{ex}\, f(W, t, U)$, i.e.,

$$f(W, t, u_t(W)) = \sum_{i=1}^{d+1} \alpha_i(W, t)h_i(W, t),$$

$h_i(W, t) \in \mathrm{ex}\, f(W, t, U)$, $\alpha_i(W, t) \geq 0$, $\sum \alpha_i(W, t) = 1$. By the "measurable" version of the Carathéodory theorem, the points $h_i(W, t)$ and weights $\alpha_i(W, t)$ can be chosen in such a way that h_i and weights α_i will be predictable processes. It follows from above that there are predictable processes α, g_1, and g_2 such that $\alpha(W, t) \in \,]0, 1]$, $g_1(W, t) \in f(W, t, U)$, $g_2(W, t) \in \mathrm{ex}\, f(W, t, U)$, and

$$f(W, t, u_t(W)) = \alpha(W, t)g_1(W, t) + (1 - \alpha(W, t))g_2(W, t).$$

Put

$$\Gamma := \{(W, t) \in C[0, T] \times [0, T] :\ f(W, t, u_t(W)) \notin \mathrm{ex}\, f(W, t, U)\}.$$

Clearly, $\Gamma = \{\alpha < 1, g_1 \neq g_2\}$. Assume that $m(\Gamma) > 0$. Then for some $\varepsilon > 0$ the set

$$\Gamma_\varepsilon = \Gamma \cap \{(W, t) :\ |f(W, t, u_t(W)) - g_1(W, t)| \wedge |f(W, t, u_t(W)) - g_2(W, t)| \geq 2\varepsilon\}$$

also is of strictly positive measure m. Let

$$a := \varepsilon\frac{(g_1 - g_2)}{|g_1 - g_2|}I_{\Gamma_\varepsilon}.$$

We introduce the process

$$X_t := \exp\left\{\int_0^t a_s^\star dW_s - \frac{1}{2}\int_0^t |a|^2 ds - \int_0^t a_s^\star f_s^u ds\right\}$$

which satisfies the linear equation

$$dX_t = X_t a_t^\star(dW_t - f_t^u dt), \quad X_0 = 1.$$

Put $\sigma = \inf\{t : X_t/(2 - X_t) \geq 2\} \wedge T$. The set $\Gamma_\varepsilon \cap [0, \sigma]$ is again of strictly positive measure m (otherwise $X_\sigma = 1$ m-a.e., that is, $\sigma = T$ m-a.e., and, consequently, $m(\Gamma_\varepsilon) = m(\Gamma_\varepsilon \cap [0, \sigma]) = 0$).

Let us consider the processes

$$f^1 := f^u I_{\bar{\Gamma}_\varepsilon} + (f^u + aI_{[0,\sigma]})I_{\Gamma_\varepsilon},$$
$$f^2 := f^u I_{\bar{\Gamma}_\varepsilon} + (f^u + bI_{[0,\sigma]})I_{\Gamma_\varepsilon},$$

where $b := -aX_t/(2 - X_t)$.

By construction, $f^i(W, t) \in f(W, t, U)$. Thus, by the Filippov lemma, there exist $u^i \in \mathcal{U}$ such that $f^i(W, t) = f((W, t, u^i(W, t)), i = 1, 2$. The process

$$Y_t := \exp\left\{\int_0^t b_s^\star dW_s - \frac{1}{2}\int_0^t |b|^2 ds - \int_0^t b_s^\star f_s^u ds\right\}$$

satisfies the equation

$$dY_t = Y_t b_t^\star(dW_t - f_t^u dt), \quad Y_0 = 1.$$

We obtain from the equations for X and Y and the definition of b that the process $Z = X + Y - 2$ is the solution of the linear equation

$$dZ_t = -Z_t \frac{a_t X_t}{2 - X_t}(dW_t - f_t^u dt), \quad Z_0 = 0.$$

Hence, $Z = 0$. But the equality $X + Y = 2$ implies that $\rho_T^u = (\rho_T^{u^1} + \rho_T^{u^2})/2$. Notice that $P(\rho_T^{u^1} \neq \rho_T^{u^2}) > 0$: otherwise, by the same arguments as used in the first part of the proof, we would get the identity $f^{u^1} = f^{u^2}$ m-a.e. which is impossible because, by construction, in the set $\Gamma_\varepsilon \cap [0, \sigma]$ the functions f^1 and f^2 are different. Thus, $P^u \notin \operatorname{ex}\mathcal{K}$ and the assumption $m(\Gamma) > 0$ leads to a contradiction. Certainly, it is possible to modify u in the predictable set of m-measure zero and find an admissible control from $E(\mathcal{U})$ generating the same probability measure P^u. $\square$

5.2.5 On the Existence of Optimal Control

The result on a structure of the set $\operatorname{ex}\mathcal{K}$ has an important application allowing us to avoid the convexity assumption for existence of optimal control in the following optimization problem:

$$E^u \xi \to \min, \quad u \in \mathcal{U}, \tag{5.2.22}$$

where ξ is a bounded random variable. Under the Roxin condition this problem obviously has a solution because

$$\inf_{u \in \mathcal{U}} E^u \xi = \inf_{\rho_T \in \mathcal{D}_T} E\rho_T \xi$$

and a continuous function in a compact set attains its minimum (recall that by Lemma 5.2.5 the set $\mathcal{D}_T$ is weakly compact in $L^1(P)$).

Let us prove the existence of the optimal solution for the problem (5.2.22) without the Roxin condition and the convexity assumption for U.

To this aim we consider a slightly more general optimization problem than that formulated above. Namely, assume that we are given a measurable set-valued mapping Γ from $(C[0,T] \times [0,T], \mathcal{P})$ into the set of convex compact subsets of $\mathbf{R}^d$. Measurability means that its graph

$$\{(W, t, v) : v \in \Gamma(W, t)\}$$

is a $\mathcal{P} \otimes \mathcal{B}^d$-measurable set. Assume that $|\Gamma(W, t)| \leq k(1 + |W_t|)$ for any (W, t). Let us denote by $\mathcal{V}$ the set of all $\mathcal{P}$-measurable selectors of Γ. For any $v \in \mathcal{V}$ we define the measure $P^v := \rho_T^v P$ where the density ρ_T^v is given by the formula (5.2.12) with $v(W, s)$ instead of $f(W, s, u_s)$. It is easy to see that the set $\mathcal{K} := \{P^v : v \in \mathcal{V}\}$ is a convex $\sigma(L^1, L^\infty)$-compact set and hence the problem

$$J(v) := E^v \xi \to \min, \quad v \in \mathcal{V},$$

(ξ is a bounded random variable) has a solution. The important observation is that there exists an optimal solution v^o with $P^{v^o} \in \mathrm{ex}\,\mathcal{K}$. To show this, we consider the convex $\sigma(L^1, L^\infty)$-compact set $K := \{\rho_T^v : J(v) = J_*\}$ where $J_* = \min_{v \in \mathcal{V}} J(v)$. By the classic Krein–Mil'man theorem it contains an extreme point $\rho_T = \rho_T^{v^o}$ and this point belongs to $\mathrm{ex}\,\mathcal{K}$. Indeed, if it is not the case, then $\rho_T = (\rho_T^{v^1} + \rho_T^{v^2})/2$ where $\rho_T^{v^1} \neq \rho_T^{v^2}$. Since at least one of the points $\rho_T^{v^i}$ does not belong to K, we would have that

$$J(v^o) = (J(v^1) + J(v^2))/2 > J_*$$

in contradiction with the definition of J_*.

Assume now that the set-valued mapping Γ satisfies all the conditions above but it values may be not convex. In this case we can consider the relaxed ("convexized") problem for the set-valued mapping

$$\Gamma' : (W, t) \mapsto \mathrm{conv}\,\Gamma(W, t)$$

with the corresponding set of $\mathcal{P}$-measurable selectors $\mathcal{V}'$. The optimization problem

$$E^v \xi \to \min, \quad v \in \mathcal{V}',$$

has a solution and in the set of all optimal controls there exists a point v^o such that the corresponding measure is in ex $\mathcal{K}'$ where $\mathcal{K}'$ is the "tube" corresponding to the relaxed problem. It is clear that the assertion of Proposition 5.2.7 can be extended and

$$\text{ex } \mathcal{K}' = \{P^v : v(W,t) \in \text{ex conv } \Gamma(W,t) \ \forall \ (W,t)\}.$$

Since

$$\text{ex conv } \Gamma(W,t) \subseteq \Gamma(W,t),$$

the control v^o takes values in $\Gamma(W,t)$ and, hence, it is the optimal solution for the original problem. $\square$

5.2.6 Comparison of Attainability Sets

We introduce the following Lipschitz condition for f:

H.5.2.4 There is a constant L such that

$$|f(W^1,t,u^1) - f(W^2,t,u^2)| \le L(\|W^1 - W^2\|_t + |u^1 - u^2|)$$

for all $W^i \in C[0,T]$, $u^i \in U$, and $t \in [0,T]$.

The main result relating the attainability sets for the model with feedback controls and the model with stochastic open loop controls is the following:

Theorem 5.2.8 (a) *Let f be a function satisfying the Roxin condition* **H.5.2.3** *such that for any $u \in \mathcal{U}$ the equation (5.2.17) has a unique strong solution. Then $\tilde{\mathcal{K}} \subseteq \mathcal{K}$ and, hence, $\tilde{\mathcal{K}}(T) \subseteq \mathcal{K}(T)$.*

(b) *Assume that f satisfies* **H.5.2.3** *and* **H.5.2.4**. *Then the imbeddings $\tilde{\mathcal{K}} \subseteq \mathcal{K}$ and $\tilde{\mathcal{K}}(T) \subseteq \mathcal{K}(T)$ are dense in the topology of total variation.*

(c) *Assume that f satisfies the Roxin condition* **H.5.2.3** *and the admissible control $u = (u_t(W))$ is such that the equation*

$$dx_t = f(x,t,u_t(x))dt + dW_t, \quad x_0 = 0, \tag{5.2.23}$$

has no strong solution. Then $P^u \notin \tilde{\mathcal{K}}$.

Notice that $\mathcal{K} \ne \tilde{\mathcal{K}}$ even in the simplest cases. For instance, let

$$f(W,t,u) := u, \quad u \in [0,1].$$

According to the well-known Tsyrel'son example, there is a bounded process $u \in \mathcal{U}$ such that the equation

$$dx_t = u_t(x)dt + dW_t, \quad y_0 = 0,$$

has no strong solution. Hence, in accordance with (c), $P^u \notin \tilde{\mathcal{K}}$.

Proof. (a) Let us consider the solution $x := x^u$ of (5.2.17) corresponding to a fix $u \in \mathcal{U}$. Let $\mathbf{F}^x = (\mathcal{F}^x_t)$ be the filtration generated by the process x and let $\pi(f^u)$ be the $\mathbf{F}^x$-predictable projection of the process $f(W, t, u_t)$. According to Th. 5 Ch. 4 of [66], the difference

$$x_t - \int_0^t \pi_s(f^u)ds \qquad (5.2.24)$$

is a Wiener process. By virtue of Lemma 5.2.9 given below there is a $\mathcal{P}$-measurable process $\alpha = (\alpha(W, t))$ satisfying the equality $\pi(f^u) = \alpha(x, t)$ up to a P-negligible set. Obviously, one can choose a version of α such that $\alpha(W, t) \in f(W, t, U)$ for all W, t. By the Filippov implicit function lemma there exists a control $v \in \mathcal{U}$ such that $\alpha(W, t) = f(W, t, v_t(W))$. Thus,

$$x_t - \int_0^t f(x, s, v_s(x))ds = \tilde{W}_t \qquad (5.2.25)$$

where $\tilde{W}$ is a Wiener process. Due to the uniqueness of the weak solution of (5.2.17) we have $P^u_x = P^v$ and (a) holds.

(b) Assume that f satisfies the Lipschitz condition **H.5.2.4**. Suppose that a given control $u \in \mathcal{U}$ also satisfies the Lipschitz condition, i.e. for all $W^i \in C[0, T]$, $t \in [0, T]$ we have

$$|u_t(W^1) - u_t(W^2)| \le K||W^1 - W^2||_t.$$

Then the equation

$$dz_t = f(z, t, u_t(z))dt + dW_t, \quad x_0 = 0, \qquad (5.2.26)$$

on the probability space $(C[0, T], \mathcal{C}, \mathbf{C}, P)$ has a unique strong solution. According to Ch.7 of [66] the distribution of the process $z = z^u$ coincides with the measure P^u.

Put $v_t = u_t(z)$. Obviously, $v \in \mathcal{U}$. Thus, the process z is a strong solution of the equation

$$dz_t = f(z, t, v_t)dt + dW_t, \quad x_0 = 0, \qquad (5.2.27)$$

and its distribution $P^v_z = P^u$ belongs to $\tilde{\mathcal{K}}$.

So, to prove (b) it is sufficient to show that for any $u \in \mathcal{U}$ there exists a sequence of admissible controls u^n satisfying the Lipschitz condition such that

$$\mathrm{Var}\,(P^u - P^n) \to 0, \quad n \to \infty, \qquad (5.2.28)$$

where P^n is the measure corresponding to u^n in the model with feedback controls.

By virtue of the criteria for the strong convergence of probability measures on a filtered space (see Appendix A.3) the relation (5.2.28) holds if and only if

$$P^n\text{-}\lim h_T(P^u, P^n) = 0 \qquad (5.2.29)$$

where $h(P^u, P^n)$ is the Hellinger process. In our case

$$h_T(P^u, P^n) = \frac{1}{8} \int_0^T |f(W, t, u_t) - f(W, t, u_t^n)|^2 \, dt.$$

The Lipschitz condition **H.5.2.4** for f implies that

$$h_T(P^u, P^n) \leq L \int_0^T |u_t - u_t^n|^2 \, dt$$

where L is a constant. Since

$$E^n h_T(P^u, P^n) \leq L E \rho_T^n \int_0^T |u_t - u_t^n|^2 \, dt, \qquad (5.2.30)$$

$E(\rho_T^n)^\alpha \leq L$ for all n for some $\alpha > 1$ (see (5.2.18)), and U is bounded, it is sufficient to show that the controls satisfying the Lipschitz condition are dense in $\mathcal{U}$ as elements of the space $L^0(C[0, T] \times [0, T], \mathcal{P}, m)$, i.e., in the sense of convergence in measure $m(dW, dt) = P(dW)dt$.

To prove this, notice that if $(\Omega, \mathcal{F}, m)$ is a space with finite measure and an algebra $\mathcal{A}$ generates $\mathcal{F}$ then the linear combinations of indicator functions of sets from $\mathcal{A}$ are dense in $L^0(\Omega, \mathcal{F}, m)$. In our case the σ-algebra $\mathcal{P}$ is generated by the algebra of all finite unions of sets $F \times]s, t]$ where F is either a null set or a union of finite number of sets of the form

$$A_c := \{W_{s(1)} \in \Gamma_1, \ldots, W_{s(k)} \in \Gamma_k\},$$

$s(i) \leq s$, U_i is rectangular in $\mathbf{R}^n$. It remains to notice that the indicator function of such a set A_c can be approximated in probability by random variables of the form $g(W_{s(1)}, \ldots, W_{s(k)})$ where g is a smooth function with compact support, thus satisfying the Lipschitz condition.

(c) Let $u \in \mathcal{U}$ be a control such that the equation

$$dy_t = f(W, t, u_t(y))dt + dW_t, \quad y_0 = 0, \qquad (5.2.31)$$

has no strong solution on $(C[0, T], \mathcal{C}, \mathbf{C}, P)$. Suppose that $P^u \in \tilde{\mathcal{K}}$. Then there exists $r \in \mathcal{U}$ such that the equation

$$dx_t = f(x, t, r_t(W))dt + dW_t, \quad x_0 = 0, \qquad (5.2.32)$$

has the strong solution on $(C[0, T], \mathcal{C}, \mathbf{C}, P)$. As in the proof of (a), we can find a process $v \in \mathcal{U}$ such that

$$dx_t = f(x, t, v_t(x))dt + d\tilde{W}_t, \quad x_0 = 0, \qquad (5.2.33)$$

where $\tilde{W}$ is a Wiener process. The distribution of this diffusion-type process is $P^v := \rho_T^v P$. As $P^v = P^u$, we have identity $\rho_T^v = \rho_T^u$ and, thus, the density processes ρ^v and ρ^u coincide (up to a P-null set). This implies that a.s. for all t we have

$$\int_0^t f_s^{u\star} dW_s - \frac{1}{2} \int_0^t |f_s^u|^2 ds = \int_0^t f_s^{v\star} dW_s - \frac{1}{2} \int_0^t |f_s^v|^2 ds,$$

where $f_t^u = f(W, t, u_t(W))$ and $f_t^v = f(W, t, v_t(W))$.

Since a continuous martingale of locally bounded variation starting from zero is zero, we obtain that

$$\int_0^T |f(W, t, u_t(W)) - f(W, t, v_t(W))|^2 dt = 0.$$

This implies that it is possible to substitute u for v in (5.2.33). Thus, we have from (5.2.32) and (5.2.33) the relation

$$dW_t = \beta_s ds + d\tilde{W}_t, \quad W_0 = 0, \tag{5.2.34}$$

where $\beta_s := f(x, s, u_s(x)) - f(x, s, r_s)$. The process β is $\mathcal{P}$-measurable and the function f satisfies the linear growth condition. Hence, the distribution of a diffusion-type process W given by the representation (5.2.34) can be obtain by a multiplication of the standard Wiener measure by density

$$\exp \left\{ \int_0^T \beta_s^\star dW_s - \frac{1}{2} \int_0^T |\beta_s|^2 ds \right\}.$$

But W is the Wiener process and this density should be equal to unity; as above we can show that $\beta = 0$ (up to P-null set). Thus, the Wiener processes W and $\tilde{W}$ coincide. But this means that the equation (5.2.31), which has a strong solution, coincides with the equation (5.2.8) which has no such solution by our assumption. This contradiction shows that $P^u \notin \tilde{\mathcal{K}}$. $\square$

Lemma 5.2.9 *Assume that X is a continuous process on some probability space $(\Omega, \mathcal{F}, P)$. Let $\mathcal{F}_t^X := \sigma\{X_s, \ s \leq t\}$, $\mathbf{F}^X := (\mathcal{F}_t^X)$. Let Y be an $\mathbf{F}^X$-predictable process. Then there exists a $\mathbf{C}$-predictable process $\alpha = (\alpha(W, t))$ on $C[0, T]$ such that $Y_t = \alpha(X, t)$ for all t.*

Proof. Let $Y = \eta I_{]u,v]}$ where random variable η is measurable with respect to $\sigma\{X_{t_1}, \ldots X_{t_k}\}$, $t_1 < \ldots < t_n \leq u < v$. By the Doob theorem there exists a measurable function f of n arguments such that

$$\eta = f(X_{t_1}, \ldots X_{t_k}).$$

This means that the assertion of the lemma holds for processes Y of the above special type. Since such processes generate $\mathcal{P}(\mathbf{F}^X)$, the assertion of Lemma 5.2.9 can be obtained by the monotone class theorem. $\square$

We finish this section with a remark that the above results holds not only for the standard Wiener process but also for the Wiener process with any non-degenerate covariance matrix.

5.3 Convergence of the Attainability Sets, I

5.3.1 The Dontchev–Veliov Theorem

We consider the controlled system of ordinary differential equations

$$\dot{x}_t = A_1(t)x_t + A_2(t)y_t + B_1(t)u_t, \quad x_0 = 0, \tag{5.3.1}$$

$$\varepsilon\dot{y}_t = A_3(t)x_t + A_4(t)y_t + B(t)u_t, \quad y_0 = 0, \tag{5.3.2}$$

where ε is a small positive number, u is any measurable function with values in a convex compact subset U of $\mathbf{R}^d$, the matrix-valued functions A_i, B_1, B are continuous and all eigenvalues of $A_4(t)$ have strictly negative real parts.

Let $K_\varepsilon(t)$ be the attainability set of the system (5.3.1), (5.3.2), i.e. the set of all end points (x_T, y_T) corresponding to various admissible controls, and $K_0^x(T)$ be the attainability set of the reduced system

$$\dot{x}_t = A_0(t)x_t + B_0(t)u_t, \quad x_0 = 0, \tag{5.3.3}$$

with the coefficients $A_0 := A_1 - A_2 A_4^{-1} A_3$, $B_0 := B_1 - A_2 A_4^{-1} B$.

We define the set

$$K_0(T) := \{(x, y) : \ x \in K_0^x(T), \ y \in R(T, x)\}$$

where $R(T, x) := -A_4^{-1}(T)A_3(T)x + Y$,

$$Y := \left\{ y : \ y = \int_0^\infty e^{A_4(T)s} B(T)v_s ds, \ v \in \mathcal{B}_U \right\}, \tag{5.3.4}$$

$\mathcal{B}_U$ is the set of all U-valued Borel functions. In other words, if we put

$$F(x, y) := (x, -A_4^{-1}(T)A_3(T)x + y)$$

then $K_0(T)$ is the image of $K_0^x(T) \times Y$ under the mapping F.

Dontchev and Veliov proved the following result (which can be extended readily to semilinear controlled systems):

Theorem 5.3.1 *The sets $K_\varepsilon(T)$ tend to $K_0(T)$ in the Hausdorff metric as $\varepsilon \to 0$.*

Let us consider for the system (5.3.1), (5.3.2) the Mayer problem

$$g(x_T, y_T) \to \min,$$

where g is a continuous function. Then for the optimal value of the singularly perturbed problem we have

$$J_\varepsilon^* = \min_{K_\varepsilon(T)} g(x, y).$$

It follows immediately from the above theorem that

$$\lim_{\varepsilon \to 0} J_\varepsilon^* = \min_{K_0(T)} g(x, y).$$

5.3.2 The First Stochastic Generalization

Our aim is to extend the Dontchev–Veliov theorem to the stochastic setting and establish the convergence of the attainability sets for the stochastic differential equations of the form

$$dx_t^\varepsilon = A_1(t, x_t^\varepsilon, u_t)dt + A_2(t)y_t^\varepsilon dt + dw_t^x, \quad x_0 = 0, \tag{5.3.5}$$

$$\varepsilon dy_t^\varepsilon = (A_3(t)x_t^\varepsilon + A_4(t)y_t^\varepsilon + B(t)u_t)dt + \beta\sqrt{\varepsilon}dw_t^y, \quad y_0 = 0, \tag{5.3.6}$$

where w^x and w^y are standard independent Wiener processes with values in $\mathbf{R}^k$ and $\mathbf{R}^n$, $0 \le t \le T < \infty$, $\varepsilon \in]0, 1]$.

We retain here the assumptions **H.5.1.1**–**H.5.1.4** of Section 5.1. For simplicity we suppose that β is continuous. Recall that **H.5.1.4** means that $\beta = o(1/|\ln\varepsilon|^{1/2})$ as $\varepsilon \to 0$. The more difficult case where $\beta^\varepsilon = \varepsilon^{1/2}$ will be studied in Section 5.4.

We understand the problem (5.3.5), (5.3.6) basically in the weak sense when a Wiener process $W = (w^x, w^y)$ is not given in advance and u is a feedback control. In fact, as explained in Section 5.2, we could avoid the above representation and use only the Girsanov transform.

The class of admissible controls $\mathcal{U}$ is the set of all U-valued predictable processes $u = (u_t)_{t \in [0,T]}$.

To adjust the above model with those considered in Section 5.2, we introduce the following notations:

$$f_\varepsilon(W, t, u) = \begin{pmatrix} A_1(t, w_t^x, u) + A_2(t)w_t^y \\ \varepsilon^{-1}A_3(t)w_t^x + \varepsilon^{-1}A_4(t)w_t^y + B(t)u_t \end{pmatrix}, \tag{5.3.7}$$

$$D_\varepsilon := \begin{pmatrix} I_k & 0 \\ 0 & \varepsilon^{-1/2}\beta I_n \end{pmatrix}, \tag{5.3.8}$$

where I_k, I_n are the identity matrices of corresponding dimensions.

Assume that the function $A_1(t, w_t^x, u)$ satisfies the Lipschitz condition **H.5.2.4** as well as the Roxin condition **H.5.2.3** introduced in the previous section. The latter means that $A_1(t, w_t^x, U)$ is a convex set for all (t, W). Thus, $f_\varepsilon(W, t, u)$ also satisfies both the Roxin and the Lipschitz conditions.

Let us consider on $(C[0, T], \mathcal{C})$ the probability measure P^ε such that the coordinate process W with respect to P^ε is the Wiener process with the correlation matrix D_ε^2.

For any admissible control u we define the measure $P^{\varepsilon,u} := \rho_T^\varepsilon(u)P^\varepsilon$ with

$$\rho_T^\varepsilon(u) = \exp\left\{\int_0^T f_\varepsilon(W, s, u_s)^\star D_\varepsilon^{-2}dW_s - \frac{1}{2}\int_0^T |f_\varepsilon(W, s, u_s)^\star D_\varepsilon^{-1}|^2 ds\right\}. \tag{5.3.9}$$

The function $f_\varepsilon(W, t, u)$ given by the formula (5.3.7) satisfies, clearly, the assumptions **H.5.2.1** and **H.5.2.2** of Section 5.2 and we can apply the general theory.

For every $\varepsilon \in]0,1]$ we consider the "tube" $\mathcal{K}_\varepsilon := \{P^{\varepsilon,u} : u \in \mathcal{U}\}$ and the attainability set $\mathcal{K}_\varepsilon(T) := \{P^{\varepsilon,u} W_T^{-1} : u \in \mathcal{U}\}$. These sets are convex compact subsets of the corresponding spaces of probability measures $\mathbf{P}(C)$ and $\mathbf{P}(\mathbf{R}^m)$ equipped with the Prohorov metric.

Let $\mathcal{K}_0^x(T)$ be the attainability set of the stochastic differential equation

$$d\bar{x}_t = A_0(t, \bar{x}_t, u_t)dt + dw_t^x, \quad \bar{x}_0 = 0, \tag{5.3.10}$$

where

$$A_0(t, x, u) := A_1(t, x, u) - A_2(t)A_4^{-1}(t)[A_3(t)x + B(t)u]. \tag{5.3.11}$$

Let $\mathbf{P}(Y)$ be the set of probability measures on the convex compact set Y given by (5.3.4).

Define the linear mapping $F(x, y) := (x, -A_4^{-1}(T)A_3(T)x + y)$ of $\mathbf{R}^m$ into itself. We put

$$S := \{\mu \in \mathbf{P}(\mathbf{R}^{k+n}) : \mu\pi_X^{-1} \in \mathcal{K}_0^x(T), \ \mu\pi_Y^{-1} \in \mathbf{P}(Y)\}$$

where $\pi_X : \mathbf{R}^{k+n} \to \mathbf{R}^k$ and $\pi_Y : \mathbf{R}^{k+n} \to \mathbf{R}^n$ are the natural projections.

Let $\mathcal{K}_0(T) := \{\mu F^{-1} : \mu \in S\}$. Clearly, $\mathcal{K}_0(T)$ is a convex compact set in $\mathbf{P}(\mathbf{R}^{k+n})$. Notice that the set S is the set of all measures

$$\mu(dx, dy) = m(x, dy)\nu(dx)$$

such that $\nu \in \mathcal{K}_0^x(T)$ and $m(x, .) \in \mathbf{P}(Y)$ for every $x \in \mathbf{R}^k$.

Also, if

$$S' := \{\mu \in \mathbf{P}(\mathbf{R}^{k+n}) : \mu\pi_X^{-1} \in \mathcal{K}', \ \mu\pi_Y^{-1} \in \mathbf{P}(Y)\}$$

where $\mathcal{K}'$ is a strongly dense subset of $\mathcal{K}_0^x(T)$, then S' is also strongly dense in S and the same holds for the sets $\{\mu F^{-1} : \mu \in S'\}$ and $\mathcal{K}_0(T)$.

Theorem 5.3.2 *The set $\cup_{\varepsilon \in [0,1]} \mathcal{K}_\varepsilon(T)$ is compact, $\mathcal{K}_\varepsilon(T)$ tend to $\mathcal{K}_0(T)$ as $\varepsilon \to 0$ in the Hausdorff metric in the space of compact subsets of $\mathbf{P}(\mathbf{R}^m)$.*

Proof. It follows from Proposition 5.1.2(a) that the family of random variables $\{(x_T^{\varepsilon,u}, y_T^{\varepsilon,u}) : u \in \mathcal{U}, \ \varepsilon \in]0,1]\}$ which are the end points of the strong solutions of the system (5.3.5), (5.3.6) are bounded in probability. It is equivalent to say that the family of their distribution is tight, i.e. relatively compact in $\mathbf{P}(\mathbf{R}^m)$. Thus, the set $\cup_{\varepsilon > 0} \tilde{\mathcal{K}}_\varepsilon(T)$ is contained in some compact set. Hence, $\cup_{\varepsilon \geq 0} \mathcal{K}_\varepsilon(T)$ is contained also in some compact set. Now, to check that $\mathcal{K}_\varepsilon(T)$ converge in the Hausdorff metric to $\mathcal{K}_0(T)$ as $\varepsilon \to 0$, it is sufficient to verify the following properties:

(1) any convergent sequence $\mu_n \in \mathcal{K}_{\varepsilon_n}(T)$ where $\varepsilon_n \to 0$ has a limit $\mu \in \mathcal{K}_0(T)$;

(2) for any $\mu \in \mathcal{K}_0(T)$ where is a sequence $\mu_n \in \mathcal{K}_{\varepsilon_n}(T)$, $\varepsilon_n \to 0$, such that μ is the limit of μ_n.

By Theorem 5.2.8(b) we can find $\tilde{\mu}_n \in \tilde{\mathcal{K}}_{\varepsilon_n}(T)$ such that the total variation distance (hence the Prohorov distance) between $\tilde{\mu}_n$ and μ_n tends to zero. This means that the sequence $\tilde{\mu}_n$ also has μ as the limit. But Proposition 5.1.2(b) and (c) assert that for the random variable $\tilde{\xi}_n := (x_T^{\varepsilon_n, u_n}, y_T^{\varepsilon_n, u_n})$ which has the distribution $\tilde{\mu}_n$ there is a random variable $\bar{\xi}_n := (\bar{x}_T^{\varepsilon_n, u_n}, \bar{y}_T^{\varepsilon_n, u_n})$ with a distribution $\bar{\mu}_n \in \mathcal{K}_0(T)$ such that the difference $\tilde{\xi}_n - \bar{\xi}_n$ tends to zero in probability. Thus, $\bar{\mu}_n$ also tends to μ. Since $\mathcal{K}_0(T)$ is closed, $\mu \in \mathcal{K}_0(T)$ and (1) holds.

Now the compactness of $\cup_{\varepsilon \in [0,1]} \mathcal{K}_\varepsilon(T)$ is clear. Indeed, let us consider an arbitrary sequence $\mu_n \in \mathcal{K}_{\varepsilon_n}(T)$. Without loss of generality we can assume that $\lim \varepsilon_n = \varepsilon_0$. In the case $\varepsilon_0 = 0$ the limit points of μ_n, as was shown above, are in $\mathcal{K}_0(T)$. If $\varepsilon_0 > 0$ then we can take $\tilde{\mu}_n \in \tilde{\mathcal{K}}_{\varepsilon_n}(T)$ such that distance between $\tilde{\mu}_n$ and μ_n tends to zero. By definition, $\tilde{\mu}_n = \mathcal{L}(\xi_n)$ for $\xi_n = (x_T^{\varepsilon_n, u_n}, y_T^{\varepsilon_n, u_n})$ where $(x^{\varepsilon_n, u_n}, y^{\varepsilon_n, u_n})$ is the solution of (5.3.5), (5.3.6). Put $\xi'_n = (x_T^{\varepsilon_0, u_n}, y_T^{\varepsilon_0, u_n})$. It is easy to show that $\xi'_n - \xi_n \to 0$ in L^2. Thus, the limit points of the sequence μ_n coincide with the limit points of the sequence $\mathcal{L}(\xi'_n) \in \mathcal{K}_{\varepsilon_0}(T)$ and it remains to recall that $\mathcal{K}_{\varepsilon_0}(T)$ is a compact set.

To prove (2) we proceed as follows.

Let $\mathcal{V}_0(T)$ be the set of random variables $(\bar{x}_T, \bar{y}_T)$ where $\bar{x}$ is the solution of the SDE (5.3.10),

$$\bar{y}_T = -A_4^{-1}(T)A_3(T)\bar{x}_T + \int_0^\infty e^{A_4(T)s} B(T)v_s ds, \qquad (5.3.12)$$

v is a $\mathcal{C} \otimes \mathcal{B}_+$-measurable function with values in U.

Let $\mathcal{U}'$ be the set of admissible controls u which has the form

$$u_t = u_t I_{[0, T-\theta[}(t) + u^o I_{[T-\theta, T]}(t) \qquad (5.3.13)$$

for some $\theta > 0$ and $u^o \in U$. Let consider the set $\tilde{\mathcal{K}}'_0(T)$ of measures $\mu F^{-1} \in \tilde{\mathcal{K}}_0(T)$ such that $\mu \pi_x^{-1}$ is the distribution of the random variable $\bar{x}_T$ corresponding to the solution of (5.3.10) with a control $u \in \mathcal{U}'$.

Clearly, for the control u given by (5.3.13) the random variables $\bar{x}_T$ and $w_T^y - w_{T-\theta}^y$ are independent.

Representing the solution of (5.3.10) as a diffusion-type process and using the inequality for the total variation distance in terms of the Hellinger process as in (5.2.30), it is easy to show that $\tilde{\mathcal{K}}'_0(T)$ is dense in $\tilde{\mathcal{K}}_0(T)$ in the total variation topology.

Lemma 5.3.3 *Let $\mathcal{L}(\mathcal{V}^0(T))$ be the set of distributions of random variables from $\mathcal{V}^0(T)$. Then $\tilde{\mathcal{K}}'_0(T) \subseteq \mathcal{L}(\mathcal{V}^0(T)) \subseteq \tilde{\mathcal{K}}_0(T)$ and these imbeddings are dense in the total variation topology.*

Proof. The inclusion $\mathcal{L}(\mathcal{V}^0(T)) \subseteq \tilde{\mathcal{K}}_0(T)$ is evident.

Let us consider an arbitrary measure μ on $\mathbf{R}^{k+n}$ such that $\mu F^{-1} \in \tilde{\mathcal{K}}_0'(T)$. Thus, $\mu\pi_X^{-1}$ is the distribution of the random variable $\bar{x}_T$, corresponding to the solution of (5.3.10) with $u \in \mathcal{U}'$ given by (5.3.13). By virtue of Lemma 5.3.4, given below the proof of Theorem 5.3.2, there is a random variable ζ on $(C[0,T], \mathcal{C}, P)$ such that the pair $(\bar{x}_T, \zeta)$ has the distribution μ.

We introduce a set-valued mapping $\Gamma : W \mapsto \Gamma_W$ of $(C[0,T], \mathcal{C})$ into the Hilbert space $\mathcal{H} = (L^2(\mathbf{R}_+)^d, e^{-\gamma t}dt)$, putting

$$\Gamma_W := \left\{ b \in \mathcal{B}_U : \zeta(W) = \int_0^\infty e^{A_4(T)s} B(T) b_s ds \right\}.$$

The graph of Γ is measurable. Hence, by the Aumann theorem it has a measurable a.s. selector, i.e., there is a measurable mapping $b : (C[0,T], \mathcal{C}) \to L^2(\mathbf{R}_+)^d$ such that $b(W) \in \Gamma_W$ for almost all W. Recall that $b(W)$ is a class of equivalence of functions coinciding up to sets of dt-measure zero. Let $\{h^i\}$ be a countable family of functions which is dense in $\mathcal{B}_U$ with respect to the convergence in $(L^2(\mathbf{R}_+)^d, e^{-\gamma t}dt)$, $j(W, n) := \min\{i : \| b(W) - h^i \|_{\mathcal{H}} \leq 1/n\}$. Then $h^{j(n)} = (h_t^{j(W,n)})$ is a $\mathcal{C} \otimes \mathcal{B}_+$-measurable function with values in U. The sequence $h_t^{j(W,n)}$ converges to $b(W)$ in the norm $\| . \|_{\mathcal{H}}$ for almost all W. As U is bounded, the sequence $h^{j(n)}$ converges to b in $(L^2(\Omega \times \mathbf{R}_+)^d, P \times e^{\gamma t}dt)$. Hence, it contains a subsequence converging $P \times dt$-a.e. to some $\mathcal{C} \otimes \mathcal{B}_+$-measurable function $v = (v_t(W))$ taking values in U.

Notice that

$$\zeta'(W) := \int_0^\infty e^{A_4(T)s} B(T) v_s(W) ds = \zeta(W) \quad P\text{-a.s.}$$

Thus, the distributions $(\bar{x}_T, \zeta)$ and $(\bar{x}_T, \zeta')$ coincide.

We checked that $\mu F^{-1} \in \mathcal{L}(\mathcal{V}^0(T))$ and, hence, the inclusion

$$\tilde{\mathcal{K}}_0'(T) \subseteq \mathcal{L}(\mathcal{V}^0(T))$$

holds.

The proof is finished because, as was mentioned, the subset $\tilde{\mathcal{K}}_0'(T)$ is dense in $\tilde{\mathcal{K}}_0(T)$ in the total variation topology. $\square$

Now the proof of (2) is obvious. Indeed, we need to find an approximation only for an arbitrary point μ from $\tilde{\mathcal{K}}_0'(T)$ since this set is dense in $\mathcal{K}_0(T)$. By the above lemma $\mu = \mathcal{L}((\bar{x}_T, \bar{y}_T))$ where $\bar{x} = \bar{x}^u$ is a solution of (5.3.10) and $\bar{y}_T$ is given by (5.3.13). But Proposition 5.1.3 asserts the existence of a sequence of admissible controls u^ε such that for the corresponding solutions of (5.3.5), (5.3.6) we have the convergence $(x_T^{\varepsilon,u^\varepsilon}, y_T^{\varepsilon,u^\varepsilon}) \to (\bar{x}_T, \bar{y}_T)$ in probability, hence, in law. $\square$

Lemma 5.3.4 *Let ξ and η be independent random variables defined on a probability space $(\Omega, \mathcal{F}, P)$ and taking values in uncountable Polish spaces $(X, \mathcal{X})$ and $(Y, \mathcal{Y})$ with the Borel σ-algebras. Assume that the distribution of* *η has no atoms. Let μ be a measure on* $(X \times Y, \mathcal{X} \otimes \mathcal{Y})$ *such that the measure* $\rho := \mu \pi_x^{-1}$ *is the distribution of the random variable* ξ.

Then there exists a random variable ζ *on* $(\Omega, \mathcal{F}, P)$ *with values in* $(Y, \mathcal{Y})$ *such that μ is the distribution of the pair* (ξ, ζ).

Proof. Assume, at first, that both spaces $(X, \mathcal{X})$ and $(Y, \mathcal{Y})$ coincide with the unit interval [0,1] equipped with the Borel σ-algebra. For this case the statement is well-known: $\zeta = g(\xi, \eta)$ where g is defined as follows. Let $F(y)$ be the distribution function of η and let $\nu(x, dy)$ be a kernel such that

$$\mu(dx, dy) = \rho(dx)\nu(x, dy).$$

Put $G_x(s) := \nu(x, [0, s])$ and $H_x(t) := \inf\{s : G_x(s) \geq t\}$. Then $g(x, y) := H_x(F(y))$. Indeed, the random variable $F(\eta)$ has the uniform distribution, and the distribution of the random variable $H_x(F(\eta))$ coincides with $\nu(x, dy)$.

The general case is reduced to that considered above due to the well-known isomorphism theorem for uncountable Polish spaces.□

5.4 Convergence of the Attainability Sets, II

5.4.1 Formulation of the Result

In this section we prove the result on convergence of the attainability sets for the stochastic differential equations

$$dx_t^\varepsilon = A_1(t, x_t^\varepsilon, u_t)dt + A_2(t)y_t^\varepsilon dt + dw_t^x, \quad x_0 = 0, \quad (5.4.1)$$

$$\varepsilon dy_t^\varepsilon = (A_3(t)x_t^\varepsilon + A_4(t)y_t^\varepsilon + B(t)u_t)dt + \sqrt{\varepsilon}dw_t^y, \quad y_0 = 0, \quad (5.4.2)$$

where w^x and w^y are standard independent Wiener processes with values in $\mathbf{R}^k$ and $\mathbf{R}^n$, $t \in [0, T]$, $\varepsilon \in]0, 1]$, and the controls $u = (u_t)$ are predictable processes with values in a compact convex subset U of $\mathbf{R}^d$.

We again use the assumptions **H.5.1.1–H.5.3** of Section 5.1, the Roxin and Lipschitz conditions $A_1(t, w_t^x, u)$, and the notations of the previous section with obvious changes: e.g., now

$$D_\varepsilon := \begin{pmatrix} I_k & 0 \\ 0 & \varepsilon^{-1/2}I_n \end{pmatrix}. \quad (5.4.3)$$

Let $\mathcal{K}_0^x(T)$ be the attainability set of the stochastic differential equation

$$d\bar{x}_t = A_0(t, \bar{x}_t, u_t)dt + dw_t^x, \quad x_0 = 0, \quad (5.4.4)$$

where

$$A_0(t, x, u) := A_1(t, x, u) - A_2(t)A_4^{-1}(t)[A_3(t)x + B(t)u]. \quad (5.4.5)$$

Let ξ be the (strong) solution of the following stochastic differential equation with constant coefficients on some filtered probability space $(\Omega, \mathcal{F}, \mathbf{F} = (\mathcal{F}_t), P)$

$$d\xi_t = A_4(T)\xi_t dt + db_t, \quad \xi_0 = \xi^o, \tag{5.4.6}$$

where b is a standard Wiener process in $\mathbf{R}^n$ independent of a Gaussian random variable ξ^o with zero mean and covariance matrix

$$\Xi := \int_0^\infty e^{A_4(T)s} e^{A_4^\star(T)s} ds. \tag{5.4.7}$$

In other words, ξ is the stationary Gaussian Markov process with the zero mean and covariance function

$$K(s,t) := E\xi_s \xi_t^\star = \Xi e^{A_4^\star(T)(t-s)}, \quad t \geq s. \tag{5.4.8}$$

Let $\mathcal{V}_U$ be the set of all U-valued processes $v = (v_t)_{t \geq 0}$ such that $v_{1/t}$ is a predictable process with respect to the filtration generated by the process $\xi_{1/t}$ and let $S_Y^o := \{\mathcal{L}(\xi_0 + I(v)) : v \in \mathcal{V}_U\}$ where

$$I(v) := \int_0^\infty e^{A_4(T)s} B(T) v_s ds. \tag{5.4.9}$$

The set S_Y^o is compact in $\mathbf{P}(\mathbf{R}^n)$, see Lemma A.6.5.

Put $S_Y := \overline{\text{conv}}\, S_Y^o$, the convex closure of S_Y^o in $\mathbf{P}(\mathbf{R}^n)$.

Let S be the set of all probability measures $\mu = \mu(dx, dy)$ on the space $\mathbf{R}^m = \mathbf{R}^k \times \mathbf{R}^n$ such that

(1) $\mu(., dy) \in S_Y$;

(2) $\mu(dx, \mathbf{R}^n) \in \mathcal{K}_0^x(T)$.

It follows from Proposition A.6.6 that S is a compact subset of $\mathbf{P}(\mathbf{R}^m)$.

Define the linear mapping $F(x, y) := (x, -A_4^{-1}(T)A_3(T)x + y)$ of $\mathbf{R}^m$ into itself. Put $\mathcal{K}_0(T) := \{\mu F^{-1} : \mu \in S\}$.

The main result of the section is

Theorem 5.4.1 *The set $\cup_{\varepsilon \in [0,1]} \mathcal{K}_\varepsilon(T)$ is compact, and $\mathcal{K}_\varepsilon(T)$ tends to $\mathcal{K}_0(T)$ as $\varepsilon \to 0$ in the Hausdorff metric in the space of compact subsets of $\mathbf{P}(\mathbf{R}^m)$.*

The above theorem immediately implies

Corollary 5.4.2 *For the optimal value*

$$J_\varepsilon^* := \inf_{u \in \mathcal{U}} E^{\varepsilon, u} g(W_T) = \inf_{\mu \in \mathcal{K}_\varepsilon(T)} \int g(x, y)\mu(dx, dy)$$

of the Mayer problem for a system (5.4.1), (5.4.2) where g is a continuous bounded function in $\mathbf{R}^m$ we have

$$\lim_{\varepsilon \to 0} J_\varepsilon^* = \inf_{\mu \in \mathcal{K}_0(T)} \int g(x, y)\mu(dx, dy).$$

Obviously, Theorem 5.4.1 can be applied to a more general optimization problem

$$J^\varepsilon(u) = F(P^{\varepsilon, u} W_T^{-1}) \to \min$$

where F is any continuous function in $\mathbf{P}(\mathbf{R}^m)$.

Remark 1. The definition of the set $\mathcal{V}_U$ seems a bit sophisticated. Essentially, $\mathcal{V}_U$ contains measurable processes v such that for any t the random variable v_t is measurable with respect to the σ-algebra $\mathcal{F}_{\geq t}^\xi := \sigma\{\xi_s, \ s \geq t\}$. To avoid a discussion of the measurable structures related to a decreasing family of σ-algebras we prefer to consider the processes in reversed time.

Remark 2. One can give an alternative description of the set S. Specifically, let α be a random variable independent of ξ, taking values in some Polish space, and having a non-atomic distribution. Define the set $\mathcal{V}_U^\alpha$ as the set of all U-valued processes $v = (v_t)_{t \geq 0}$ such that $v_{1/t}$ is a predictable process with respect to the filtration generated by the process $\xi_{1/t}$ and the random variable α. Then the set $S = \{\mathcal{L}(\xi_0 + I(v)) : v \in \mathcal{V}_U^\alpha\}$; see Lemma A.6.7.

5.4.2 The Fast Variable Model

The proof of Theorem 5.4.1 follows the same scheme as the proof of Theorem 5.3.1. Since the general case requires rather long arguments, we clarify the main idea on the simplest example of a one-dimensional model with constant coefficients containing only the fast variable.

Let us consider the controlled stochastic differential equation

$$\varepsilon dy_t^{\varepsilon, u} = (-\gamma y_t^{\varepsilon, u} + u_t)dt + \varepsilon^{1/2} dw_t^y, \quad y_0 = 0, \tag{5.4.10}$$

where u is a predictable process which takes values in $U = [0, 1]$. Now the set $\mathcal{K}_0(T)$ is the convex closure of the set $\{\mathcal{L}(\xi_0 + I(v)), \ v \in \mathcal{V}_U\}$ where

$$I(v) := \int_0^\infty e^{-\gamma s} v_s ds,$$

ξ is the Ornstein–Uhlenbeck process on some probability space $(\Omega, \mathcal{F}, P)$ with correlation function $K(s, t) = (2\gamma)^{-1} e^{-\gamma|t-s|}$, and $\mathcal{V}_U$ is the set of all U-valued processes v such that $v_{1/t}$ is a predictable process with respect to the filtration generated by the process $\xi_{1/t}$. For our purpose it is more convenient to use the alternative description of $\mathcal{K}_0(T)$ as the set $\{\mathcal{L}(\xi_0 + I(v)), \ v \in \mathcal{V}_U^\alpha\}$ where α is a random variable independent of ξ with values in a Polish space and non-atomic distribution, $\mathcal{V}_U^\alpha$ is the set of all U-valued processes v such that $v_{1/t}$ is a predictable process with respect to the filtration generated by the process $\xi_{1/t}$ and the random variable α. The equation (5.4.10) we understand in the strong sense. Its solution can be represented in the following way:

$$y_t^{\varepsilon, u} = \varepsilon^{-1} \int_0^t e^{-\gamma(t-s)/\varepsilon} u_s ds + \eta_t^\varepsilon \tag{5.4.11}$$

where

$$\eta_t^\varepsilon := \varepsilon^{-1/2} \int_0^t e^{-\gamma(t-s)/\varepsilon} dw_s^y. \tag{5.4.12}$$

Put $T_\varepsilon := T(1 - \varepsilon^{1/2})$. Let us consider on the interval $[T_\varepsilon, T]$ the Gaussian stationary process

$$\tilde{\xi}_t^\varepsilon := (2\gamma)^{-1/2} \exp\{-\gamma(t - T_\varepsilon)/\varepsilon\}\beta + \varepsilon^{-1/2} \int_{T_\varepsilon}^t e^{-\gamma(t-s)/\varepsilon} dw_s^y,$$

where β is a standard normal random variable independent of the Wiener process w^y (to define β we can extend our canonical coordinate probability space). The process $\tilde{\xi}^\varepsilon$ is the solution of the linear equation

$$\varepsilon d\tilde{\xi}_t^\varepsilon = -\gamma\tilde{\xi}_t^\varepsilon dt + \varepsilon^{1/2} dw_t^y, \quad \tilde{\xi}_{T_\varepsilon}^\varepsilon = \beta.$$

Let us consider the Ornstein–Uhlenbeck process $\xi_t^\varepsilon = \tilde{\xi}_{T-\varepsilon t}^\varepsilon$, $t \in [0, T/\sqrt{\varepsilon}]$.

Obviously, $\eta_T^\varepsilon - \xi_0^\varepsilon = \eta_T^\varepsilon - \tilde{\xi}_T^\varepsilon \to 0$ in L^2 as $\varepsilon \to 0$.

For $u \in \mathcal{U}$ we define the process $v_s = v_s^\varepsilon := u_{T-\varepsilon s} I_{[0,T/\sqrt{\varepsilon}[}$.

Now we can write that

$$y_T^{\varepsilon,u} = \eta_T^\varepsilon + \int_0^{T/\sqrt{\varepsilon}} e^{-\gamma s} u_{T-\varepsilon s} ds + \int_{T/\sqrt{\varepsilon}}^{T/\varepsilon} e^{-\gamma s} u_{T-\varepsilon s} ds = \bar{y}_T^{\varepsilon,u} + R^\varepsilon(u)$$

where $\bar{y}_T^{\varepsilon,u} = \xi_0^\varepsilon + I(v)$ and

$$R^\varepsilon(u) := \int_{T/\sqrt{\varepsilon}}^{T/\varepsilon} e^{-\gamma s} u_{T-\varepsilon s} ds + \eta_T^\varepsilon - \xi_0^\varepsilon.$$

Since $\sup_{u \in \mathcal{U}} |R^\varepsilon(u)| \to 0$ in probability, to accomplish the first step we need to check only that $\mathcal{L}(\xi_0^\varepsilon + I(v)) \in \mathcal{K}_0(T)$. Indeed, let us associate with ξ the process ξ^ε defined above. For any $s \leq T/\sqrt{\varepsilon}$ the random variable v_s is measurable with respect to the σ-algebra $\mathcal{C}_{T-\varepsilon s}$. But

$$\mathcal{C}_{T-\varepsilon s} = \sigma\{w_r, \ r \leq T_\varepsilon\} \vee \sigma\{w_r, \ T_\varepsilon \leq r \leq T - \varepsilon s\}$$
$$\subseteq \sigma\{w_r, \ r \leq T_\varepsilon\} \vee \sigma\{\tilde{\xi}_r^\varepsilon, \ T_\varepsilon \leq r \leq T - \varepsilon s\}$$
$$= \sigma\{w_r, \ r \leq T_\varepsilon\} \vee \sigma\{\xi_r^\varepsilon, \ s \leq r \leq T/\sqrt{\varepsilon}\}$$

and we see that $v \in \mathcal{V}_U^\alpha$ where the random variable α is defined as the projection mapping of $C[0,T]$ onto $C[0,T_\varepsilon]$. The above considerations show that for any convergent sequence $\mu^n \in \mathcal{K}_{\varepsilon_n}(T)$ its limit is an element of $\mathcal{K}_0(T)$.

Now we introduce the set $\mathcal{V}_U^{\alpha'}$ consisting of all processes

$$v_s = \sum_{i=1}^N \varphi_i I_{]s_i, s_{i+1}]}(s) + u^0 I_{]s_{N+1}, \infty[}(s), \tag{5.4.13}$$

where $0 = s_1 < \ldots < s_{N+1}$, $u^0 \in U$, and random U-valued random variables φ_i have the form

$$\varphi_i = f_i(\alpha, \xi(r_1^i), \ldots, \xi(r_{M_i}^i)), \quad s_{i+1} < r_j^i \leq s_N. \tag{5.4.14}$$

Let $\mathcal{K}_0'(T) := \{\mathcal{L}(\xi_0 + I(v)), \ v \in \mathcal{V}_U^{\alpha'}\}$. It is not difficult to show that the set $\{I(v), \ v \in \mathcal{V}_U^{\alpha'}\}$ is dense in $\{I(v), \ v \in \mathcal{V}_U\}$ in probability. Thus, $\mathcal{K}'(T)$ is dense in $\mathcal{K}_0(T)$ in $\mathbf{P}(\mathbf{R})$.

Let $\mu \in \mathcal{K}'(T)$. This means that μ is the distribution of a random variable $\chi := \xi_0 + I(v)$ where v is of the form (5.4.13). The result will be proved if we construct a random variable χ^ε and a control u^ε such that $\mathcal{L}(\chi^\varepsilon) = \mathcal{L}(\chi)$ and $\chi^\varepsilon - y_T^{u^\varepsilon, \varepsilon} \to 0$ in probability. To this aim it is enough to find on the coordinate probability space $(C[0, T], \mathcal{C}, P)$ a stationary Gaussian Markov process ξ^ε with correlation function $K(s, t)$, a standard normal random variable α^ε independent of ξ^ε, and an admissible control $u^\varepsilon \in \mathcal{U}$ such that $\xi_0^\varepsilon - \eta_T^\varepsilon \to 0$ in probability (η_T^ε is defined by (5.4.12)) and

$$\int_0^\infty e^{-\gamma s} v_s^\varepsilon ds - \frac{1}{\varepsilon} \int_0^T e^{-\gamma(T-s)/\varepsilon} u_s^\varepsilon ds \to 0.,$$

where v^ε is the process given by the formula (5.4.13) if we substitute ξ, φ, and α by ξ^ε, φ^ε, and α^ε. Indeed, in this case the random variable $\chi^\varepsilon := \xi_0^\varepsilon + I(v^\varepsilon)$ meets the required properties.

The process ξ^ε can be constructed by the following way. For sufficiently small ε let $T_\varepsilon^k := T(1 - k\varepsilon^{1/2})$, $k = 1, 2, 3$. Put

$$\alpha^\varepsilon := (w_{T_\varepsilon^2} - w_{T_\varepsilon^3})/(T_\varepsilon^2 - T_\varepsilon^3)^{1/2},$$

$$\rho^\varepsilon := (2\gamma)^{-1/2}(w_{T_\varepsilon^1} - w_{T_\varepsilon^2})/(T_\varepsilon^1 - T_\varepsilon^2)^{1/2},$$

$$\tilde{\xi}_t^\varepsilon := e^{-\gamma(t - T_\varepsilon^1)/\varepsilon} \rho^\varepsilon + \frac{1}{\varepsilon^{1/2}} \int_{T_\varepsilon^1}^t e^{-\gamma(t-s)/\varepsilon} dw_s, \quad t \geq T_\varepsilon^1.$$

Define the process ξ^ε on $[0, \varepsilon^{-1/2}T]$ by the equality $\xi_t^\varepsilon := \tilde{\xi}_{T-\varepsilon t}^\varepsilon$.

Obviously,

$$\xi_0^\varepsilon - \eta_T^\varepsilon = \exp\{-\gamma(T - T_\varepsilon^1)/\varepsilon\}\rho^\varepsilon - \frac{1}{\varepsilon^{1/2}} \int_0^{T_\varepsilon^1} e^{-\gamma(T-s)/\varepsilon} dw_s \to 0 \text{ in } L^2.$$

For sufficiently small ε we put

$$u^\varepsilon := u^0 I_{[0, t_{N+1}[} + \sum_{i=1}^{N+1} \varphi_i^\varepsilon I_{[t_{i+1}, t_i[}$$

where $t_i := T - \varepsilon s_i$, $i \leq N + 1$.

The random variables φ_i^ε are $\mathcal{C}_{t_{i+1}}$-measurable. Thus, $u^\varepsilon \in \mathcal{U}$. It follows that

$$\int_0^\infty e^{-\gamma s} v_s^\varepsilon ds - \frac{1}{\varepsilon} \int_0^T e^{-\gamma(T-s)/\varepsilon} u_s^\varepsilon ds = \int_0^\infty e^{-\gamma s} v_s^\varepsilon ds - \int_0^{T/\varepsilon} e^{-\gamma s} u_{T-\varepsilon s}^\varepsilon ds$$

$$= \int_{T/\varepsilon}^\infty e^{-\gamma s} v_s^\varepsilon ds \to 0.$$

The proof of the result for this particular case is finished.

5.4.3 General Case

For the proof of Theorem 5.4.1 in the general case we need the following:

Proposition 5.4.3 *Let $(x^{\varepsilon,u}, y^{\varepsilon,u})$ be the solution of (5.4.1), (5.4.2) corresponding to some $u \in \mathcal{U}$ and let $\bar{x}^u$ be the solution of the reduced equation (5.4.4). Let the random variable $\bar{y}_T^{\varepsilon,u}$ be defined by*

$$\bar{y}_T^{\varepsilon,u} := -A_4^{-1}(T)A_3(T)\bar{x}_T^u + \int_0^\infty e^{A_4(T)r} B(T) v_r^\varepsilon dr + \tilde{\xi}_T^\varepsilon \tag{5.4.15}$$

where $v_r^\varepsilon := u_{T-r\varepsilon} I_{[0,T/\sqrt{\varepsilon}]}(r) + u^0 I_{]T/\sqrt{\varepsilon},\infty[}(r)$, u^0 is an arbitrary point in U,

$$\tilde{\xi}_T^\varepsilon := e^{\varepsilon^{-1} A_4(T)(T-T^\varepsilon)} \rho^\varepsilon + \frac{1}{\sqrt{\varepsilon}} \int_{T_\varepsilon}^T e^{\varepsilon^{-1} A_4(T)(T-s)} dw_s^y, \tag{5.4.16}$$

$T_\varepsilon := (1 - \sqrt{\varepsilon})T$, and ρ^ε is a Gaussian random variable with zero mean and covariance Ξ; the matrix Ξ is defined in (5.4.7).

Then for any $p \in [1, \infty[$ the sets

$$M^x := \{\| x^{\varepsilon,u} \|_T : u \in \mathcal{U}, \ \varepsilon \in]0,1]\},$$

are bounded in $L^p(\Omega)$,

$$\sup_\varepsilon \sup_{u \in \mathcal{U}} \sup_{t \leq T} E|y_t^{\varepsilon,u}|^p < \infty, \tag{5.4.17}$$

$$\limsup_{\varepsilon \to 0} \sup_{u \in \mathcal{U}} E \| x^{\varepsilon,u} - \bar{x}^u \|_T^p = 0, \tag{5.4.18}$$

$$\limsup_{\varepsilon \to 0} \sup_{u \in \mathcal{U}} E|y_T^{\varepsilon,u} - \bar{y}_T^{\varepsilon,u}|^p = 0. \tag{5.4.19}$$

Proof. The boundedness of M^x in L^p as well as (5.4.18) has been established in the proof of Proposition 5.1.2. The property (5.4.17) follows easily from the representation

$$y_t^\varepsilon := \varepsilon^{-1} \int_0^t \Psi^\varepsilon(t,s)[A_3(s)x_s^\varepsilon + B(s)u_s^\varepsilon] ds + \eta_t^\varepsilon \tag{5.4.20}$$

where

$$dn_t^\varepsilon = \varepsilon^{-1} A_4(t)\eta_t^\varepsilon dt + \varepsilon^{-1/2} dw_t^y, \quad \eta_0^\varepsilon = 0,$$

and

$$\sup_{t \geq 0} E|\eta_t^\varepsilon|^p \leq C_p.$$

Let $\tilde{y}^{\varepsilon,u}$ be the solution of the stochastic differential equation

$$\varepsilon d\tilde{y}_t^{\varepsilon,u} = (A_3(T)\bar{x}_T^u + A_4(T)\tilde{y}_t^{\varepsilon,u} + B(T)u_t)dt + \sqrt{\varepsilon}dw_t^y \quad \tilde{y}_0^{\varepsilon,u} = 0. \quad (5.4.21)$$

Put

$$\tilde{\Delta}_t^{y,\varepsilon,u} := y_t^{\varepsilon,u} - \tilde{y}_t^{\varepsilon,u},$$
$$\hat{x}_t^{\varepsilon,u} := x_t^{\varepsilon,u} - x_T^{\varepsilon,u},$$
$$\hat{A}_i(t) := A_i(t) - A_i(T),$$
$$\hat{B}(t) := B(t) - B(T).$$

The process $\tilde{\Delta}^{y,\varepsilon,u}$ is the solution of the ordinary differential equation

$$d\tilde{\Delta}_t^{y,\varepsilon,u} = (A_4(T)\tilde{\Delta}_t^{y,\varepsilon,u} + \varphi_t^{\varepsilon,u})dt, \quad \tilde{\Delta}_0^{y,\varepsilon,u} = 0,$$

where

$$\varphi_t^{\varepsilon,u} := \hat{A}_4(t)y_t^{\varepsilon,u} + \hat{A}_3(t)x_t^{\varepsilon,u} + A_3(T)\hat{x}_t^{\varepsilon,u} + A_3(T)\bar{\Delta}_T^{x,\varepsilon,u} + \hat{B}(t)u_t.$$

Thus,

$$\tilde{\Delta}_T^{y,\varepsilon,u} = \frac{1}{\varepsilon}\int_0^T e^{\varepsilon^{-1}A_4(T)(T-s)}\varphi_s^{\varepsilon,u}ds. \quad (5.4.22)$$

By virtue of **H.5.1.3** for all $t \geq 0$ we have that

$$|e^{\varepsilon^{-1}A_4(T)t}| \leq Ce^{-2\kappa t/\varepsilon}. \quad (5.4.23)$$

Taking into account the L^p-boundedness of M^x we get from (5.4.22) that the L^p-norm of $\tilde{\Delta}_T^{y,\varepsilon,u}$ is bounded by

$$\cdot \quad C\frac{1}{\varepsilon}\int_0^T e^{-2\kappa(T-s)/\varepsilon}(|\hat{A}_4(s)| + |\hat{A}_3(s)| + f_s^\varepsilon + \bar{g}^\varepsilon + |\hat{B}(s)|)ds \quad (5.4.24)$$

where

$$f_s^\varepsilon := \sup_{u \in \mathcal{U}}(E|x_s^{\varepsilon,u} - x_T^{\varepsilon,u}|^p)^{1/p}, \quad \bar{g}^\varepsilon := \sup_{u \in \mathcal{U}}(E|\bar{\Delta}_T^{x,\varepsilon,u}|^p)^{1/p}.$$

Let $\bar{f}_s$ be the function similar to f_s^ε but defined for $\bar{x}^u$. It follows from (5.4.17) that for any $\delta > 0$ we have $f_s^\varepsilon \leq \bar{f}_s + \delta$ for all sufficiently small ε. But it is clear from the equation (5.4.4) that $\lim_{s \to T} \bar{f}_s = 0$. Taking into account the above remarks we check easily that the expression (5.4.24) tends to zero as $\varepsilon \to 0$ and, hence,

$$\lim_{\substack{\varepsilon \to 0 \\ u \in \mathcal{U}}} \sup E |y_T^{\varepsilon,u} - \tilde{y}_T^{\varepsilon,u}|^p = 0. \tag{5.4.25}$$

Now we show that

$$\lim_{\substack{\varepsilon \to 0 \\ u \in \mathcal{U}}} \sup E |\bar{y}_T^{\varepsilon,u} - \tilde{y}_T^{\varepsilon,u}|^p = 0. \tag{5.4.26}$$

Indeed,

$$\bar{y}_T^{\varepsilon,u} - \tilde{y}_T^{\varepsilon,u} = \left(-A_4^{-1}(T) - \frac{1}{\varepsilon} \int_0^T e^{\varepsilon^{-1} A_4(T)(T-s)} ds \right) A_3(T) \bar{x}_T^u$$

$$+ \int_{T/\varepsilon}^\infty e^{A_4(T)r} B(T) u_r^0 dr - \int_{T/\sqrt{\varepsilon}}^{T/\varepsilon} e^{A_4(T)r} B(T) u_{T-\varepsilon r} dr$$

$$+ e^{\varepsilon^{-1/2} A_4(T)T} \rho^\varepsilon - \frac{1}{\sqrt{\varepsilon}} \int_0^{T_\varepsilon} e^{\varepsilon^{-1} A_4(T)(T-s)} dw_s^y.$$

Obviously, the L^p-norms of all terms on the right-hand side of this identity tend to zero and the convergence of the first one is uniform in $u \in \mathcal{U}$ because the set $\{\| \bar{x}^u \|_T : u \in \mathcal{U}\}$ is L^p-bounded. Thus, (5.4.26) holds. The relations (5.4.25), (5.4.26) imply (5.4.20).

Proposition 5.4.3 is proved. □

5.4.4 Proof of Theorem 5.4.1

Assume that the sequence $\mathcal{L}(x_T^{\varepsilon_n,u_n}, y_T^{\varepsilon_n,u_n})$ converges in $\mathbf{P}(\mathbf{R}^m)$ to some μ. We choose in the representation (5.4.16) the random variable ρ^ε independent of W. It follows from Proposition 5.4.3 that the sequence $\mathcal{L}(\bar{x}_T^{u_n}, \bar{y}_T^{\varepsilon_n,u_n})$ converges to the same limit. Let us introduce the modified controls

$$\hat{u}_n = u_n I_{[0,T_{\varepsilon_n}]} + u^0 I_{]T_{\varepsilon_n},T]}$$

where u^0 is a fixed point from U. Since $\bar{x}_T^{u_n} - \bar{x}_T^{\hat{u}_n}$ tends to zero in probability, the sequence $\mathcal{L}(\bar{x}_T^{\hat{u}_n}, \bar{y}_T^{\varepsilon_n,u_n})$ converges to μ and we need to check only that $\mathcal{L}(\bar{x}_T^{\hat{u}_n}, \bar{y}_T^{\varepsilon_n,u_n}) \in \mathcal{K}_0(T)$. To show this we notice that $\bar{x}_T^{\hat{u}_n}$ is a function of the natural projection

$$i^{\varepsilon_n} : \{w_t^x, w_t^y, \ t \in [0,T]\} \mapsto (\{w_t^x, \ t \in [0,T]\}, \{w_t^y, \ t \in [0,T_{\varepsilon_n}]\}).$$

As in the previously treated one-dimensional case it can be shown that the regular conditional distribution of the random variable $\xi_0^{\varepsilon_n} + I(v^{\varepsilon_n})$ for a fixed value i^{ε_n} belongs to S. Since S is a convex closed set and $\bar{x}_T^{\hat{u}_n}$ is a measurable function of i^{ε_n}, it follows from Lemma A.6.3 that the regular conditional distribution of $\xi_0^{\varepsilon_n} + I(v^{\varepsilon_n})$ for a fixed value $\bar{x}_T^{\hat{u}_n}$ also belongs to S. Thus, the limit of any converging sequence $\mu_n \in \mathcal{K}_{\varepsilon_n}(T)$ is an element of $\mathcal{K}_0(T)$.

Now we must show that for any measure $\mu F^{-1} \in \mathcal{K}_0(T)$ there exists a sequence $\mu_n \in \mathcal{K}_{\varepsilon_n}(T)$ which converges to μF^{-1} in $\mathbf{P}(\mathbf{R}^n)$. It is sufficient to find such a sequence for an arbitrary μF^{-1} from the set $\tilde{\mathcal{K}}_0(T)$ which is dense in $\mathcal{K}_0(T)$ in the total variation topology. The latter property holds since the attainability set $\tilde{\mathcal{K}}_0^x$ corresponding to the strong solutions of (2.1) is dense in $\mathcal{K}_0^x$ in the total variation topology. Thus, there are dense imbeddings $\tilde{\mathcal{K}}_0 \subseteq \mathcal{K}_0$ and $\tilde{\mathcal{K}}_0(T) \subseteq \mathcal{K}_0(T)$.

Let us fix $\delta > 0$ and a measure $\mu = m(x, dy)\nu(dx)$ with $\mu F^{-1} \in \mathcal{K}_0(T)$. By definition, $\nu = \mathcal{L}(\bar{x}_T^u)$ where $\bar{x}^u$ solves the reduced equation (5.4.4) corresponding to some admissible control u. Let $\nu_h := \mathcal{L}(\bar{x}_{T-h}^u)$ and let $\mu_h(dx, dy) := m(x, dy)\nu_h(dx)$, $h \in [0, T]$. Then there exists $h_0 > 0$ such that

$$d(\mu F^{-1}, \mu_h F^{-1}) \le \delta \tag{5.4.27}$$

for all $h \le h_0$.

To prove (5.4.27) we use the following:

Lemma 5.4.4 *Let $\bar{x}^u$ be the solution of (5.4.4). Then*

$$\lim_{s \to 0} \sup_{u \in \mathcal{U}} \text{Var}(\mathcal{L}(\bar{x}_{T-s}^u) - \mathcal{L}(\bar{x}_T^u)) = 0. \tag{5.4.28}$$

Proof. For any $u \in \mathcal{U}$ let

$$\theta_t^r := \bar{x}_t^u I_{[0, T-r]}(t) + (\bar{x}_{T-r}^u w_t - w_{T-r}) I_{[T-r, T]}(t).$$

It follows from the bound for the total variation distance in terms of the Hellinger process h_t (see Appendix A.3) that

$$\text{Var}(\mathcal{L}(\bar{x}^u) - \mathcal{L}(\theta^r)) \le Cr^{1/2}. \tag{5.4.29}$$

This bound holds since in the considered situation the Hellinger process for the pair $(\mathcal{L}(\bar{x}^u), \mathcal{L}(\theta^r))$ has the form

$$h_t = \int_0^t I_{[T-r, T]}(\tau)|A_0(\tau, W, \hat{u}_\tau)|^2 d\tau.$$

Fix $\gamma > 0$ and $r > 0$ such that $Cr^{1/2} \le \gamma$. For any $s \in [0, r[$ we have

$$\mathcal{L}(\theta_{T-s}^r) = \mathcal{L}(\bar{x}_{T-r}^u) * \mathcal{N}(0, (r-s)I)$$

where $*$ denotes the convolution, and $\mathcal{N}(0, (r-s)I)$ is the Gaussian distribution with zero mean and covariance matrix $(r-s)I$. In particular,

$$\mathcal{L}(\theta_T^r) = \mathcal{L}(\bar{x}_{T-r}^u) * \mathcal{N}(0, rI).$$

The well-known inequality

$$\text{Var}(F * G - F * \tilde{G}) \le \text{Var}(G - \tilde{G})$$

implies that

$$\mathrm{Var}(\mathcal{L}(\theta^r_{T-s}) - \mathcal{L}(\theta^r_T)) \leq \mathrm{Var}(\mathcal{N}(0, (r-s)I) - \mathcal{N}(0, rI))$$

where the right-hand side tends to zero as $s \to 0$.

Thus, for sufficiently small s we have

$$\sup_{u \in \mathcal{U}} \mathrm{Var}(\mathcal{L}(\theta^r_{T-s}) - \mathcal{L}(\theta^r_T)) \leq \gamma. \tag{5.4.30}$$

It follows from (5.4.29) and (5.4.30) that

$$\sup_{u \in \mathcal{U}} \mathrm{Var}(\mathcal{L}(\bar{x}^u_{T-s}) - \mathcal{L}(\bar{x}^u_T)) \leq 3\gamma$$

and the lemma is proved. $\square$

Since

$$\mathrm{Var}(\mu F^{-1} - \mu_h F^{-1}) = \mathrm{Var}(\mu - \mu_h) \leq \mathrm{Var}(\nu - \nu_h) \to 0$$

by virtue of the above lemma, the relation (5.4.27) holds.

Furthermore, there exists $h_1 > 0$ such that for $h \leq h_1$ we have

$$\sup_{\varepsilon} \sup_{z \in \mathcal{U}_h(u)} d(\mathcal{L}(x^{\varepsilon,z}_{T-h}, y^{\varepsilon,z}_T), \mathcal{L}(x^{\varepsilon,z}_T, y^{\varepsilon,z}_T)) \leq \delta \tag{5.4.31}$$

where $\mathcal{U}_h(u)$ is the set consisting of all $z \in \mathcal{U}$ such that

$$z I_{[0,T-h]} = u I_{[0,T-h]}. \tag{5.4.32}$$

The relation (5.4.31) is an obvious corollary of Proposition 5.4.3 and the following:

Lemma 5.4.5 Let $(\xi^{(i)}_{\iota,h})$, $\iota \in I(h)$, $h \in [0, T]$, $i = 1, 2$, be two families of random variables with values in $\mathbf{R}^m$ such that

$$\sup_h \sup_{\iota \in I(h)} E|\xi^{(i)}_{\iota,h}|^p < \infty, \quad i = 1, 2,$$

$$\lim_{h \to 0} \sup_{\iota \in I(h)} E|\xi^{(1)}_{\iota,h} - \xi^{(2)}_{\iota,h}|^p = 0$$

for some $p \geq 1$. Then for any bounded continuous function f in $\mathbf{R}^m$

$$\lim_{h \to 0} \sup_{\iota \in I(h)} |Ef(\xi^{(1)}_{\iota,h}) - Ef(\xi^{(2)}_{\iota,h})| = 0.$$

The easy proof of Lemma 5.4.5 is omitted.

Lemma 5.4.5 implies also the existence of $h_2 > 0$ such that for all $h \leq h_2$

$$\sup_\iota d(\mathcal{L}(\bar{x}^u_{T-h}, -A_4(T)A_3(T)\bar{x}^u_{T-h}+\eta_\iota), \mathcal{L}(\bar{x}^u_{T-h}, -A_4(T)A_3(T)\bar{x}^u_T+\eta_\iota)) \leq \delta$$

$$(5.4.33)$$

where the family (η_ι) consists of all random variables with distribution from the set S_Y.

Let us consider some $h \leq h_0 \wedge h_1 \wedge h_2$. The desired result will be proved if we find for any sufficiently small ε an admissible control $z = z^\varepsilon$ satisfying (5.4.32) such that

$$d(\mathcal{L}(x^{\varepsilon,z}_{T-h}, y^{\varepsilon,z}_T), \mu_h F^{-1}) \leq 2\delta. \qquad (5.4.34)$$

Indeed, it follows from (5.4.27), (5.4.31), and (5.4.34) that

$$d(\mathcal{L}(x^{\varepsilon,z}_T, y^{\varepsilon,z}_T), \mu_h F^{-1}) \leq 4\delta$$

and this means that any point in $\mathcal{K}_0(T)$ can be approximated by points from $\mathcal{K}_\varepsilon(T)$.

Let $(\Omega, \mathcal{F}, P)$ be a probability space with a countably generated σ-algebra and independent random elements ζ, α, ξ where ζ has the distribution ν_h, i.e. the same distribution as $\bar{x}^u_{T-h}$, α has the standard normal distribution, ξ is a stationary Gaussian Markov process with zero mean and covariance function given by (5.4.7), (5.4.8). Let us consider the set $\mathcal{V}^\alpha_U$ of U-valued processes which are predictable with respect to the filtration generated by $\xi_{1/t}$ and α (we denote by $\mathcal{P}$ the corresponding predictable σ-algebra in $\Omega \times \mathbf{R}_+$).

Lemma 5.4.6 *There is a function $v : \Omega \times \mathbf{R}_+ \times \mathbf{R}^m \to U$ which is measurable with respect to $\mathcal{P} \otimes \mathcal{B}(\mathbf{R}^m)$ such that $v(.,x) \in \mathcal{V}^\alpha$ for all $x \in \mathbf{R}^m$ and the law $\mathcal{L}(\xi_0 + I(v(.,x)))$ is equal to $m(x,dy)$ for ν_h-almost all $x \in \mathbf{R}^m$.*

Proof. Evidently, $v \mapsto \mathcal{L}(\xi_0 + I(v))$ is a continuous, hence, measurable mapping from the space $\mathbf{V} := L^1(\Omega \times \mathbf{R}_+, \mathcal{P}, \rho)^d$ into $\mathbf{P}(\mathbf{R}^n)$ where

$$\rho(d\omega, dt) = e^{-2\kappa t} P(d\omega)dt.$$

Thus, the multivalued mapping

$$\Gamma : x \mapsto \{v \in \mathbf{V} : v(\omega, t) \in U \ \rho\text{-a.e.}, \ \mathcal{L}(\xi_0 + I(v)) = m(x,.)\}$$

has the measurable graph. Hence, it admits a measurable selector $x \mapsto V(x)$. Notice that $V(x)$, as an element of $\mathbf{V}$, is a class of ρ-equivalent functions. To choose from $V(x)$ a representative in a measurable way we proceed as follows. Let (v^i) be a sequence of elements from the set $\mathcal{V}^\alpha_U$ which is dense in $\mathcal{V}^\alpha_U \cap \mathbf{V}$ and let

$$j(x,l) := \min\{i : \| v(x) - v^i \| \leq 1/l\}.$$

Then $v^{j(l)} = v^{j(x,l)}(\omega, t)$ is a $\mathcal{P} \otimes \mathcal{B}(\mathbf{R}^m)$-measurable function with values in U. The sequence $v^{j(x,l)}$ converges to $V(x)$ in $\mathbf{V}$. Since U is bounded, the sequence $v^{j(l)}$ converges to V in $L^1(\Omega \times \mathbf{R}_+ \times \mathbf{R}^m, \mathcal{P} \otimes \mathcal{B}(\mathbf{R}^m), \rho \times \nu_h)^d$. Hence, there exists a subsequence which converges $\rho \times \nu_h$-a.e. to some $\mathcal{P} \otimes \mathcal{B}(\mathbf{R}^m)$-measurable function $v = v(\omega, t, x)$. For ν_h-almost all x we have the inclusion $v(.,x) \in V(x)$ implying that $\mathcal{L}(\xi_0 + I(v(.,x))) = \mu(x, dy)$ for such x. $\square$

It follows from the above lemma that the measure μ_h is the distribution of the random variable $(\zeta, \xi_0 + I(v(.,\zeta)))$, i.e.

$$\mu_h = \mathcal{L}(\zeta, \xi_0 + I(v(.,\zeta))). \tag{5.4.35}$$

Generalizing the arguments of the one-dimensional case, we introduce the set $V_U^{(\alpha,\zeta)'}$ consisting of all functions

$$v(s,x) = \sum_{i=1}^{N} \varphi_i(x) I_{]s_i, s_{i+1}]}(s) + u^0 I_{]s_{N+1}, \infty[}(s), \tag{5.4.36}$$

where $0 = s_1 < \ldots < s_{N+1}$, $u^0 \in U$, and $\varphi_i(x)$ have the form

$$\varphi_i(x) = f_i(\alpha, \xi(r_1^i), \ldots, \xi(r_{M_i}^i), x), \quad s_{i+1} < r_j^i \leq s_N, \tag{5.4.37}$$

functions f_i are measurable with respect to their arguments and take values in U.

Assume that the representation (5.4.9) holds with $v \in V_U^{(\alpha,\zeta)'}$. There is a freedom in the choice of ζ, α, and ξ which we use in the following constructions.

Put $T_\varepsilon^k := T(1 - k\varepsilon^{1/2})$, $k = 1, 2, 3$, $\zeta := \bar{x}_{T-h}^u$.

Define

$$\alpha^\varepsilon := (w_{T_\varepsilon^2}^{y,1} - w_{T_\varepsilon^3}^{y,1})/(T_\varepsilon^2 - T_\varepsilon^3)^{1/2}$$

where $w^{y,1}$ is the first component of the vector process w^y,

$$\rho^\varepsilon := \Xi^{1/2}(w_{T_\varepsilon^1}^y - w_{T_\varepsilon^2}^y)/(T_\varepsilon^1 - T_\varepsilon^2)^{1/2}.$$

Let us consider on $[T_\varepsilon^1, T]$ the linear stochastic differential equation

$$\varepsilon d\tilde{\xi}_t^\varepsilon = A_4(T)\tilde{\xi}_t^\varepsilon dt + \varepsilon^{1/2} dw_t^y, \quad \tilde{\xi}_{T_\varepsilon^1}^\varepsilon = \rho^\varepsilon.$$

Put $\xi_t^\varepsilon := \tilde{\xi}_{T-\varepsilon t}^\varepsilon$, $t \in [0, \varepsilon^{-1/2}T]$. For sufficiently small ε we define the admissible control

$$z^\varepsilon := u I_{[0, t_{N+1}[} + \sum_{i=1}^{N+1} \varphi_i^\varepsilon(\bar{x}_{T-h}^u) I_{[t_{i+1}, t_i[}$$

where $t_i := T - \varepsilon s_i$, $i \leq N+1$, and φ_i^ε is constructed by in accordance with (5.4.37).

It follows from Proposition 5.4.3 that

$$(x_{T-h}^{\varepsilon, z^\varepsilon}, y_T^{\varepsilon, z^\varepsilon}) - (\bar{x}_{T-h}^u, -A_4(T)A_3(T)\bar{x}_T^u + \xi_0^\varepsilon + I(v(., \bar{x}_{T-h}^u))) \to 0$$

in probability as $\varepsilon \to 0$. Thus,

$$d(\mathcal{L}(x_{T-h}^{\varepsilon,z^\varepsilon}, y_T^{\varepsilon,z^\varepsilon}), \mathcal{L}(\bar{x}_{T-h}^u, -A_4(T)A_3(T)\bar{x}_T^u + \xi_0^\varepsilon + I(v(., \bar{x}_{T-h}^u)))) \leq \delta$$

for all sufficiently small ε. Taking into account (5.4.33) we get from here the desired inequality (5.4.34).

The second step of Theorem 5.4.1, consisting in approximation of points of the limiting set, is done now for the case when μ_h is given by (5.4.35) with $v \in \mathcal{V}_U^{(\alpha,\varsigma)'}$. Since the set $\{I(v) : v \in \mathcal{V}_U^{(\alpha,\varsigma)'}\}$ is dense in probability in the set $\{I(v) : v \in \mathcal{V}_U^{\alpha,\varsigma}\}$, the result holds for the general case as well. $\square$

6 Applications

6.1 Applications to PDEs

It is well-known that various results for stochastic differential equations can be translated, via probabilistic representations, to results for PDEs and vice versa, enriching both theories. In this section we apply our stochastic Tikhonov theorem to a study of asymptotics of boundary-value problems for the second-order PDEs with small parameters.

Cauchy problem. Let us consider the initial-value problem in $\mathbf{R}^{k+n}$

$$\frac{\partial v^\varepsilon}{\partial t}(t, z) = \mathcal{L}^\varepsilon v^\varepsilon(t, z) + r(t, z)v^\varepsilon(t, z) + h(t, z), \quad t > 0, \qquad (6.1.1)$$

$$v^\varepsilon(0, z) = b(z), \qquad (6.1.2)$$

with the elliptic differential operator

$$\mathcal{L}^\varepsilon := \sum_{i,j=1}^{k+n} d_\varepsilon^{ij}(t, z)\frac{\partial^2}{\partial z_i \partial z_j} + \sum_{i=1}^{k} f(t, z)\frac{\partial}{\partial z_i} + \frac{1}{\varepsilon}\sum_{i=k+1}^{n+k} F_i(t, z)\frac{\partial}{\partial z_i}, \qquad (6.1.3)$$

where

$$D_\varepsilon(t, z) := (d_\varepsilon^{ij}(t, z)) = \begin{bmatrix} g(t, z) \\ \sigma(\varepsilon)\varepsilon^{-1}G(t, z) \end{bmatrix} [g^\star(t, z), \sigma(\varepsilon)\varepsilon^{-1}G^\star(t, z)].$$

We shall exploit here the probabilistic representation of the above problem using the family of solutions $Z^{z,\varepsilon} = (X^{z,\varepsilon}, Y^{z,\varepsilon})$ of the two-scale stochastic system

$$dX_t^{z,\varepsilon} = f(t, X_t^{z,\varepsilon}, Y_t^{z,\varepsilon})dt + g(t, Y_t^{z,\varepsilon})dw_t^X, \qquad (6.1.4)$$

$$\varepsilon dY_t^{z,\varepsilon} = F(t, X_t^{z,\varepsilon}, Y_t^{z,\varepsilon})dt + \sigma(\varepsilon)G(t, Y_t^{z,\varepsilon})dw_t^Y, \qquad (6.1.5)$$

where the superscript z reveals the dependence of a solution of the initial point $Z_0^{z,\varepsilon} = z = (x, y)$.

We impose on the coefficients the assumptions **H.2.1.1–H.2.5**; in particular, we suppose that $\sigma(\varepsilon) = o(\sqrt{\varepsilon}/|\ln\varepsilon|)$. Moreover, we modify **H.2.1.3** by suggesting that for any x all points y (in the notation of this section) belong to the domain of the influence of the unique root $\varphi(t, x)$ of the equation $F(t, x, \varphi(t, x)) = 0$.

Regarding the coefficients of the problem (6.1.1), (6.1.2) we assume that
the scalar functions r and h are uniformly continuous and bounded in $\mathbf{R}^n$, b
is continuous and bounded, and there is a positive constant C_ε such that

$$C_\varepsilon^{-1}||\lambda||^2 \leq \lambda^\star D(t,z)\lambda \leq C_\varepsilon ||\lambda||^2 \quad \forall\, z, \lambda \in \mathbf{R}^{k+n}. \tag{6.1.6}$$

Recall that by **H.2.1.1** the functions F and G satisfy the linear growth and
local Lipschitz conditions.

It is well-known that under such assumptions the solution $v^\varepsilon(t,x)$ of the
Cauchy problem (6.1.1), (6.1.2) exists, is unique, and admits the probabilistic
representation

$$v^\varepsilon(t,z) = Eb(Z_t^{z,\varepsilon})\exp\left\{\int_0^t r(s, Z_s^{z,\varepsilon})ds\right\}$$

$$+ E\int_0^t h(s, Z_s^{z,\varepsilon})\exp\left\{\int_0^s r(u, Z_u^{z,\varepsilon})du\right\}ds. \tag{6.1.7}$$

We consider also the initial-value problem

$$\frac{\partial v}{\partial t}(t,x) = \mathcal{L}v(t,z) + r(t,\varphi(t,x))v(t,z) + h(t,\varphi(t,x)), \quad t > 0, \tag{6.1.8}$$

$$v(0,x) = b(x, \varphi(0,x)), \tag{6.1.9}$$

with

$$\mathcal{L} := \sum_{i,j=1}^k d^{ij}(t,x,\varphi(t,x))\frac{\partial^2}{\partial x_i \partial x_j} + \sum_{i=1}^k f_i(t,x,\varphi(t,x))\frac{\partial}{\partial x_i}. \tag{6.1.10}$$

Its probabilistic representation is given by

$$v(t,x) = Eb(X_t^x)\exp\left\{R_t^x\right\} + E\int_0^t h(s, X_s^x, \varphi(s, X_s^x))\exp\left\{R_s^x\right\}ds, \tag{6.1.11}$$

where

$$dX_t^x = f(t, X_t^x, \varphi(t, X^x, t))dt + g(t, X_t^x, \varphi(t, X^x, t))dw^X, \quad X_0^x = x, \tag{6.1.12}$$

and

$$R_t^x := \int_0^t r(s, X_s^x, \varphi(s, X_s^x))ds.$$

Theorem 6.1.1 *Under the above assumptions for any $t > 0$*

$$\lim_{\varepsilon \to 0} v^\varepsilon(t,x) = v(t,x). \tag{6.1.13}$$

Proof. The claim follows from the probabilistic representations by a straight-forward application of Theorem 2.1.1. □

The above result has an especially simple form for the problem with time-homogeneous coefficients and $k = 0$, i.e. when there is no term corresponding to slow variables in the probabilistic representation of the initial value problem in $\mathbf{R}^n$:

$$\frac{\partial v^\varepsilon}{\partial t}(t, y) = \mathcal{L}^\varepsilon v^\varepsilon(t, y) + r(y)v^\varepsilon(t, y) + h(y), \quad t > 0, \tag{6.1.14}$$

$$v^\varepsilon(0, y) = b(y), \tag{6.1.15}$$

where

$$\mathcal{L}^\varepsilon := \frac{\sigma^2}{\varepsilon^2} \sum_{i,j}^n d^{ij}(y) \frac{\partial^2}{\partial y_i \partial y_j} + \frac{1}{\varepsilon} \sum_{i=1}^n F_i(y) \frac{\partial}{\partial y_i}, \tag{6.1.16}$$

and $D(y) := G(y)G^\star(y)$. The corresponding "fast" variables follow the SDEs

$$\varepsilon dY_t^{y,\varepsilon} = F(Y_t^{y,\varepsilon})dt + \sigma(\varepsilon)G(Y_t^{y,\varepsilon})dw_t^Y, \quad Y_0^{y,\varepsilon} = y. \tag{6.1.17}$$

In this case we have

Theorem 6.1.2 *Under the above assumptions, for any $t > 0$*

$$\lim_{\varepsilon \to 0} v^\varepsilon(t, x) = \begin{cases} b_\infty e^{r_\infty t} + \frac{h_\infty}{r_\infty}(e^{r_\infty t} - 1), & r_\infty \neq 0, \\ b_\infty + h_\infty t, & r_\infty = 0, \end{cases} \tag{6.1.18}$$

where $b_\infty := r(\tilde{Y}_\infty)$, $r_\infty := r(\tilde{Y}_\infty)$, and $h_\infty := h(\tilde{Y}_\infty)$ with $\tilde{Y}_\infty$ denoting the unique root of the equation $F(y) = 0$.

Dirichlet problem. Let us consider in a bounded region V of $\mathbf{R}^n$ with the smooth boundary ∂V the Dirichlet problem for the elliptic equation

$$\mathcal{L}^\varepsilon u^\varepsilon(y) + r(y)u^\varepsilon(y) = h(y), \quad u^\varepsilon(y)|_{\partial V} = b(y), \tag{6.1.19}$$

where $\mathcal{L}^\varepsilon$ is given by (6.1.16).

We say that the trajectory $\tilde{Y}^y$ of the equation $d\tilde{Y}_s^y = F(\tilde{Y}_s^y)ds$ with $\tilde{Y}_0^y = y \in V$ leaves V in a *regular manner* if $T(y) := \inf\{s : \tilde{Y}_s^y \notin V\} < \infty$ and $\tilde{Y}_{T(y)+\delta}^y \notin V$ for all sufficiently small $\delta > 0$.

Theorem 6.1.3 *In addition to the previous assumptions, suppose that $\tilde{Y}_\infty \notin \partial V$ and $r(y) < 0$ for all $y \in V \cup \partial V$. If the trajectory $\tilde{Y}^y$ does not leave V, then*

$$\lim_{\varepsilon \to 0} u^\varepsilon(y) = -\frac{h_\infty}{r_\infty}. \tag{6.1.20}$$

If the trajectory $\tilde{Y}^y$ leaves V in a regular manner, then

$$\lim_{\varepsilon \to 0} u^\varepsilon(y) = b(\tilde{Y}_{T(y)}^y). \tag{6.1.21}$$

Proof. Notice that Theorem 2.2.1 implies that for any $t > 0$

$$P\text{-}\lim_{\varepsilon \to 0} \|\tilde{Y}^{y,\varepsilon} - \tilde{Y}^y\|_{t/\varepsilon} = 0 \tag{6.1.22}$$

where

$$d\tilde{Y}_t^{y,\varepsilon} = F(\tilde{Y}^{y,\varepsilon})dt + \sigma(\varepsilon)\varepsilon^{-1/2}G(\tilde{Y}^{y,\varepsilon})d\tilde{w}_t^Y, \qquad \tilde{Y}_0^{y,\varepsilon} = 0,$$

and $\tilde{w}_t^Y = \varepsilon^{-1/2}w_{t\varepsilon}^Y$ is a Wiener process.

Our assumptions ensure that for any $\varepsilon > 0$ there exists a unique solution of the boundary value problem (6.1.19) and, moreover, it admits the probabilistic representation

$$u^\varepsilon(y) = Eb(Y_{\tau_\varepsilon}^{y,\varepsilon})e^{R_{\tau_\varepsilon}^{y,\varepsilon}} - E\int_0^{\tau_\varepsilon} h(Y_t^{y,\varepsilon})e^{R_t^{y,\varepsilon}}dt \tag{6.1.23}$$

where

$$R_t^{y,\varepsilon} := \int_0^t r(Y_s^{y,\varepsilon})ds,$$

$y \in V$ and $\tau_\varepsilon := \tau_\varepsilon(y) := \inf\{t \geq 0 : Y_t^{y,\varepsilon} \notin V\}$ is the exit time.

At first, we consider the case where $T(y) = \infty$, i.e. the trajectory does not leave V. Since $\tilde{Y}_\infty \notin \partial V$, we have $\tilde{Y}_\infty \in V$. Using **H.2.2.3** it is easy to show that

$$\mu := \inf_{t \geq 0} \rho(\tilde{Y}_t^y, \partial V) > 0,$$

where ρ is the Euclidean distance.

Furthermore, for any $t > 0$ we have

$$P(\tau_\varepsilon < t) = P(Y_{\tau_\varepsilon}^{y,\varepsilon} \notin V, \ \tau_\varepsilon \in \,]0,t[) \leq P(|\tilde{Y}_{\tau_\varepsilon/\varepsilon}^{y,\varepsilon} - \tilde{Y}_{\tau_\varepsilon/\varepsilon}^y| > \mu/2, \ \tau_\varepsilon \in \,]0,t[)$$

$$\leq P(\|\tilde{Y}^{y,\varepsilon} - \tilde{Y}^y\|_{t/\varepsilon} > \mu/2).$$

Applying (6.1.22) we get from here for any $t > 0$ that

$$\lim_{\varepsilon \to 0} P(\tau_\varepsilon < t) = 0. \tag{6.1.24}$$

Let $\Delta^{y,\varepsilon} := u^\varepsilon(y) + h_\infty/r_\infty$. The probabilistic representation (6.1.23) yields:

$$|\Delta^{y,\varepsilon}| \leq E|b(Y_{\tau_\varepsilon}^{y,\varepsilon})|e^{R_{\tau_\varepsilon}^{y,\varepsilon}} + \left| E\int_0^{\tau_\varepsilon} h(Y_t^{y,\varepsilon})e^{R_t^{y,\varepsilon}}dt - \int_0^\infty h_\infty e^{r_\infty t}dt \right|$$

$$\leq b_0 Ee^{-r_0\tau_\varepsilon} + 2h_0 P(\tau_\varepsilon < S)\int_0^\infty e^{-r_0 t}dt + 2h_0 \int_S^\infty e^{-r_0 t}dt$$

$$+ \left| E\int_0^S h(Y_t^{y,\varepsilon})e^{R_t^{y,\varepsilon}}dt - \int_0^S h_\infty e^{r_\infty t}dt \right|$$

where

$$h_0 := \sup_{z \in V \cup \partial V} |h(z)|, \quad b_0 := \sup_{z \in V \cup \partial V} |b(z)|, \quad r_0 := -\sup_{z \in V \cup \partial V} r(z).$$

Taking in the resulting inequality successively the limits with $\varepsilon \to 0$ and $S \to \infty$ and using (2.2.3), (2.2.6), and (6.1.24) we obtain (6.1.20).

Before the proof of the second part of Theorem 6.1.3 we check that

$$P\text{-}\lim_{\varepsilon \to 0} \tau_\varepsilon / \varepsilon = T(y) \tag{6.1.25}$$

for any $y \in V$. Indeed, for sufficiently small $\nu > 0$ we have

$$\tilde{Y}^y_{T(y)-\nu} \in V, \quad \tilde{Y}^y_{T(y)+\nu} \notin V \cup \partial V. \tag{6.1.26}$$

Let $\tilde{\tau}_\varepsilon := \tau_\varepsilon / \varepsilon$,

$$\lambda_1 := \inf_{t \le T(y)-\nu} \rho(\tilde{Y}^y_t, \partial V), \quad \lambda_2 := \rho(\tilde{Y}^y_{T(y)+\nu}, \partial V);$$

then $\lambda_0 := \lambda_1 \wedge \lambda_2 > 0$. Put also $\Gamma^\varepsilon := \{\|\tilde{Y}^{y,\varepsilon} - \tilde{Y}^y\|_{T(y)+\nu} > \lambda_0/2\}$. It follows from the definitions that

$$\{\tilde{\tau}_\varepsilon < T(y) - \nu\} = \{\tilde{Y}^y_{\tilde{\tau}_\varepsilon} \in V, \ \tilde{Y}^{y,\varepsilon}_{\tilde{\tau}_\varepsilon} \notin V, \ \tilde{\tau}_\varepsilon < T(y) - \nu\}$$

$$\subseteq \{|\tilde{Y}^{y,\varepsilon}_{\tilde{\tau}_\varepsilon} - \tilde{Y}^y_{\tilde{\tau}_\varepsilon}| > \lambda_0/2, \ \tilde{\tau}_\varepsilon < T(y) - \nu\} \subseteq \Gamma^\varepsilon,$$

$$\{\tilde{\tau}_\varepsilon > T(y) + \nu\} \subseteq \{\tilde{Y}^{y,\varepsilon}_{T(y)+\nu} \in V\} \subseteq \Gamma^\varepsilon.$$

Thus, for sufficiently small ν

$$P(|\tau_\varepsilon - T(y)| > \nu) \le 2P(\Gamma^\varepsilon)$$

and (6.1.25) follows by virtue of (6.1.22).

The result follows from (6.1.23) by (6.1.22), (6.1.25), and the Lebesgue theorem on dominated convergence. $\square$

Let us consider the case where $r(y)$ is an arbitrary continuous function not necessary strictly negative. In general, a solution of the problem (6.1.9) may not exist. However, the following condition provides its existence and uniqueness as well as the probabilistic representation (6.1.9):

$$Ee^{r_1 \tau_\varepsilon} < \infty \tag{6.1.27}$$

where $r_1 := \sup_{y \in V \cup \partial V} |r(y)|$; see [27].

To show that τ_ε has exponential moments for sufficiently small ε we consider the random variable

$$\tilde{\tau}_\varepsilon := \tilde{\tau}_\varepsilon(y) := \inf\{t \ge 0 : \ \tilde{Y}^{y,\varepsilon}_t \notin V\}.$$

which is a stopping time with respect to the filtration generated by the Wiener process $\tilde{w}^Y_t = \varepsilon^{-1/2} w^Y_{t\varepsilon}$ and $\tilde{\tau}_\varepsilon = \tau_\varepsilon / \varepsilon$.

Lemma 6.1.4 *Let us suppose that for all $y \in V$ the trajectory $\tilde{Y}^y$ leaves V in a regular manner, $\sup_{y \in V} T() \leq T_0/2$ for some $T_0 < \infty$ and there is $\nu > 0$ such that*

$$\sup_{T(y) \leq t \leq T(y)+\nu} \rho(\tilde{Y}_t^y, V \cup \partial V) \geq \mu > 0. \qquad (6.1.28)$$

Then for any $\lambda > 0$ there exist $A = A(\lambda)$ and $\varepsilon_0 = \varepsilon_0(\lambda)$ such that

$$\sup_{y \in V} E e^{\lambda \tilde{\tau}_\varepsilon(y)} \leq A(\lambda) < \infty \qquad (6.1.29)$$

for all $\varepsilon \leq \varepsilon_0$.

Proof. By the continuity theorem for stochastic differential equations for arbitrary numbers T and $\mu > 0$

$$\lim_{\varepsilon \to 0} \sup_{y \in V} P(\|\tilde{Y}^{y,\varepsilon} - \tilde{Y}^y\|_T > \mu) = 0 \qquad (6.1.30)$$

(in fact, this relation can be easily derived directly by using the Gronwall–Bellman lemma). Choose a number $\theta \in]0,1[$. Let us show that there is $\varepsilon_0 > 0$ such that

$$\sup_{y \in V} P(\tilde{\tau}_\varepsilon(y) > T_0) \leq \theta \qquad (6.1.31)$$

for all $\varepsilon \in]0, \varepsilon_0]$. Indeed, it follows from the hypothesis (6.1.28) that there is a point $t_1 = t_1(y)$ in the interval $[T(y), T(y) + \nu]$ such that

$$\rho(\tilde{Y}_{t_1}^y, V \cup \partial V) > \mu/2.$$

Without loss of generality we can assume that $T_0 > \nu/2$. Then

$$P(\tilde{\tau}_\varepsilon(y) > T_0) \leq P(\tilde{\tau}_\varepsilon(y) > t_1(y)) \leq P(\tilde{Y}_{t_1}^{y,\varepsilon} \in V, \ \tilde{Y}_{t_1}^{y,\varepsilon} \in V).$$

Therefore, by virtue of (6.1.31) we have

$$P(\tilde{\tau}_\varepsilon(y) > T_0) \leq P(\|\tilde{Y}^{y,\varepsilon} - \tilde{Y}^y\|_{T_0} > \mu/2)$$

and the assertion (6.1.22) follows from (6.1.21).

Since $Y^{y,\varepsilon}$ is a homogeneous Markov process we have that

$$\sup_{y \in V} P(\tilde{\tau}_\varepsilon(y) > nT_0 \mid \tilde{\tau}_\varepsilon(y) > (n-1)T_0) \leq \sup_{y \in V} P(\tilde{\tau}_\varepsilon(y) > T_0) \leq \theta.$$

Hence,

$$\sup_{y \in V} P(\tilde{\tau}_\varepsilon(y) > nT_0) \leq \theta \sup_{y \in V} P(\tilde{\tau}_\varepsilon(y) > (n-1)T_0) \leq \theta^n.$$

If we choose $\theta < e^{-\lambda T_0}$ then for all $y \in V$ and $\varepsilon \in]0, \varepsilon_0]$

$$\sup_{y \in V} E e^{\lambda \tilde{\tau}_\varepsilon(y)} \leq \sum_{n=0}^\infty e^{(n+1)\lambda T_0} P(\tilde{\tau}_\varepsilon(y) > nT_0)$$

$$\leq e^{\lambda T_0} \sum_{n=0}^\infty \left(e^{\lambda T_0} \theta \right)^n = A(\lambda) < \infty$$

and the lemma is proved. $\square$

Notice that the bound (6.1.29) implies that for any $m \geq 1$ there is a constant $L(m)$ such that

$$E \tilde{\tau}_\varepsilon^m \leq L(m)$$

for sufficiently small ε (see [27]). Therefore,

$$E \tau_\varepsilon^m \leq \varepsilon^m L(m). \tag{6.1.32}$$

Theorem 6.1.5 *Assume that the coefficients of problem (6.1.19) satisfy the conditions* **H.2.2.1** *and* **H.2.1.5**, *(6.1.6) holds, the domain V is bounded and has a smooth boundary ∂V. Assume also that for all $y \in V$ the trajectories $\tilde{Y}^y$ leave V in a regular manner, $\sup_{y \in V} T(y) \leq T_0 < \infty$ and for some $\nu > 0$*

$$\inf_{y \in V} \sup_{T(y) \leq t \leq T(y)+\nu} \rho(\tilde{Y}_t^y, V \cup \partial V) > 0.$$

Then for any continuous function $r(y)$ on $y \in V \cup \partial V$ for sufficiently small $\varepsilon > 0$ the problem (6.1.19) has the unique solution and

$$\lim_{\varepsilon \to 0} u^\varepsilon(y) = b(\tilde{Y}_{T(y)}^y). \tag{6.1.33}$$

Proof. By Lemma 6.1.4 for

$$E e^{r_1 \tau_\varepsilon} \leq E e^{r_1 \varepsilon \tilde{\tau}_\varepsilon} \leq E e^{r_1 \tilde{\tau}_\varepsilon} < \infty$$

for sufficiently small ε. Hence, (6.1.27) holds and the result follows from the probabilistic representation (6.1.19) and (6.1.25). $\square$

6.2 Fast Markov Modulations Revisited

6.2.1 Main Result

We return here to the model with fast Markov switchings introduced in Section 0.1 and show that $\sqrt{\varepsilon}$ is the exact rate of convergence for the total variation distance as $\varepsilon \to 0$ in the case of smooth Q. More precisely, we prove the following result.

Theorem 6.2.1 *Assume that $Q = (Q_t)$ is a continuously differentiable function and the initial distribution does not depend on ε, i.e. $p^\varepsilon = p$. Then*

$$\lim_{\varepsilon \to 0} \varepsilon^{-1/2} \mathrm{Var} \left(P_T^\varepsilon - R_T \right) = \sigma \sqrt{2/\pi} \qquad (6.2.1)$$

where

$$\sigma^2 := \int_0^\infty \left(\lambda^\star e^{Q_0^\star r} (p - \pi_0) \right)^2 dr + \int_0^T \lambda^\star M_t \lambda \, dt; \qquad (6.2.2)$$

the matrix M_t (depending on λ) is given by

$$M_t := \int_0^\infty e^{Q_t^\star r} \phi(\pi_t) \phi^\star(\pi_t) e^{Q_t r} \, dr \qquad (6.2.3)$$

with $\phi(x) := \mathrm{diag}\, \lambda\, x - x(\lambda^\star x)$.

Clearly, the right-hand side of (6.2.1) is nothing but $E|\eta|$ where $\eta \sim N(0, \sigma^2)$. The first term on the right-hand side of (6.2.2), related to the boundary layer behavior, depends only on the value of Q at zero and it vanishes if $p = \pi_0$.

Our strategy is the following. First, we recall some simple facts about the convergence in distributions. Afterwards, we prove the result postulating hypotheses sufficient for the desired convergence of the distributions of diffusion-type processes. The concluding part of the section contains rather tedious calculations and estimates to verify the properties needed; it relies on the moment bound for solutions of singularly perturbed stochastic equations given by Proposition 1.2.4.

6.2.2 Preliminaries from Weak Convergence

We begin with a lemma about the weak convergence of one-dimensional distributions of continuous local martingales.

Lemma 6.2.2 *Let $Y^\varepsilon = (Y_t^\varepsilon)$, $\varepsilon \in]0, 1]$, be a family of continuous real-valued martingales starting from zero. Assume that $\langle Y^\varepsilon \rangle_T \to \sigma^2$ in probability as $\varepsilon \to 0$, where $\sigma^2 > 0$ is a constant. Then the distribution of Y_T^ε converges weakly to $N(0, \sigma^2)$.*

Proof. Assume for a moment that the family of random variables $\langle Y^\varepsilon \rangle_T$ is bounded by a constant. Then for every $u \in \mathbf{R}$ we have

$$E e^{iu Y_T^\varepsilon + (1/2)u^2 \langle Y^\varepsilon \rangle_T} = 1$$

and, using this "representation of unit",

$$|E e^{iu Y_T^\varepsilon + (1/2)u^2 \sigma^2} - 1| \leq E |e^{iu Y_T^\varepsilon + (1/2)u^2 \sigma^2} - e^{iu Y_T^\varepsilon + (1/2)u^2 \langle Y^\varepsilon \rangle_T}|$$

$$\leq E |e^{(1/2)u^2 \sigma^2} - e^{(1/2)u^2 \langle Y^\varepsilon \rangle_T}| \to 0, \qquad \varepsilon \to 0.$$

Thus, $Ee^{iuY_T^\varepsilon} \to e^{-(1/2)u^2\sigma^2}$ and the result follows. In the general case we define the truncated process $\tilde{Y}_t^\varepsilon = Y_{t\wedge\tau^\varepsilon}^\varepsilon$ with

$$\tau^\varepsilon := \inf\{t : \langle Y^\varepsilon\rangle_t \geq \sigma^2 + 1\} \wedge T.$$

It remains to notice that $Y_T^\varepsilon = \tilde{Y}_T^\varepsilon + (Y_T^\varepsilon - \tilde{Y}_T^\varepsilon)$ where the first summand converges in distribution to $N(0, \sigma^2)$ as we just proved while the second one converges to zero in probability since $P(\tau^\varepsilon < T) \to 0$. $\square$

Of course, the above assertion is a very particular case of the central limit theorem for semimartingales. The proof of the latter in full generality is rather involved but for continuous processes the method of stochastic exponentials yields the result immediately: we prefer to provide arguments instead of a reference.

Lemma 6.2.3 *Let*

$$\rho_T^\varepsilon := \exp\{\varepsilon^{1/2}Y_T^\varepsilon - (1/2)\varepsilon\langle Y^\varepsilon\rangle_T\} \tag{6.2.4}$$

where Y^ε satisfy the hypotheses of Lemma 6.2.2 and let $\vartheta^\varepsilon := \varepsilon^{-1/2}(\rho_T^\varepsilon - 1)$. Assume that

$$\lim_{c\to\infty} \limsup_{\varepsilon\to 0} E|\vartheta^\varepsilon|I_{\{|\vartheta^\varepsilon|\geq c\}} = 0. \tag{6.2.5}$$

Then

$$\lim_{\varepsilon\to 0} \varepsilon^{-1/2}E|\rho_T^\varepsilon - 1| = E|\eta|. \tag{6.2.6}$$

Proof. Using Lemma 6.2.2 we conclude that the random variables

$$\eta^\varepsilon := Y_T^\varepsilon - (1/2)\varepsilon^{1/2}\langle Y^\varepsilon\rangle_T$$

converge in distribution to η (because $\varepsilon^{1/2}\langle Y^\varepsilon\rangle_T \to 0$ in probability). The same limit in distributions have the random variables

$$\vartheta^\varepsilon := \varepsilon^{-1/2}(\rho_T^\varepsilon - 1) = \eta^\varepsilon + \varepsilon^{1/2}\frac{e^{\varepsilon^{1/2}\eta^\varepsilon} - 1 - \varepsilon^{1/2}\eta^\varepsilon}{\varepsilon}.$$

Indeed, the convergence in distribution implies that the family η^ε with ε varying near zero is bounded in probability. This implies that the second summand on the right-hand side of the above decomposition tends to zero in probability.

The result follows since the assumed condition (6.2.5) (close to the uniform integrability) guarantees that the function $|x|$ (continuous but unbounded) can be added to the set of "test" functions (bounded continuous) in the definition of convergence in distribution. $\square$

A tractable sufficient condition ensuring (6.2.5) is given by

Lemma 6.2.4 *Assume that*

$$\limsup_{\varepsilon \to 0} E\langle Y^\varepsilon \rangle_T^2 < \infty \qquad (6.2.7)$$

and

$$\limsup_{\varepsilon \to 0} E e^{p\varepsilon \langle Y^\varepsilon \rangle_T} < \infty \qquad \forall p > 0. \qquad (6.2.8)$$

Then (6.2.5) holds.

Proof. As for the usual uniform integrability, it is sufficient to check that

$$\limsup_{\varepsilon \to 0} \varepsilon^{-1} E(\rho_T^\varepsilon - 1)^2 < \infty. \qquad (6.2.9)$$

Since

$$d\rho_t^\varepsilon = \varepsilon^{1/2} \rho_t^\varepsilon dY_t^\varepsilon, \quad \rho_0^\varepsilon = 1,$$

we have, using the Cauchy–Schwartz and Doob inequalities, that

$$\varepsilon^{-1} E(\rho_T^\varepsilon - 1)^2 = E \int_0^T \rho_t^{\varepsilon 2} \, d\langle Y^\varepsilon \rangle_t \leq E \sup_{t \leq T} \rho_t^{\varepsilon 2} \langle Y^\varepsilon \rangle_T$$

$$\leq (E \sup_{t \leq T} \rho_t^{\varepsilon 4})^{1/2} (E \langle Y^\varepsilon \rangle_T^2)^{1/2} \leq (4/3)(E\rho_T^{\varepsilon 4})^{1/2} (E \langle Y^\varepsilon \rangle_T^2)^{1/2}.$$

Now (6.2.9) is obviously implied by (6.2.7) and the relation

$$\limsup_{\varepsilon \to 0} E\rho_T^{\varepsilon m} < \infty \qquad \forall m > 0. \qquad (6.2.10)$$

The latter follows from (6.2.8) because

$$E\rho_T^{\varepsilon m} = E e^{m\varepsilon^{1/2} Y_T^\varepsilon - m^2 \varepsilon \langle Y^\varepsilon \rangle_T} e^{(1/2)(2m^2 - m)\langle Y^\varepsilon \rangle_T}$$

$$\leq (E e^{2m\varepsilon^{1/2} Y_T^\varepsilon - 2m^2 \varepsilon \langle Y^\varepsilon \rangle_T})^{1/2} (E e^{(2m^2 - m)\varepsilon \langle Y^\varepsilon \rangle_T})^{1/2}$$

$$\leq (E e^{(2m^2 - m)\varepsilon \langle Y^\varepsilon \rangle_T})^{1/2},$$

where we used the standard trick to "hide" the martingale component into the stochastic exponential. $\square$

6.2.3 Proof of Theorem 6.3.1

Taking P_T^ε as a dominating measure we can write that

$$\text{Var}\,(P_T^\varepsilon - R_T) = \int \left| \frac{dR_T}{dP_T^\varepsilon}(x) - 1 \right| P_T^\varepsilon(dx) = E \left| \frac{dR_T}{dP_T^\varepsilon}(X^\varepsilon) - 1 \right|. \qquad (6.2.11)$$

According to [66]

$$\rho_T^\varepsilon := \frac{dR_T}{dP_T^\varepsilon}(X^\varepsilon) = \exp\left\{ -\int_0^T \lambda^\star z_s^\varepsilon \, d\tilde{w}_s - \frac{1}{2}\int_0^T (\lambda^\star z_s^\varepsilon)^2 \, ds \right\}. \quad (6.2.12)$$

where z^ε is the solution of (0.1.17) with $z_0^\varepsilon = z_0 = p - \pi_0$. Setting

$$Y_t^\varepsilon := -\varepsilon^{-1/2} \int_0^t \lambda^\star z_s^\varepsilon \, d\tilde{w}_s, \quad (6.2.13)$$

we come to the notations of Lemma 6.2.3. It remains to verify that

$$\langle Y^\varepsilon \rangle_T := \varepsilon^{-1} \int_0^T (\lambda^\star z_s^\varepsilon)^2 \, ds \to \sigma^2 \quad (6.2.14)$$

in probability with σ^2 given by (6.2.2) and that the hypotheses (6.2.7) and (6.2.8) are fulfilled. This will be done in the next subsection. $\square$

6.2.4 Calculations and Estimates

Let us denote by λ_1 the projection of λ on $\mathcal{L}$. To work with our usual matrix notation we fix a basis in $\mathcal{L}$; all vectors and operators in this subspace can be written in this basis. In particular, A_t is the matrix of the operator Q_t restricted to $\mathcal{L}$. We shall study properties of the solution of (0.1.17) with $z_0^\varepsilon = z_0 = p - \pi_0$. It is convenient to rewrite the representation of z^ε given by the Cauchy formula (0.1.18) as

$$z_t^\varepsilon = v_t^\varepsilon + \varepsilon^{1/2}\xi_t^\varepsilon + \varrho_t^\varepsilon, \quad (6.2.15)$$

where

$$v_t^\varepsilon := \Phi^\varepsilon(t, s)z_0, \quad (6.2.16)$$

$$\xi_t^\varepsilon := \varepsilon^{-1/2} \int_0^t \Phi^\varepsilon(t, s)\phi(\widehat{J}_s^\varepsilon)d\tilde{w}_s, \quad (6.2.17)$$

$$\varrho_t^\varepsilon := \int_0^t \Phi^\varepsilon(t, s)\dot{\pi}_s \, ds. \quad (6.2.18)$$

By virtue of (0.1.20)–(0.1.22), we have for some constant C, the following bounds:

$$|v_t^\varepsilon| \le Ce^{-2\kappa t/\varepsilon}, \quad (6.2.19)$$
$$E|\xi_t^\varepsilon|^2 \le C, \quad (6.2.20)$$
$$|\varrho_t^\varepsilon| \le C\varepsilon, \quad (6.2.21)$$

implying, in particular, that

$$E|z_t^\varepsilon| \le C(e^{-2\kappa t/\varepsilon} + \varepsilon^{1/2}). \quad (6.2.22)$$

The bound (6.2.20) can be extended to the moments of order $2m$, of course, with a constant depending on m. Indeed, the inequality (1.1.13) applied to the process ξ^ε with

$$\varepsilon d\xi_t^\varepsilon = A_t \xi_t^\varepsilon dt + \sqrt{\varepsilon}\phi(\widehat{J}_t^\varepsilon)d\widetilde{w}_t, \quad \xi_0^\varepsilon = 0,$$

ensures the existence of a constant L, independent on $\varepsilon > 0$ and $m \geq 1$, such that

$$E|\xi_t^\varepsilon|^{2m} \leq m! L^{2m}. \tag{6.2.23}$$

The structure of this estimate is will be used below.

The bounds (6.2.18)–(6.2.20) yield that

$$P\text{-}\lim_{\varepsilon \to 0} \frac{1}{\varepsilon} \int_0^T (\lambda^\star z_s^\varepsilon)^2 \, ds = \lim_{\varepsilon \to 0} \frac{1}{\varepsilon} \int_0^T (\lambda^\star v_s^\varepsilon)^2 \, ds + P\text{-}\lim_{\varepsilon \to 0} \int_0^T (\lambda^\star \xi_s^\varepsilon)^2 \, ds.$$

We summarize the remaining calculations in the following two lemmas.

Lemma 6.2.5 *We have*

$$\lim_{\varepsilon \to 0} \varepsilon^{-1} \int_0^T (\lambda_1^\star v_t^\varepsilon)^2 \, dt = \int_0^\infty \left(\lambda_1^\star e^{A_0^\star r}(p - \pi_0)\right)^2 \, dr. \tag{6.2.24}$$

The proof is standard, see Appendix A.2.

Lemma 6.2.6 *We have*

$$\lim_{\varepsilon \to 0} E \left| \int_0^T \xi_t^\varepsilon \xi_t^{\varepsilon\star} \, dt - \int_0^T M_t \, dt \right| = 0. \tag{6.2.25}$$

Proof. Let us define the process ζ^ε and the function M^ε (both matrix-valued) with

$$\zeta_t^\varepsilon := \xi_t^\varepsilon \xi_t^{\varepsilon\star} - E\xi_t^\varepsilon \xi_t^{\varepsilon\star}, \tag{6.2.26}$$

$$M_t^\varepsilon := \frac{1}{\varepsilon} \int_0^t \varPhi^\varepsilon(t,s)\phi(\pi_s)\phi^\star(\pi_s)\varPhi^{\varepsilon\star}(t,s) \, ds. \tag{6.2.27}$$

Obviously,

$$E \left| \int_0^T \xi_t^\varepsilon \xi_t^{\varepsilon\star} \, dt - \int_0^T M_t \, dt \right| \leq E \left| \int_0^T \zeta_t^\varepsilon \, dt \right| + \int_0^T |E\xi_t^\varepsilon \xi_t^{\varepsilon\star} - M_t^\varepsilon| \, dt$$

$$+ \int_0^T |M_t^\varepsilon - M_t| \, dt = I_1^\varepsilon + I_2^\varepsilon + I_3^\varepsilon.$$

Noticing that

$$E\xi_t^\varepsilon \xi_t^{\varepsilon\star} = \frac{1}{\varepsilon} \int_0^t \varPhi^\varepsilon(t,s) E[\phi(\widehat{J}_s^\varepsilon)\phi^\star(\widehat{J}_s^\varepsilon)]\varPhi^{\varepsilon\star}(t,s) \, ds, \tag{6.2.28}$$

and using the exponential bound for the fundamental matrix, the Lipschitz property of the function $\phi(x)\phi^*(x)$ (considered on the set of vectors which are probability distributions), and the inequality (6.2.22) we get that

$$|E\xi_t^\varepsilon \xi_t^{\varepsilon\star} - M_t^\varepsilon| \leq \frac{1}{\varepsilon}\int_0^t |\Phi^\varepsilon(t,s)|E|\phi(\widehat{J}_s^\varepsilon)\phi^\star(\widehat{J}_s^\varepsilon) - \phi(\pi_s)\phi^\star(\pi_s)||\Phi^{\varepsilon\star}(t,s)|\,ds$$

$$\leq C\frac{1}{\varepsilon}\int_0^t e^{-2\kappa(t-s)/\varepsilon}E|z_s^\varepsilon|\,ds \leq C(e^{-\kappa t/\varepsilon} + \varepsilon^{1/2}).$$

Thus, $I_2^\varepsilon \to 0$. The convergence to zero of I_3^ε can be proved by similar calculations as for the claim of Lemma 6.2.5, see Appendix A.2.

It remains to check that $I_1^\varepsilon \to 0$. It is easier to prove a stronger assertion, namely, that

$$\lim_{\varepsilon \to 0} E\left|\int_0^T \zeta_t^\varepsilon \, dt\right|^2 = 0. \tag{6.2.29}$$

We have:

$$E\left|\int_0^T \zeta_t^\varepsilon \, dt\right|^2 = 2\int_0^T \left(\int_s^T \operatorname{tr} E\zeta_t^\varepsilon \zeta_s^{\varepsilon\star} \, dt\right) ds. \tag{6.2.30}$$

From the representation

$$\xi_t^\varepsilon = \Phi^\varepsilon(t,s)\xi_s^\varepsilon + \frac{1}{\varepsilon^{1/2}}\int_s^t \Phi^\varepsilon(t,u)\phi(\widehat{J}_u^\varepsilon)\,du$$

and the formula (6.2.28) we easily derive that

$$E(\zeta_t^\varepsilon|\mathcal{F}_s) = E(\xi_s^\varepsilon\xi_s^{\varepsilon\star}|\mathcal{F}_s) - \frac{1}{\varepsilon}\int_0^t \Phi^\varepsilon(t,u)E[\phi(\widehat{J}_u^\varepsilon)\phi^\star(\widehat{J}_u^\varepsilon)]\Phi^{\varepsilon\star}(t,u)\,du$$

$$= \Phi^\varepsilon(t,s)\xi_t^\varepsilon\xi_t^{\varepsilon\star}\Phi^{\varepsilon\star}(t,s) - \frac{1}{\varepsilon}\int_0^s \Phi^\varepsilon(t,u)E[\phi(\widehat{J}_u^\varepsilon)\phi^\star(\widehat{J}_u^\varepsilon)]\Phi^{\varepsilon\star}(t,s)\,du$$

and, hence, for some constant C

$$|E(\zeta_t^\varepsilon|\mathcal{F}_s)| \leq C(1 + |\xi_s^\varepsilon|^2)e^{-2\kappa(t-s)/\varepsilon}.$$

Using the moment bound (6.2.23) and the definition (6.2.26) we easily infer that

$$\operatorname{tr} E\zeta_t^\varepsilon \zeta_s^{\varepsilon\star} \leq E|E(\zeta_t^\varepsilon|\mathcal{F}_s)||\zeta_s^{\varepsilon\star}| \leq Ce^{-2\kappa(t-s)/\varepsilon}. \tag{6.2.31}$$

Therefore, by virtue of (6.2.29)–(6.2.31), we have

$$E\left|\int_0^T \zeta_t^\varepsilon \, dt\right|^2 \leq C\int_0^T \left(\int_s^T e^{-2\kappa(t-s)/\varepsilon}\,dt\right) \leq C'\varepsilon$$

and the result follows. $\square$

At last, we check the condition (6.2.8) of Lemma 6.2.4 leaving a verification of (6.2.7) to the reader as a simple exercise.

Lemma 6.2.7 *For any $p > 0$*

$$\limsup_{\varepsilon \to 0} E \exp \left\{ p \int_0^T (\lambda^\star z_t^\varepsilon)^2 \, dt \right\} < \infty. \tag{6.2.32}$$

Proof. The decomposition (6.2.15) of z^ε contains three components but only one, ξ^ε, is random. Isolating it using the elementary inequality

$$(a + b + c)^2 \leq 3(a^2 + b^2 + c^2)$$

and applying for the deterministic components the bounds (6.2.19) and (6.2.21) we easily reduce the claim to the following statement:

For any $p > 0$ there is ε_p such that

$$E \exp \left\{ p\varepsilon \int_0^T |\xi_t^\varepsilon|^2 \, dt \right\} < \infty. \tag{6.2.33}$$

for all $\varepsilon \in \,]0, \varepsilon_p[$.

By virtue of the Jensen inequality and the bound (6.2.23) we have

$$E \left(\int_0^T |\xi_t^\varepsilon|^2 \, dt \right)^m \leq T^{m-1} \int_0^T E|\xi_t^\varepsilon|^{2m} \, dt \leq m! L^{2m} T^m. \tag{6.2.34}$$

Using this fact to estimate the terms of the Taylor expansion for the exponential we conclude that

$$E \exp \left\{ p\varepsilon \int_0^T |\xi_t^\varepsilon|^2 \, dt \right\} \leq \sum_{m=0}^\infty (pL^2 T)^m \varepsilon^m < \infty \tag{6.2.35}$$

for $\varepsilon < 1/(pL^2 T)$. $\square$

6.2.5 Cox Processes with Fast Markov Modulations

The considered model has a "twin" in the theory of counting processes where $\lambda^T J^\varepsilon$ denotes the intensity of a Cox process X^ε (in other terminology, conditionally Poisson, or doubly stochastic Poisson process), P_T^ε is the distribution of X^ε, and R is the distribution of a Poisson process with the intensity $\lambda^T \pi$. The model of such a type arises in the reliability theory as a process of failures of a highly reliable complex system with states changing according to finite state Markov processes with the transition intensity matrix Q. The usual rescaling allows us to consider, instead of time intervals growing to infinity, a fixed one, but switchings become "fast". It is interesting to notice that for such a model the rate of convergence of $\text{Var}\,(P_T^\varepsilon - R_T)$ to zero is of order ε; see [19], [20]. Moreover, the signed measures $\varepsilon^{-1}(P_T^\varepsilon - R_T)$ converge in a certain sense. The mentioned papers use a similar approach based on a study of singularly perturbed stochastic equations but involving a discontinuous martingale.

6.3 Accuracy of Approximate Filters

Let us consider the model of Section 0.3 with the unobservable process x and observations y^ε given by

$$dx_t = f_t dt + \sigma dw^x, \quad x_0 = x^0, \tag{6.3.1}$$

$$dy_t^\varepsilon = x_t dt + \varepsilon dw^y, \quad y_0^\varepsilon = y^0, \tag{6.3.2}$$

assuming that σ is a matrix which does not depend on t and

$$f_t = b_t + D_t x_t \tag{6.3.3}$$

where b and D are continuous functions taking values, respectively, in $\mathbf{R}^n$ and in the set of $n \times n$-matrices.

We specify the approximate filter for x by the linear equation

$$d\widehat{x}_t^\varepsilon = D_t \widehat{x}_t^\varepsilon dt - \varepsilon^{-1} A(dy_t^\varepsilon - \widehat{x}_t^\varepsilon dt), \quad \widehat{x}_0^\varepsilon = \widehat{x}^0, \tag{6.3.4}$$

i.e. we take $\widehat{f}_t^\varepsilon = D_t \widehat{x}_t^\varepsilon$. We assume that A is a symmetric negative definite matrix which commutes with $\sigma\sigma^\star$. For instance, one can take $A = -\gamma I_n$ where $\gamma > 0$ but, as we observe in Section 0.3, if σ is known and nondegenerate, $A = -(\sigma\sigma^\star)^{1/2}$ is a more reasonable choice.

It is worth noticing that, for the assumed structure, our filter requires only the knowledge of D and it tracks x_t for any value of b. We show that large deviation results provide useful information about the filter performance.

For the discrepancy process $\Delta^\varepsilon := \widehat{x}^\varepsilon - x$ of the filter we have the equation

$$d\Delta_t^\varepsilon = \varepsilon^{-1} A \Delta_t^\varepsilon dt + (D_t \Delta_t^\varepsilon - b_t)dt + G d\tilde{w}_t, \quad \Delta_0^\varepsilon = \widehat{x}^0 - x^0, \tag{6.3.5}$$

where $G := (A^2 + \sigma\sigma^\star)^{1/2}$ and $\tilde{w} := -G^{-1}Aw^y - G^{-1}\sigma dw^x$ is a Wiener process in $\mathbf{R}^n$. Now we shall use as a measure of filter accuracy the norm $|||\cdot|||_T$ of $L^2[0, T]$.

Theorem 6.3.1 *For every $\eta > 0$*

$$\lim_{\varepsilon \to 0} \varepsilon^2 \ln P(|||\Delta^\varepsilon|||_T > \eta) = -\frac{1}{2}\eta^2 \lambda_{\min}(A^2(A^2 + \sigma\sigma^\star)^{-1}. \tag{6.3.6}$$

In particular, for the scalar case when $A = -\gamma < 0$, we have

$$\lim_{\varepsilon \to 0} \varepsilon^2 \ln P(|||\Delta^\varepsilon|||_T > \eta) = -\frac{\gamma^2\eta^2}{2(\gamma^2 + \sigma^2)}. \tag{6.3.7}$$

Proof. Let us introduce the process ζ^ε with

$$\varepsilon d\zeta_t^\varepsilon = A\zeta_t^\varepsilon dt + \varepsilon G d\tilde{w}_t, \quad \zeta_0^\varepsilon = \widehat{x}^0 - x^0. \tag{6.3.8}$$

According to Theorem 3.2.1 on large deviations in the L^2-metric applied to the linear equation (see (3.2.4)–(3.2.6)) we have

$$\lim_{\varepsilon \to 0} \varepsilon^2 \ln P(|||\zeta^\varepsilon|||_T > \eta) = -\frac{1}{2}\eta^2 \lambda_{\min}(A(GG^\star)^{-1}A). \tag{6.3.9}$$

Put $r^\varepsilon := \Delta^\varepsilon - \zeta^\varepsilon$. Then

$$\varepsilon dr_t^\varepsilon := (A_t + \varepsilon D_t)r_t^\varepsilon dt + \varepsilon(D_t\zeta_t^\varepsilon - b_t)dt, \quad r_0^\varepsilon = 0. \tag{6.3.10}$$

Using Proposition A.2.5 we easily obtain for sufficiently small ε the bound

$$|||r^\varepsilon|||_T \le C\varepsilon(1 + |||\zeta^\varepsilon|||_T). \tag{6.3.11}$$

Thus,

$$(1 - C\varepsilon)|||\zeta^\varepsilon|||_T - C\varepsilon \le |||\Delta^\varepsilon|||_T \le (1 + C\varepsilon)|||\zeta^\varepsilon|||_T + C\varepsilon. \tag{6.3.12}$$

Using these bounds and the continuity in η of the right-hand side of (6.3.9) we conclude that

$$\lim_{\varepsilon \to 0} \varepsilon^2 \ln P(|||\Delta^\varepsilon|||_T > \eta) = \lim_{\varepsilon \to 0} \varepsilon^2 \ln P(|||\zeta^\varepsilon|||_T > \eta) \tag{6.3.13}$$

and the result follows. $\Box$

Corollary 6.3.2 *In the model where the nondegenerate matrix σ is known and the filter parameter $A = -(\sigma\sigma^\star)^{1/2}$ we have*

$$\lim_{\varepsilon \to 0} \varepsilon^2 \ln P(|||\Delta^\varepsilon|||_T > \eta) = -\eta^2/4. \tag{6.3.14}$$

6.4 Signal Estimation

Let us consider the system

$$dx_t = b_t dt, \quad x_0 = 0, \tag{6.4.1}$$
$$dy_t^\varepsilon = x_t dt + \varepsilon dw, \quad y_0^\varepsilon = y^0, \tag{6.4.2}$$

where x is an unknown function with continuous derivative $\dot{x} = b$ observed via the process y^ε. Obviously, this is a particular case of the model studied in Sections 0.3 and 6.3 but now $\sigma = 0$ and the filtering problem for deterministic signals becomes a statistical estimation problem (the first equation does not carry specific information and it is usually omitted in this setting).

Though x is an element of $C^1[0, T]$ we define for every $\varepsilon \in]0, 1]$ an estimator as a continuous map

$$F^\varepsilon : [0, T] \times C[0, T] \to C[0, T]$$

assuming that for every $t \in [0, T]$ the map $x \mapsto F^\varepsilon(t, x)$ is measurable with respect to the σ-algebra $\mathcal{C}_t$ in $C[0, T]$.

For an estimator F^ε we put $x_t^\varepsilon := F(t, y^\varepsilon)$ getting a continuous process adapted with respect to the filtration generated by y^ε.

In this section we are interested in the family of estimators $\widehat{F} = \{\widehat{F}^\varepsilon\}_{\varepsilon \in [0,1]}$ with

$$\widehat{F}^\varepsilon(t, y) := e^{tA/\varepsilon}\widehat{x}^0 - \frac{1}{\varepsilon}Ay_t + \frac{1}{\varepsilon}Ae^{tA/\varepsilon}y_0 - \frac{1}{\varepsilon^2}A^2\int_0^t e^{(t-s)A/\varepsilon}y_s ds \quad (6.4.3)$$

where A is a symmetric negative definite matrix. The corresponding process $\widehat{x}^\varepsilon$ is the solutions of singularly perturbed stochastic equation

$$d\widehat{x}_t^\varepsilon = -\varepsilon^{-1}A(dy_t^\varepsilon - \widehat{x}_t^\varepsilon dt), \quad \widehat{x}_0^\varepsilon = \widehat{x}^0. \quad (6.4.4)$$

Indeed, using the Cauchy formula and integrating by parts we get from (6.4.4) that for arbitrary process y^ε given by (6.4.2)

$$\widehat{x}_t^\varepsilon = e^{tA/\varepsilon}\widehat{x}^0 - \frac{1}{\varepsilon}A\int_0^t e^{(t-s)A/\varepsilon}y_s^\varepsilon ds$$

$$= e^{tA/\varepsilon}\widehat{x}^0 - \frac{1}{\varepsilon}Ay_t^\varepsilon + \frac{1}{\varepsilon}Ae^{tA/\varepsilon}y_0^\varepsilon - \frac{1}{\varepsilon^2}A^2\int_0^t e^{(t-s)A/\varepsilon}y_s^\varepsilon ds$$

$$= F^\varepsilon(t, y^\varepsilon).$$

Notice that (6.4.4) is exactly the equation (6.3.4) with $D = 0$.

Our aim is to show that the family $\widehat{F}$ is optimal with respect to a quite natural optimality criterion when one wants to minimize the probability of deviation of $\widehat{x}^\varepsilon$ from x in the L^2-norm $|||.|||_T$ above a certain threshold, say, η. When the parameter ε is small, it is natural to expect that estimators do the work better. We formalize this idea by introducing the risk of the family as follows.

Fix functions $\lambda :]0, 1] \to \mathbf{R}_+$ and $\nu :]0, 1] \to \mathbf{R}_+$ such that

$$\lim_{\varepsilon \to 0} \lambda_\varepsilon = \infty, \qquad \lim_{\varepsilon \to 0} \varepsilon^{1/2}\lambda_\varepsilon = 0, \quad (6.4.5)$$

and

$$\lim_{\varepsilon \to 0} \varepsilon\nu_\varepsilon = 0. \quad (6.4.6)$$

Let

$$K^\varepsilon := \{x \in C^1[0, T] : \|x\|_T \le \lambda_\varepsilon, \|\dot{x}\|_T \le \nu_\varepsilon\}.$$

For a chosen $\eta > 0$ we define the risk of a family $F = \{F^\varepsilon\}$ of estimators as

$$R(F) := \limsup_{\varepsilon \to 0} \varepsilon^2 \sup_{x \in K^\varepsilon} \ln P(|||x^\varepsilon - x|||_T \ge \eta). \quad (6.4.7)$$

Theorem 6.4.1 *For any $\eta > 0$*

$$R(\widehat{F}) = \inf_F R(F) = -\eta^2/2. \quad (6.4.8)$$

Proof. We establish first the lower bound for the risk and then show, using the large deviation result, that this bound is attained on $\widehat{F}$.

Proposition 6.4.2 *For any $\eta > 0$ and every family of estimators F we have*

$$R(F) \geq -\eta^2/2. \tag{6.4.9}$$

Proof. Put $U_N := [-N, N]^n$. Identifying $\theta \in \mathbf{R}^n$ with the constant function $x_t = \theta$ we consider the set of signals of the form $\theta = \varepsilon h/\sqrt{T}$ where $h \in U_{2N/\varepsilon}$. Since $\lambda_\varepsilon \to \infty$ this set is in K^ε for all sufficiently small ε. Thus, it suffices to show that for some $N > 0$ we have

$$\limsup_{\varepsilon \to 0} \varepsilon^2 \sup_{h \in U_{2N/\varepsilon}} \ln P(|||x^\varepsilon - \varepsilon h/\sqrt{T}|||_T \geq \eta) \geq -\eta^2/2 \tag{6.4.10}$$

where $x_t^\varepsilon = F^\varepsilon(t, y^\varepsilon)$ and the observation process is $y_t^\varepsilon = \varepsilon h/\sqrt{T}t + \varepsilon w_t$.

By the Girsanov theorem $w^h := h/\sqrt{T}t + w_t$ is the Wiener process under the probability

$$P^h = e^{-h^\star/\sqrt{T}w_T - |h|^2/2}P.$$

Notice also that

$$|||x^\varepsilon - \varepsilon h/\sqrt{T}|||_T \geq \frac{1}{\sqrt{T}} \int_0^T |x_t^\varepsilon - \varepsilon h/\sqrt{T}| \, dt \geq |\bar{x}^\varepsilon - \varepsilon h|$$

where

$$\bar{x}^\varepsilon := \frac{1}{\sqrt{T}} \int_0^T x_t^\varepsilon \, dt = \frac{1}{\sqrt{T}} \int_0^T F^\varepsilon(t, \varepsilon w^h) \, dt.$$

Thus,

$$\begin{aligned} P(|||x^\varepsilon - \varepsilon h|||_T \geq \eta) &\geq P(|\bar{x}^\varepsilon - \varepsilon h| \geq \eta) \\ &= E^h I_{\{|\bar{x}^\varepsilon - \varepsilon h| \geq \eta\}} e^{h^\star/\sqrt{T}w_T^h - |h|^2/2} \\ &= E I_{\{|\zeta^\varepsilon - \varepsilon h| \geq \eta\}} e^{h^\star/\sqrt{T}w_T - |h|^2/2} \end{aligned}$$

where

$$\zeta^\varepsilon := \frac{1}{\sqrt{T}} \int_0^T F^\varepsilon(t, \varepsilon w) \, dt.$$

The claim follows now from Lemma 6.4.3 below. $\square$

Lemma 6.4.3 *Let $\xi \sim N(0, I_n)$ and let $\{\zeta^\varepsilon\}$ be a family of random variables with values in $\mathbf{R}^n$. Then for any sufficiently large N*

$$\limsup_{\varepsilon \to 0} \varepsilon^2 \sup_{h \in U_{2N/\varepsilon}} \ln E I_{\{|\zeta^\varepsilon - \varepsilon h| \geq \eta\}} e^{h^\star \xi - |h|^2/2} \geq -\eta^2/2. \tag{6.4.11}$$

Proof. Let $N = 2\eta$. Obviously,

$$EI_{\{|\zeta^\varepsilon - \varepsilon h| \geq \eta\}} e^{h^* \xi - |h|^2/2} = EI_{\{|\zeta^\varepsilon - \varepsilon h| \geq \eta\}} e^{-|h-\xi|^2/2} e^{|\xi|^2/2}$$

$$\geq EI_{\{|\zeta^\varepsilon - \varepsilon h| \geq \eta\}} e^{-|h-\xi|^2/2} e^{|\xi|^2/2} I_{\{\xi \in U_{N/\varepsilon}\}}. \quad (6.4.12)$$

We have:

$$\sup_{h \in U_{2N/\varepsilon}} EI_{\{|\zeta^\varepsilon - \varepsilon h| \geq \eta\}} e^{-|h-\xi|^2/2} e^{|\xi|^2/2} I_{\{\xi \in U_{N/\varepsilon}\}}$$

$$\geq (4N/\varepsilon)^{-n} \int_{U_{2N/\varepsilon}} EI_{\{|\zeta^\varepsilon - \varepsilon h| \geq \eta\}} e^{-|h-\xi|^2/2} e^{|\xi|^2/2} I_{\{\xi \in U_{N/\varepsilon}\}} dh$$

$$= (4N/\varepsilon)^{-n} E e^{|\xi|^2/2} I_{\{\xi \in U_{N/\varepsilon}\}} \int_{U_{2N/\varepsilon-\xi}} I_{\{|\zeta^\varepsilon - \varepsilon(\xi+z)| \geq \eta\}} e^{-|z|^2/2} dz$$

$$\geq (4N/\varepsilon)^{-n} E e^{|\xi|^2/2} I_{\{\xi \in U_{N/\varepsilon}\}} \int_{U_{N/\varepsilon}} I_{\{|\varepsilon z + (\varepsilon\xi - \zeta^\varepsilon)| \geq \eta\}} e^{-|z|^2/2} dz \quad (6.4.13)$$

where the last inequality holds because $U_{2N/\varepsilon} - \xi \supseteq U_{N/\varepsilon}$ for $\xi \in U_{N/\varepsilon}$. Using the Anderson inequality (see Lemma 6.4.4 below) we get that

$$\int_{U_{N/\varepsilon}} I_{\{|\varepsilon z + (\varepsilon\xi - \zeta^\varepsilon)| \geq \eta\}} e^{-|z|^2/2} dz \geq \int_{U_{N/\varepsilon}} I_{\{|\varepsilon z| \geq \eta\}} e^{-|z|^2/2} dz$$

$$\geq \int I_{\{\eta/\varepsilon \leq |z| \leq N/\varepsilon\}} e^{-|z|^2/2} dz$$

$$\geq \int I_{\{\eta/\varepsilon \leq |z| \leq (1+\delta)\eta/\varepsilon\}} e^{-|z|^2/2} dz$$

$$\geq e^{-(1+\delta)\eta^2/2} C_n (\eta/\varepsilon)^n [(1+\delta)^n - 1] \quad (6.4.14)$$

where the constant C_n is the volume of the unit ball in $\mathbf{R}^n$ and $\delta \in \,]0,1[$ is arbitrary. At last,

$$E e^{|\xi|^2/2} I_{\{\xi \in U_{N/\varepsilon}\}} = (2\pi)^{-n/2} (2N/\varepsilon)^n. \quad (6.4.15)$$

It follows from (6.4.12)–(6.4.15) that

$$\limsup_{\varepsilon \to 0} \varepsilon^2 \sup_{h \in U_{2N/\varepsilon}} \ln EI_{\{|\zeta^\varepsilon - \varepsilon h| \geq \eta\}} e^{h^* \xi - |h|^2/2} \geq -(1+\delta)\eta^2/2$$

implying the claim.

Lemma 6.4.4 *Let $X = (X_1, \ldots, X_n)$ be a random vector with independent components. Assume that for each i the distribution of X_i has the density f_i which is an even function decreasing on $\mathbf{R}_+$. Then for any $\eta > 0$ and $a \in \mathbf{R}^n$*

$$P(|X + a| \geq \eta) \geq P(|X| \geq \eta). \quad (6.4.16)$$

Proof. The case $n = 1$ is obvious. Assume that the result holds up to dimension $n - 1$. Then

$$P\Big(\sum_{i=1}^{n}(X_i + a_i)^2 \geq \eta^2\Big) = EP\Big(\sum_{i=1}^{n-1}(X_i + a_i)^2 \geq \eta^2 - (X_n + a_n)^2\Big|X_n\Big)$$

$$\geq P\Big(\sum_{i=1}^{n-1} X_i^2 + (X_n + a_n)^2 \geq \eta^2\Big).$$

Repeating the above argument with conditioning, e.g., with respect X_1 leads to the desired inequality. $\square$

To accomplish the proof of Theorem 6.4.1 it remains to verify that the lower bound is attained on $\widehat{F}$. To do this we re-examine the arguments of the previous section.

Let $x \in K^\varepsilon$. For the discrepancy process $\Delta^\varepsilon := \widehat{x}^\varepsilon - x$ we have the equation

$$d\Delta_t^\varepsilon = \varepsilon^{-1} A \Delta_t^\varepsilon dt - \dot{x}_t dt - \varepsilon^{-1} dw_t, \quad \Delta_0^\varepsilon = \widehat{x}^0 - x^0,$$

and, hence,

$$\Delta_t^\varepsilon = e^{tA/\varepsilon}\Delta_0^\varepsilon - \int_0^t e^{(t-s)A/\varepsilon}\dot{x}_s\, ds + \xi_t^\varepsilon$$

where

$$\varepsilon d\xi_t^\varepsilon = A\xi_t^\varepsilon dt - A\, dw_t, \quad \xi^\varepsilon = 0.$$

Let $\gamma > 0$ be a constant such that all eigenvalues of A are strictly greater than γ. Clearly,

$$|\Delta_t^\varepsilon| \leq Ce^{-t\gamma/\varepsilon}|\Delta_0^\varepsilon| + C\int_0^t e^{-(t-s)\gamma/\varepsilon}|\dot{x}_s|ds + |\xi_t^\varepsilon|$$

$$\leq Ce^{-t\gamma/\varepsilon}\lambda_\varepsilon + C\nu_\varepsilon\varepsilon + \xi_t^\varepsilon.$$

It follows that

$$|||\Delta^\varepsilon|||_T \leq C\varepsilon^{1/2}\lambda_\varepsilon + C\nu_\varepsilon\varepsilon + |||\xi^\varepsilon|||_T$$

$$= |||\xi^\varepsilon|||_T + o(1)$$

according to (6.4.5) and (6.4.6). Therefore,

$$R(\widehat{F}) := \limsup_{\varepsilon \to 0} \varepsilon^2 \sup_{x \in K^\varepsilon} \ln P(|||\Delta^\varepsilon|||_T \geq \eta)$$

$$\leq \limsup_{\varepsilon \to 0} \varepsilon^2 \ln P(|||\xi^\varepsilon|||_T \geq \eta) = -\eta^2/2$$

by virtue of Theorem 3.2.1. $\square$

6.5 Linear Regulator with Infinite Horizon

The main purpose of this section is to prove that the classic feedback solution of the LQG-problem has an advantage with respect to any alternative non-anticipating control not only in the mean but almost surely, being applied in a long run. A technique of singular perturbations can be used in our analysis since the dynamics under the optimal feedback is given by a stochastic differential equation exactly of the same type as has been studied throughout the whole book.

6.5.1 Sensitive Probabilistic Criteria

We start from a rather general optimal control model by describing a basic idea of optimality a.s. and in probability. We deal with strong solutions, so the stochastic basis $(\Omega, \mathcal{F}, \mathbf{F} = (\mathcal{F}_t), P)$ with an adapted Wiener process is fixed in advance. The model is described by the stochastic differential equations

$$dx_t = f(t, x_t, u_t)dt + \sigma(t, x_t, u_t)dw_t, \quad x_0 = \xi, \qquad (6.5.1)$$

which is assumed to have a strong solution, an admissible control being chosen. Sets $\mathcal{U}_T$ of admissible controls are given for any time horizon T and their elements are predictable processes $u = (u_t)_{t \leq T}$ (a value u_∞ when $T = \infty$ may be not defined). It is assumed that for $S \leq T$ the restriction $u|_S$ of $u \in \mathcal{U}_T$ to $\Omega \times [0, S]$ belongs to $\mathcal{U}_S$.

The time horizon will tend to infinity, i.e., actually, we consider the parametric family of optimal control problems. We shall use the superscript T to exhibit the dependence of a chosen control and the corresponding process on the time horizon. The costs functions $J_T : \mathcal{U}_T \to L^0(\Omega, \mathcal{F}_T, P)$ are of the form

$$J_T(u) = \int_0^T h(t, x_t, u_t)dt \qquad (6.5.2)$$

where $h \geq 0$.

Traditionally, in the stochastic setting with a fixed time horizon the objective is to minimize the mean value of (6.5.2), i.e. the expectation of the integral of the running cost

$$\tilde{J}_T(u) = E \int_0^T h(t, x_t, u_t)dt. \qquad (6.5.3)$$

However, the classic mean value approach is subject to criticism since the criterion, in fact, involves all possible states of nature described by points $\omega \in \Omega$. Practically, quite often a single dynamical experiment can be performed, though on an unboundedly large time interval. This means that only one state nature ω is available and relevant. The idea, inspired by the laws of large numbers, is to replace the expectation of mean value by its time average and consider as the most preferable a control u^o such that the limit of $(1/T)J_T(u^o)$ is less than the limit of $(1/T)J_T(u)$ for any other u. Various types of convergence lead to different definitions of optimality. Of course, the above limits may not exist and it is more reasonable to look for an asymptotic behavior of the average gain that can be obtained by using a competing control. Roughly speaking, one can agree to implement a control u^o as a standard if for every alternative u on the set where the random variable $(J_T(u^o) - J_T(u))^+$ is positive, i.e. a certain a gain is could be achieved, this gain is not excessively large relatively to the length of the time interval.

To formalize this idea of comparing alternative controls we suggest the following probabilistic criteria of optimality.

Let g be a positive decreasing function on $\mathbf{R}_+$.

We say that a control $\widehat{u} \in \mathcal{U}_\infty$ is *g-optimal a.s.* (respectively, *in probability*) if for any $u \in \mathcal{U}_\infty$

$$\lim_{T \to \infty} (J_T(\widehat{u}|_T) - J_T(u|_T))^+ g_T = 0 \quad a.s. \tag{6.5.4}$$

(respectively,

$$P\text{-}\lim_{T \to \infty} (J_T(\widehat{u}|_T) - J_T(u|_T))^+ g_T = 0). \tag{6.5.5}$$

We say that a control $\widehat{u} \in \mathcal{U}_\infty$ is *strongly optimal in probability* if for any family of controls $u^T \in \mathcal{U}_T$, $T \in \mathbf{R}_+$,

$$P\text{-}\lim_{T \to \infty} (J_T(\widehat{u}|_T) - J_T(u^T))^+ g_T = 0. \tag{6.5.6}$$

The function g in the above definition is called a *rate function*. Typical examples considered in the literature are $g_T = 1$, T^{-1}, or, more generally, T^{-q} with $q \geq 0$ etc. Criteria with rate functions which decrease to zero slower than T^{-1} sometimes are referred to as *sensitive* while the 1-optimality a.s. is called the *overtaking optimality a.s.*

Notice that in the definition of optimality in probability the control $\widehat{u}$ competes with a very particular family of controls of the form $u|_T$ for some $u \in \mathcal{U}_\infty$.

Remark. Each random variable $J_T(u^T)$ is defined up to a null set and we cannot expect, for an arbitrary family of controls, that they can be chosen to form a process with regular paths. This is the reason why we avoid to defining the strong optimality a.s.

6.5.2 Linear-Quadratic Regulator

We consider the classic model of the completely observed linear-quadratic regulator with Gaussian disturbances and time-invariant coefficients given by the stochastic differential equation

$$dx_t = (Ax_t + Bu_t)dt + Gdw_t, \quad x_0 = \xi, \tag{6.5.7}$$

where $x_t \in \mathbf{R}^n$, $u_t \in \mathbf{R}^m$, w is a p-dimensional Wiener process, A, B, G are constant matrices of appropriate dimensions, $GG^\star \neq 0$. We assume that the filtration $\mathbf{F}$ is generated by w.

The set $\mathcal{U}_T$ consists of all adapted processes $(u_t)_{t \leq T}$ such that $|u_t|^2$ is integrable on any finite interval almost surely.

The cost function $J_T(u)$ is given by

$$J_T(u) := \int_0^T (x_t^\star Q x_t + u_t^\star u_t) dt = \int_0^T (|Q^{1/2} x_t|^2 + |u_t|^2) dt \qquad (6.5.8)$$

where Q is a symmetric positive semidefinite matrix.

In the system theory a pair of matrices (A, B) is called controllable if the matrix $[B, AB, \ldots, A^{n-1}, B]$ is of the full rank. A pair (A, C) is observable if $[C^\star, A^\star C^\star, \ldots, A^{\star n-1}, C^\star]$ is of the full rank.

We adopt the following assumption which is standard.

H.6.5.1 The pair (A, B) is controllable and the pair $(A, Q^{1/2})$ is observable.

Notice that it is not assumed that A is a stable matrix.

We recall the well-known result describing the structure of the optimal solution for the LQG-problem; it is an immediate consequence of Lemma 6.5.3 below.

Proposition 6.5.1 *For a fixed T the optimal pair $(\widehat{u}^T, \widehat{x}^T)$ in the problem of minimization of the mean value cost*

$$\tilde{J}_T(u) := E \int_0^T (|Q^{1/2} x_t|^2 + |u_t|^2) dt \to \min \qquad (6.5.9)$$

is given by the following formulae:

$$\widehat{u}_t^T = -B^\star \Pi_t^T \widehat{x}_t^T, \qquad (6.5.10)$$
$$d\widehat{x}_t^T = (A - BB^\star \Pi_t^T) \widehat{x}_t^T dt + G dw_t, \quad \widehat{x}_0^T = \xi, \qquad (6.5.11)$$

where the matrix function Π_t^T solves the differential Riccati equation

$$\dot{\Pi}_t^T + \Pi_t^T A + A^\star \Pi_t^T - \Pi_t^T BB^\star \Pi_t^T + Q = 0, \quad \Pi_T^T = 0. \qquad (6.5.12)$$

Under the assumption of controllability of (A, B) and observability of $(A, Q^{1/2})$ there exists the limit

$$\Pi = \lim_{T \to \infty} \Pi_0^T$$

where the symmetric matrix Π is the positive definite solution of the algebraic Riccati equation (ARE)

$$\Pi A + A^\star \Pi - \Pi BB^\star \Pi + Q = 0, \qquad (6.5.13)$$

the matrix $A - BB^\star \Pi$ is stable, and the pair $(\widehat{u}, \widehat{x})$ with

$$\widehat{u}_t = -B^\star \Pi \widehat{x}_t, \tag{6.5.14}$$

$$d\widehat{x}_t = (A - BB^\star \Pi)\widehat{x}_t dt + Gdw_t, \quad \widehat{x}_0 = \xi, \tag{6.5.15}$$

is strongly T^{-1}-optimal for the cost functions $\tilde{J}_T$ given by (6.5.9).

We show that the feedback control $\widehat{u}$ is also optimal with respect to sensitive probabilistic criteria with very slowly decreasing rate functions.

Theorem 6.5.2 *Assume that Q is nondegenerated. Then in the infinite horizon LQG-problem with the cost functions given by the formula (6.5.8) the control $\widehat{u}$ defined in (6.5.14), (6.5.15) is*

(a) g-optimal a.s. for any rate function such that $g_T = o(1/\ln T)$ as $T \to \infty$;

(b) strongly g-optimal in probability for any g with $g_T = o(1)$.

6.5.3 Preliminaries

The study of LQG-problem is based on an appropriate transformation of the cost functional.

Lemma 6.5.3 *Let $\Pi^T = (\Pi_t^T)$ be the positive semidefinite solution on $[0, T]$ of the differential Riccati equation*

$$\dot{\Pi}_t^T + \Pi_t^T A + A^\star \Pi_t^T - \Pi_t^T BB^\star \Pi_t^T + Q = 0, \quad \Pi_T^T = N, \tag{6.5.16}$$

and x be the solution of the linear equation (6.5.7) corresponding to some admissible control $u \in \mathcal{U}_T$. Then

$$J_T(u) := \int_0^T (|Q^{1/2}x_t|^2 + |u_t|^2)dt$$

$$= \xi^\star \Pi_0^T \xi - x_t^\star N x_t + \int_0^T |u_t + B^\star \Pi_t^T x_t|^2 dt$$

$$+ \int_0^T \operatorname{tr} G^\star \Pi_t^T G dt + 2 \int_0^T x_t^\star \Pi_t^T G dw_t. \tag{6.5.17}$$

Proof. By the Ito formula for the product $x_t^\star \Pi_t^T x_t$ we have

$$x_T^\star N x_T - \xi^\star \Pi_0^T \xi = \int_0^T [(Ax_t + Bu_t)^\star \Pi_t^T x_t + x_t^\star \Pi_t^T (Ax_t + Bu_t)$$

$$+ x_t^\star \dot{\Pi}^T x_t]dt + \int_0^T \operatorname{tr} G^\star \Pi_t^T G dt + 2 \int_0^T x_t^\star \Pi_t^T G dw_t.$$

Substituting the expression for $\dot{\Pi}^T$ from (6.5.16) and making use of the identity

$$|u + B^\star \Pi^T x|^2 - |Q^{1/2} x|^2 - |u|^2 = (Ax + Bu)^\star \Pi^T x + x^\star \Pi^T (Ax + Bu)$$
$$- x^\star (\Pi^T A + A^\star \Pi^T - \Pi^T BB^\star \Pi_t^T + Q)x$$

we get the result. □

Notice that Proposition 6.5.1 asserting that $(\widehat{u}^T, \widehat{x}^T)$ given by (6.5.10), (6.5.11) is the optimal pair for $\tilde{J}_T$ follows immediately from the above lemma if we put $N = 0$ and take mathematical expectation of (6.5.17). The optimal value is, evidently,

$$\tilde{J}_T(\widehat{u}^T) = \xi^\star \Pi_0^T \xi + \int_0^T \operatorname{tr} G^\star \Pi_t^T G \, dt. \qquad (6.5.18)$$

The following result on the exponential rate of convergence of the solution of the differential Riccati equation to the solution of ARE which we give without a proof shows that the use of the family of controls $\widehat{u}^T$ instead of $\widehat{u}|_T$ in a long run does not lead to a large gain.

Lemma 6.5.4 Let P_t, $t \in \mathbf{R}_+$, be the solution of

$$-\dot{P}_t + P_t A + A^\star P_t - P_t BB^\star P_t + Q = 0, \quad P_0 = 0. \qquad (6.5.19)$$

Then for some positive constants c and κ

$$|P_t - \Pi| \le c e^{-\kappa t}. \qquad (6.5.20)$$

Obviously, the solution Π_t^T of (6.5.12) (i.e. of (6.5.16) with $N = 0$) can be written as

$$\Pi_t^T = P_{T-t}. \qquad (6.5.21)$$

It follows from Lemma 6.5.3 that

$$\sup_{T \in \mathbf{R}_+} \left| \int_0^T \operatorname{tr} G^\star \Pi_t^T G \, dt - T \operatorname{tr} G^\star \Pi G \right| < \infty. \qquad (6.5.22)$$

Lemma 6.5.5 Let $c > 0$ be a constant and let $(Y^T)_{T \ge 0}$ be a family of random variables of the form $Y^T = M_T^T - c\langle M^T \rangle_T$ where $M^T = (M_t^T)_{t \le T}$ is a continuous local martingale with $M_0^T = 0$. Then for any function g such that $g_T \to 0$ as $T \to \infty$ we have

$$P\text{-} \lim_{T \to \infty} (Y^T)^+ g_T = 0. \qquad (6.5.23)$$

Proof. We may assume that $g \ge 0$. Then for any $\varepsilon > 0$

$$P(Y^T g_T \ge \varepsilon) = P(e^{2cM_T^T - 2c^2 \langle M^T \rangle_T} \ge e^{2c\varepsilon/g_T}).$$

The result follows by the Chebyshev inequality since $E e^{\lambda M_T^T - (\lambda^2/2)\langle M^T \rangle_T} \le 1$.
□

Lemma 6.5.6 *Let M be a continuous local martingale, $M_0 = 0$, and $c > 0$ be a constant. Then*

(a) $M_T - c\langle M\rangle_T \to -\infty$ *a.s. on* $\{\langle M\rangle_\infty = \infty\}$,

(b) $M_T - c\langle M\rangle_T \to M_\infty - c\langle M\rangle_\infty$ *a.s. on* $\{\langle M\rangle_\infty < \infty\}$.

Proof. The assertion (b) holds since $\{\langle M\rangle_\infty < \infty\}$ coincides a.s with the set where M converges. To prove (a) it is sufficient to show that $M_T/\alpha_T \to 0$ a.s. on the set $\{\langle M\rangle_\infty = \infty\}$ where $\alpha := 1 + \langle M\rangle^{3/4}$. To this aim, define the local martingale X with

$$X_T := \int_0^T \alpha_s^{-1} dM_s$$

which converges a.s. to X_∞ since

$$\langle X\rangle_\infty = \int_0^\infty \alpha_s^{-2} d\langle M\rangle_s < \infty.$$

Integrating by parts we get

$$\frac{M_T}{\alpha_T} = \frac{1}{\alpha_T} \int_0^T \alpha_s dX_s = X_T - \frac{1}{\alpha_T} \int_0^T X_s d\alpha_s \to 0$$

on the set $\{\alpha_\infty = \infty\} = \{\langle M\rangle_\infty = \infty\}$. $\square$

6.5.4 Proof of Theorem 6.5.2

Let $u \in \mathcal{U}_T$ and x be the corresponding solution of (6.5.7). Put $y := \hat{x} - x$, $v := \hat{u} - u$. Evidently,

$$\dot{y}_t = Ay_t + Bv_t, \quad y_0 = 0. \tag{6.5.24}$$

It follows that

$$|\Pi^{1/2} y_T|^2 = y_T^\star \Pi y_T = 2\int_0^T \dot{y}_t^\star \Pi y_t dt = 2\int_0^T y_t^\star A^\star \Pi y_t dt + 2\int_0^T v_t^\star B^\star \Pi y_t dt$$

$$= \int_0^T y_t^\star (\Pi BB^\star \Pi - Q) y_t dt + 2\int_0^T v_t^\star B^\star \Pi y_t dt \tag{6.5.25}$$

where we use at the last step the algebraic Riccati equation (6.5.13).

For $\Delta_T := J_T(\hat{u}|_T) - J_T(u)$ we have by Lemma 6.5.3 with $N = \Pi$ the following representation:

$$\Delta_T = |\Pi^{1/2} x_T|^2 - |\Pi^{1/2} \hat{x}_T|^2 - \int_0^T |u_t + B^\star \Pi x_t|^2 dt + M_T$$

$$= |\Pi^{1/2}(\hat{x}_T - y_T)|^2 - |\Pi^{1/2} \hat{x}_T|^2 - \int_0^T |v_t + B^\star \Pi y_t|^2 dt + M_T$$

where

$$M_T := 2 \int_0^T y_t^\star \Pi G \, dw_t.$$

Let $\gamma > 0$. Notice that

$$|\Pi^{1/2}(\widehat{x}_T - y_T)|^2 - |\Pi^{1/2}\widehat{x}_T|^2$$
$$= \gamma^{-1}|\Pi^{1/2}\widehat{x}_T|^2 - \gamma^{-1}|\Pi^{1/2}(\widehat{x}_T + \gamma y_T)|^2 + (1+\gamma)|\Pi^{1/2}y_T|^2.$$

Using this representation together with the formula (6.5.25) and regrouping terms with help of the identity

$$(1+\gamma)y^\star(\Pi BB^\star \Pi - Q)y + 2(1+\gamma)v^\star B^\star \Pi y - |v + B^\star \Pi y|^2$$
$$= -|v - \gamma B^\star \Pi y|^2 - (1+\gamma)y^\star(Q - \gamma\Pi BB^\star \Pi)y,$$

we get from (6.5.26) that

$$\Delta_T = \gamma^{-1}|\Pi^{1/2}\widehat{x}_T|^2 - \gamma^{-1}|\Pi^{1/2}(\widehat{x}_T + \gamma y_T)|^2 - \int_0^T |v_t - \gamma B^\star \Pi y_t|^2 dt$$

$$- (1+\gamma)\int_0^T y_t^\star(Q - \gamma\Pi BB^\star \Pi)y_t \, dt + M_T. \tag{6.5.26}$$

Hence, we have always

$$\Delta_T \leq \gamma^{-1}|\Pi^{1/2}\widehat{x}_T|^2 - (1+\gamma)\int_0^T y_t^\star(Q - \gamma\Pi BB^\star \Pi)y_t \, dt + M_T. \tag{6.5.27}$$

The crucial observation is that for some positive constants γ and c which do not depend on y one has the inequality

$$(1+\gamma)\int_0^T y_t^\star(Q - \gamma\Pi BB^\star \Pi)y_t \, dt \geq 4c \int_0^T y_t^\star \Pi GG^\star \Pi y_t \, dt. \tag{6.5.28}$$

Indeed, the assumption that Q is nondegenerated implies the existence of $c > 0$ such that for sufficiently small $\gamma > 0$

$$(1+\gamma)z^\star(Q - \gamma\Pi BB^\star \Pi)z \geq 4c \, z^\star \Pi GG^\star \Pi z$$

for all $z \in \mathbf{R}^n$.

Thus, for such γ we have the bound

$$\Delta_T \leq \gamma^{-1}|\Pi^{1/2}\widehat{x}_T|^2 - c\langle M\rangle_T + M_T. \tag{6.5.29}$$

In particular, for arbitrary $u \in \mathcal{U}_\infty$ it follows that

$$J_T(\widehat{u}|T) - J_T(u|T) \leq \gamma^{-1}|\Pi^{1/2}\widehat{x}_T|^2 - c\langle M\rangle_T + M_T. \tag{6.5.30}$$

The martingale M here does not depend on T and the assertion (a) follows since by Proposition A.2.6 and Lemma 6.5.6 the right-hand side multiplied by any function $g_T = o(1/\ln T)$ will tend to zero a.s.

In the case of the family of admissible controls u^T the inequality

$$J_T(\widehat{u}|_T) - J_T(u^T) \leq \gamma^{-1}|\Pi^{1/2}\widehat{x}_T|^2 - c\langle M^T\rangle_T + M_T^T \qquad (6.5.31)$$

involves a martingale depending on T and the assertion (b) follows from Lemma 6.5.5 and the observation that for the solution of the $\widehat{x}$ of the linear equation (6.5.15) with the stable matrix $A - BB^\star\Pi$ we have $E|\widehat{x}_t|^2 \leq C$ by Theorem 1.2.3. $\square$

6.5.5 Example

Let us consider the scalar model

$$dx_t = (1/2)u_t dt + dw_t, \quad x_0 = 0, \qquad (6.5.32)$$

with

$$J_T(u) := \int_0^T (x_t^2 + u_t^2)dt. \qquad (6.5.33)$$

In this case (where $A = 0$, $B = 1/2$, $G = 1$, and $Q = 1$) we have by virtue of (6.5.13)–(6.5.15) that $\Pi = 2$, $\widehat{u}_t = -\widehat{x}_t$, and

$$d\widehat{x}_t = -(1/2)\widehat{x}_t dt + dw_t, \quad \widehat{x}_0 = 0. \qquad (6.5.34)$$

Theorem 6.5.2 asserts that the control $\widehat{u}$ is g-optimal a.s. for any rate function g such that $g_T = o(1/\ln T)$ as $T \to \infty$. However, $\widehat{u}$ is not $(1/\ln T)$-optimal a.s. To show this, we find a control $u = (u_t)_{t\geq 0}$ such that

$$\limsup_{T\to\infty} (J_T(\widehat{u}|_T) - J_T(u|_T))/\ln T > 0 \text{ a.s.} \qquad (6.5.35)$$

For the arbitrary control u we define $v := \widehat{u} - u$ and

$$\dot{y}_t = (1/2)v_t, \quad y_0 = 0. \qquad (6.5.36)$$

According to (6.5.26)

$$\begin{aligned}
\Delta_T &= -4\widehat{x}_T y_T + 2y_T^2 - \int_0^T (2\dot{y}_t + y_t)^2 dt + 4\int_0^T y_t dw_t \\
&= -4\widehat{x}_T y_T - \int_0^T (4\dot{y}_t^2 + y_t^2)dt + 4\int_0^T y_t dw_t \\
&\geq -4\widehat{x}_T y_T - 2\int_0^T (4\dot{y}_t^2 + y_t^2)dt + \int_0^T y_t^2 dt + 4\int_0^T y_t dw_t.
\end{aligned}$$

By Lemma 6.5.6 the sum of the two last integrals converges a.s. either to $+\infty$ or to a finite limit. Since knowing y one can restore u, we need to find an adapted process y with absolute continuous trajectories such that

$$\limsup_{T \to \infty} \frac{1}{\ln T} \left(- 2\widehat{x}_T y_T - \int_0^T (4\dot{y}_t^2 + y_t^2)dt \right) > 0 \quad \text{a.s.} \tag{6.5.37}$$

Notice that $W_t := -\widehat{x}_{\ln(t+1)}\sqrt{t+1}$ is a Wiener process. The time change $t \mapsto \ln(t+1)$ reduces the problem to the following one: find a process z adapted to the filtration generated by W, having absolute continuous trajectories, and such that

$$\limsup_{T \to \infty} R_T(z) > 0 \quad \text{a.s.} \tag{6.5.38}$$

where

$$R_t(z) := \frac{1}{\ln_2 T} \left(\frac{2W_T z_T}{\sqrt{T+1}} - \int_0^T F(t, z_t, \dot{z}_t)dt \right),$$

$$F(t, z, \dot{z}) := 4\dot{z}^2(t+1) + z^2/(t+1).$$

Take a continuous function $f : \mathbf{R}_+^2 \to [0, 1]$ such that for every θ the section $f(., \theta)$ is a function vanishing outside the interval $[\theta, 4\theta+3]$, increasing on $[\theta, 2\theta+1]$ from zero to one, decreasing from one to zero on $[2\theta+1, 4\theta+3]$, and

$$\int_\theta^{2\theta+1} F(t, f(t, \theta), \dot{f}(t, \theta))dt = \int_{2\theta+1}^{4\theta+3} F(t, f(t, \theta), \dot{f}(t, \theta)) = a$$

where $a > 0$ is a constant which does not depend on θ. For instance, for the function $f(t, \theta) := \sqrt{2}g(t, \theta)$ where $g(t, \theta)$ is the function

$$\left(\sqrt{\frac{t+1}{\theta+1}} - \sqrt{\frac{\theta+1}{t+1}} \right) I_{[\theta, 2\theta+1]}(t) + \left(2\sqrt{\frac{\theta+1}{t+1}} - \frac{1}{2}\sqrt{\frac{t+1}{\theta+1}} \right) I_{]2\theta+1, 4\theta+3]}(t)$$

we have $a = 6$.

Fix $\delta \in]0, 1/2[$ and define the following sequence of stopping times converging to infinity:

$$\tau_1 := \inf\{t \geq e^e : W_t \geq (1 - \delta)\sqrt{2t \ln_2 t}\},$$

$$\tau_n := \inf\{t \geq \phi_{n-1} + 4\tau_{n-1} + 3 : W_t \geq (1 - \delta)\sqrt{2t \ln_2 t}\}, \quad n \geq 2,$$

where

$$\ln_2 \phi_{n-1} := n^2 \sum_{i=1}^{n-1} \ln_2 \tau_i \geq n^2. \tag{6.5.39}$$

By the law of iterated logarithms $\tau_n < \infty$.

We check that the process

$$z_t := c \sum_{i=1}^{\infty} \sqrt{\ln_2 \tau_i} f(t, \tau_i) I_{\{t \geq \tau_i\}} \qquad (6.5.40)$$

with $c := (1 - 2\delta)/a$ satisfies (6.5.38).

Put $\sigma_n := 2\tau_n + 1$. By construction, $\ln_2 \tau_n > \ln_2 \phi_{n-1}$ and

$$\int_0^{\sigma_n} F(t, z_t, \dot{z}_t) dt = 2ac^2 n^{-2} \ln_2 \phi_{n-1} + ac^2 \ln_2 \tau_n \sim ac^2 \ln_2 \sigma_n.$$

The process $W - W_{\tau_n}$ is a Wiener process independent on τ_n. Thus,

$$P\left(W_{\sigma_n} - W_{\tau_n} \leq -\delta\sqrt{2\tau_n \ln_2 \tau_n} \,\Big|\, \tau_n\right) = \Phi\left(-\delta\sqrt{2\tau_n \ln_2 \tau_n}/\sqrt{\tau_n + 1}\right)$$
$$\leq \Phi(-\delta n^2).$$

Therefore,

$$\sum_{n=1}^{\infty} P\left(W_{\sigma_n} - W_{\tau_n} \leq -\delta\sqrt{\tau_n \ln_2 \tau_n}\right) \leq \sum_{n=1}^{\infty} \Phi(-\delta n^2) < \infty$$

and by the Borel–Cantelli lemma

$$\limsup_{n \to \infty} \frac{W_{\sigma_n} - W_{\tau_n}}{\sqrt{2\tau_n \ln_2 \tau_n}} \geq -\delta \quad \text{a.s.}$$

Since $z_{\sigma_n} = c\sqrt{\ln_2 \tau_n}$ and $W_{\tau_n} = (1 - \delta)\ln_2 \tau_n$ it follows that

$$\limsup_{n \to \infty} \frac{2W_{\sigma_n} z_{\sigma_n}}{\sqrt{\sigma_n + 1}\ln_2 \sigma_n} \geq 2c(1 - 2\delta) \quad \text{a.s.}$$

Hence,

$$\limsup_{n \to \infty} R_{\sigma_n}(z) \geq 2c(1 - 2\delta) - ac^2 = (1 - 2\delta)^2/a > 0 \quad \text{a.s.}$$

and we get (6.5.38).

Appendix

A.1 Basic Facts About SDEs

A.1.1 Existence and Uniqueness of Strong Solutions for SDEs with Random Coefficients

Let $(\Omega, \mathcal{F}, \mathbf{F} = (\mathcal{F}_t)_{t \geq 0}, P)$ be a filtered probability space with an adapted Wiener process $W = (W_t)$ taking values in $\mathbf{R}^d$ and let $\mathcal{P}$ be a predictable σ-algebra, i.e. the σ-algebra in $\Omega \times \mathbf{R}_+$ generated by all adapted continuous processes. We are given functions $a(\omega, t, x)$ and $b(\omega, t, x)$ defined on the set $\Omega \times [0, T] \times \mathbf{R}^k$ and take their values, respectively, in $\mathbf{R}^k$ and in the set of $k \times d$-matrices. Both functions are assumed to be $\mathcal{P} \otimes \mathcal{B}^n$-measurable and satisfying the (usual) Lipschitz condition and a kind of linear growth condition, namely:

$$|a(\omega, t, x) - a(\omega, t, y)| + |b(\omega, t, x) - b(\omega, t, y)| \leq L|x - y|, \quad (A.1.1)$$

$$|a(\omega, t, x)| + |b(\omega, t, x)| \leq L|x| + \lambda_t, \quad (A.1.2)$$

where L is a constant and $\lambda \geq 0$ is an adapted process such that for an integer $m \geq 1$ we have

$$\sup_{t \leq T} E\lambda_t^{2m} < \infty. \quad (A.1.3)$$

Theorem A.1.1 *Under the assumptions (A.1.1)–(A.1.3) there is a unique strong solution of the stochastic differential equation*

$$dX_t = a(t, X_t)dt + b(t, X_t)dW_t, \quad X_0 = \xi, \quad (A.1.4)$$

where ξ is a $\mathcal{F}_0$-measurable random variable such that $E|\xi|^{2m} < \infty$; moreover, $E\|X\|_T^{2m} < \infty$.

Recall the precise meaning of the concept of strong solution: this is a continuous adapted process $X = (X_t)$ on the space $(\Omega, \mathcal{F}, \mathbf{F} = (\mathcal{F}_t)_{t \geq 0}, P)$ such that

$$X = \xi + \int_0^{\cdot} a(s, X_s)ds + \int_0^{\cdot} b(s, X_s)dW_s.$$

The proof of Theorem A.1.1 is similar to that of Theorem 4.6 in [66].

A.1.2 Existence and Uniqueness with a Lyapunov Function

In the above theorem the linear growth condition, though more general than the usual one, is rather restrictive. We provide here another result on existence and uniqueness of the strong solution. It involves a hypothesis on existence of a Lyapunov function for SDE (A.1.4).

Put

$$\mathcal{L} := \frac{\partial}{\partial t} + \sum_{i=1}^{k} a^i(t,x)\frac{\partial}{\partial x_i} + \frac{1}{2}\sum_{i,j=1}^{d} \sigma^{ij}(t,x)\frac{\partial^2}{\partial x_i \partial x_j}$$

with $\sigma(t,x) := b(t,x)b(t,x)^{*T}$.

Definition. We say that $V = V(t,x) \in C^{1,2}([0,T] \times \mathbf{R}^k)$ is a *Lyapunov function* for (A.1.4) if $V \geq 0$, there is a constant $\gamma > 0$ such that

$$\mathcal{L}V \leq \gamma V,$$

and

$$V_*^n := \inf_{(t,x)\in[0,T]\times\bar{U}_n} V(t,x) \to \infty, \quad n \to \infty, \qquad (A.1.5)$$

where $\bar{U}_n := \{x \in \mathbf{R}^k : |x| > n\}$ is the complement of the ball U_n.

Obviously, for if $V(t,x)$ is a Lyapunov function for (A.1.4), then for the function $\tilde{V}(t,x) := e^{-\gamma t}V(t,x)$ we have the inequality $\mathcal{L}\tilde{V} \leq 0$.

Theorem A.1.2 *Assume that for any $N > 0$ there is a constant L_N such that*

$$|a(\omega,t,x) - a(\omega,t,y)| + |b(\omega,t,x) - b(\omega,t,y)| \leq L_N|x-y|, \quad (A.1.6)$$
$$|a(\omega,t,x)| + |b(\omega,t,x)| \leq L_N|x| \qquad (A.1.7)$$

for all $\omega \in \Omega$, $t \in [0,T]$, and $x,y \in U_N$. Suppose that for (A.1.4) there exists a Lyapunov function V. Then for any $\mathcal{F}_0$-measurable random variable ξ the equation (A.1.4) has a unique strong solution.

Proof. For simplicity we assume that ξ is bounded, i.e. $\xi \in U_n$ a.s. for sufficiently large n; only such n will be considered. Put

$$a_n(t,x) := a(t,x)I_{U_n}(x) + a(t,nx/|x|)I_{\bar{U}_n}(x),$$
$$b_n(t,x) := b(t,x)I_{U_n}(x) + b(t,nx/|x|)I_{\bar{U}_n}(x).$$

Let us consider the sequence of continuous processes x^n which are the solutions of the following SDEs:

$$dX_t^n = a_n(t,X_t^n)dt + b_n(t,X_t^n)dW_t, \quad X_0^n = \xi. \qquad (A.1.8)$$

Define also the stopping times $\tau_n := \inf\{t \geq 0 : |X_t^n| \geq n\} \wedge T$. It follows from Theorem A.1.1 that the process X^{n+1} coincides with X^n on $[0,\tau_n]$ and $\tau_n \leq \tau_{n+1}$. Let us show that

$$\lim_{n \to \infty} P(\tau_n < T) = 0. \tag{A.1.9}$$

Indeed, the process $(\partial/\partial x)\tilde{V}(t, X_t^n) = e^{-\gamma t}(\partial/\partial x)V(t, X_t^n)$ is bounded on $[0, \tau_n]$ and, by the Ito formula,

$$E\tilde{V}(\tau_n, X_{\tau_n}^n) - EV(0, \xi) = E\int_0^{\tau_n} \frac{\partial \tilde{V}(t, X_t^n)}{\partial x} dW_t + E\int_0^{\tau_n} \mathcal{L}\tilde{V}(t, X_t^n) dt \leq 0.$$

Thus,

$$EV(\tau_n, X_{\tau_n}^n) \leq e^{\gamma T} EV(0, \xi).$$

Using this bound and the property (A.1.5) of the Lyapunov function we get that

$$P(\tau_n < T) \leq P(V(\tau_n, x_{\tau_n}^n) \geq V_*^n) \leq \frac{e^{\gamma T} EV(0, \xi)}{V_*^n} \to 0, \quad n \to \infty,$$

and (A.1.9) holds. But this means that there exists a process $X = (X_t)_{t \leq T}$ such that $X = X^n$ on $[0, \tau^n]$. Since on $[0, \tau_n]$ the equations (A.1.4) and (A.1.8) coincide (and, in particular, any other solution must be equal to X^n on this set) we obtain the result. $\square$

A.1.3 Moment Bounds for Linear SDEs

Let us consider on a finite interval $[0, T]$ the following linear SDE:

$$dX_t = (AX_t + BY_t + R)dt + (A^1 X_t + B^1 Y_t + R^1)dW_t, \quad X_0 = \xi, \tag{A.1.10}$$

where $E|\xi|^2 < \infty$, X and Y are continuous adapted processes with values in $\mathbf{R}^k$ and $\mathbf{R}^n$, A, B, etc. are matrix-valued predictable processes of compatible dimensions,

$$\|A\|_T + \|B\|_T + \|A^1\|_T + \|B^1\|_T \leq C, \tag{A.1.11}$$

$$\int_0^T |R_s| ds + \int_0^T |R_s^1|^2 ds < \infty. \tag{A.1.12}$$

Suppose that for all $t \in [0, T]$

$$|Y_t| \leq L|X_t| + \varphi_t \tag{A.1.13}$$

where φ is a nonnegative predictable process.

Lemma A.1.3 *Under the assumptions (A.1.11)–(A.1.13) there is a constant $L = L_T$ such that*

$$E(\|X\|_T^2 + \|Y\|_T^2) \leq LE\left[|\xi|^2 + \|\varphi\|_T^2 + \left(\int_0^T |R_s| ds\right)^2 + \int_0^T |R_s^1|^2 ds\right]. \tag{A.1.14}$$

Proof. The inequality is nontrivial only when the expression in the square brackets is finite; we shall assume this. Using elementary inequalities and the assumption (A.1.12) we get that

$$|X_t|^2 \le L \left[|\xi|^2 + \int_0^t (|X_s|^2 + |Y_s|^2) ds + \left(\int_0^t |R_s| ds \right)^2 + \left| \int_0^t \Psi_s dW_s \right|^2 \right]$$

(A.1.15)

where $\Psi_s := A^1 X_s + B^1 Y_s + R^1$ and the constant L depends of T. Put

$$\tau_N := \inf\{ t \ge 0 : |X_s| + |Y_s| \ge N \} \wedge T.$$

By the Doob inequality and (A.1.12)

$$E \sup_{u \le t \wedge \tau_N} \left| \int_0^u \Psi_s dW_s \right|^2 \le 4E \int_0^{t \wedge \tau_N} |\Psi_s|^2 ds$$

$$\le LE \int_0^t (|R_s^1| + |X_{s \wedge \tau_N}^1|^2 + |Y_{s \wedge \tau_N}^1|^2) ds. \quad \text{(A.1.16)}$$

It follows from (A.1.15), (A.1.16), and (A.1.13) that

$$E\|X\|_{t \wedge \tau_N}^2 \le L_T \left(K_T + \int_0^t E\|X\|_{s \wedge \tau_N}^2 ds \right).$$

where K_T is the expression in the square brackets of the right-hand side of (A.1.14). The Gronwall–Bellman lemma implies that for a some constant L we have the inequality $E\|X\|_{T \wedge \tau_N}^2 \le L K_T$ for all N. Since τ_N tends to T almost surely, the same bound holds for $E\|X\|_T^2$. Taking into account (A.1.13) we obtain (A.1.14). $\square$

A.1.4 The Novikov Condition

Let $M = (M_t)_{t \le T}$ be a continuous local martingale. The positive local martingale $\mathcal{E}(M) := e^{M - (1/2)\langle M \rangle}$, solving the linear equation

$$dZ_t = Z_t dM_t, \quad Z_0 = 1,$$

is a local martingale; by the Fatou lemma, it is a supermartingale with $E\mathcal{E}_T(M) \le 1$. A number of sufficient conditions is known to guarantee its *true* martingale property (or, equivalently, the equality $E\mathcal{E}_T(M) = 1$). The most convenient one is the Novikov condition

$$E e^{\frac{1}{2}\langle M \rangle_T} < \infty; \quad \text{(A.1.17)}$$

it is an obvious corollary of the slightly weaker condition

$$\lim_{\varepsilon \downarrow 0} \varepsilon \ln E e^{\frac{1}{2}(1-\varepsilon)\langle M \rangle_T} = 0. \quad \text{(A.1.18)}$$

We give here a short proof of the sufficiency of the latter.

Theorem A.1.4 *If (A.1.18) holds, then $\mathcal{E}(M)$ is a martingale.*

Proof. First, notice that $\mathcal{E}(M)$ is a martingale if $Ee^{(\frac{1}{2}+\delta)\langle M\rangle_T} < \infty$ for some $\delta > 0$. Indeed, let $p := 1 + \delta$ and $q := (1+\delta)/\delta$. There is $r > 1$ such that $(r^2 p - r)q = 1 + 2\delta$. Using the Hölder inequality and the bound $E\mathcal{E}_\tau(rpM) \leq 1$, we get, for any stopping time τ with values in $[0, T]$, that

$$E\mathcal{E}_\tau^\tau(M) = E\mathcal{E}_\tau^{1/p}(rpM)e^{\frac{1}{2}(r^2 p - r)\langle M\rangle_\tau} \leq (Ee^{\frac{1}{2}(r^2 p - r)q\langle M\rangle_\tau})^{1/q}$$
$$\leq (Ee^{(\frac{1}{2}+\delta)\langle M\rangle_T})^{1/q}.$$

Thus, the family $\{\mathcal{E}_\tau(M)\}$ is uniformly integrable and $\mathcal{E}(M)$ is a martingale.

It follows from (A.1.18) that $e^{\frac{1}{2}(1-\varepsilon)\langle M\rangle_T}$ is integrable when $\varepsilon \in \,]0, 1]$. Hence, a smaller random variable $e^{\frac{1}{2}(1+\varepsilon)(1-\varepsilon)^2\langle M\rangle_T}$ is also integrable and, by virtue of the just proven sufficient condition, $\mathcal{E}((1-\varepsilon)M)$ is a martingale. Applying the Hölder inequality (now with $p := 1/(1-\varepsilon)$ and $q := 1/\varepsilon$), we have:

$$1 = E\mathcal{E}_T((1-\varepsilon)M) = E\mathcal{E}_T^{1-\varepsilon}(M)e^{\frac{1}{2}(1-\varepsilon)\varepsilon\langle M\rangle_T}$$
$$\leq (E\mathcal{E}_T(M))^{1-\varepsilon}(Ee^{\frac{1}{2}(1-\varepsilon)\langle M\rangle_T})^\varepsilon.$$

Using (A.1.18) we obtain that $E\mathcal{E}_T(M) \geq 1$ and, hence, $E\mathcal{E}_T(M) = 1$. $\square$

A.2 Exponential Bounds for Fundamental Matrices

A.2.1 Uniform Bound in the Time-Homogeneous Case

It is well-known that if B is a constant stable $q \times q$-matrix, i.e. such that

$$\operatorname{Re}\lambda(B) < -\gamma < 0 \tag{A.2.1}$$

then the corresponding fundamental matrix $\Phi(t, s)$, which is in this case $e^{(t-s)B}$, admits the bound

$$|e^{(t-s)B}| \leq Ce^{-\gamma(t-s)}, \quad 0 \leq s \leq t < \infty, \tag{A.2.2}$$

where the constant C depends only on $|B|$ and the dimension q. In the theory of Lyapunov exponents it is useful the following ramification of this result.

Proposition A.2.1 *Let $U(R, \gamma)$ be the set of $q \times q$ matrices B with $|B| \leq R$ and all eigenvalues satisfying (A.2.1). Then there exists a constant C, depending only on R, γ and q, such that*

$$\sup_{B\in U(R,\gamma)} |e^{(t-s)B}| \leq Ce^{-\gamma(t-s)} \tag{A.2.3}$$

for all $s, t \in \mathbf{R}_+$, $s \leq t$.

Proof. Let B be an arbitrary $q \times q$-matrix and $\Lambda := \max_j \operatorname{Re} \lambda_j$ where $\lambda_1, \ldots, \lambda_q$ are eigenvalues of B. Obviously, (A.2.3) is a consequence of the bound

$$|e^{tB}| \leq e^{t\Lambda} \left(1 + 2|B| \sum_{j=1}^{q-1} \frac{1}{j!} (2t|B|)^j \right). \tag{A.2.4}$$

To show (A.2.4), we recall that for any analytic function $f(\lambda)$ the matrix $f(B)$ is equal to the matrix $P(B)$ where $P(\lambda)$ is any polynomial of degree less or equal to $q - 1$ which has the same values as $f(\lambda)$ on the spectrum of B (this means the coincidence of $m - 1$ derivatives at the eigenvalue of multiplicity m). Assume that all λ_j are distinct and consider the Newton interpolation polynomial

$$P_{q-1}(\lambda) = b_1 + \sum_{j=2}^{q} b_j (\lambda - \lambda_1) \ldots (\lambda - \lambda_j). \tag{A.2.5}$$

The equalities $f_j := f(\lambda_j) = P_{q-1}(\lambda_j)$, $1 \leq j \leq q$, hold if and only if

$$\begin{aligned} f_1 &= b_1, \\ f_2 &= b_1 + b_2(\lambda_2 - \lambda_1), \\ &\cdots \\ f_q &= b_1 + b_2(\lambda_q - \lambda_1) + \ldots + b_q(\lambda_q - \lambda_1) \ldots (\lambda_q - \lambda_{q-1}). \end{aligned}$$

It is easy to show that $b_{j+1} = u_j(\lambda_{j+1})$, $1 \leq j \leq q - 1$, where $u_j(\lambda)$ denotes the integral

$$\int_0^1 \ldots \int_0^{t_{j-1}} f^{(j)}(\lambda_1 + (\lambda_2 - \lambda_1)t_1 + \ldots + (\lambda_j - \lambda_{j-1})t_{j-1} + (\lambda - \lambda_j)t_j) dt_j \ldots dt_1.$$

Notice that on the complex plane the point

$$\lambda_1 + (\lambda_2 - \lambda_1)t_1 + \ldots + (\lambda_{j+1} - \lambda_j)t_j = \lambda_1(1 - t_1) + \lambda_2(t_1 - t_2) + \ldots + \lambda_{j+1}t_j$$

with $0 \leq t_j \leq \ldots \leq t_1 \leq 1$ belongs to the convex hull of the set $\{\lambda_1, \ldots, \lambda_{j+1}\}$ which lies within the circle $\{z : |z| \leq |B|\}$. Hence for the function $f(\lambda) := e^{t\lambda}$ and $1 \leq j \leq q - 1$ we have from the above integral representation that

$$|b_{j+1}| \leq |u_j(\lambda_{j+1})| \leq \frac{1}{j!} \sup_{\{z : |z| \leq |B|\}} f^{(j)}(z) \leq \frac{1}{j!} e^{t\Lambda} (t|B|)^j. \tag{A.2.6}$$

The inequality (A.2.4) follows obviously from (A.2.5) and (A.2.6). By continuity, it holds also in the case of multiple eigenvalues. $\square$

A.2.2 Nonhomogeneous Case

If $A = (A_t)$ is time-dependent then assumptions on eigenvalues do not imply that the corresponding fundamental matrix $\Phi(t, s)$ has an exponential decay. In the example given in Section 1.2 the eigenvalues are negative and constant but the fundamental matrix increases at infinity exponentially fast. In fact, the behavior of the latter is controlled not only by the spectrum but also by the amplitude of changes of A_t at infinity. The following simple lemma shows that the fundamental matrix admits an exponential bound if A_t is bounded and converges to a stable matrix.

Lemma A.2.2 *Assume that $|A_t| \leq c$ for all t and $A_t \to A_\infty$ as $t \to \infty$ where the matrix A_∞ is such that*

$$\operatorname{Re} \lambda(A_\infty) < -2\gamma < 0$$

Then for some constant C we have

$$|\Phi(t, s)| \leq C e^{-\gamma(t-s)} \qquad \forall s, t \in \mathbf{R}_+, \ s \leq t.$$

Proof. Writing the equation for the fundamental matrix as

$$\frac{\partial \Phi(t, s)}{\partial t} = A_\infty \Phi(t, s) + (A_t - A_\infty)\Phi(t, s), \quad \Phi(s, s) = I,$$

and "solving" it, we get the representation

$$\Phi(t, s) = e^{A_\infty(t-s)} + \int_s^t e^{A_\infty(t-u)}(A_u - A_\infty)\Phi(u, s)du.$$

Multiplying both sides by $e^{\gamma(t-s)}$ and denoting

$$\phi(t, s) = e^{\gamma(t-s)}|\Phi(t, s)|$$

we obtain easily, making use the inequality

$$|e^{A_\infty(t-u)}| \leq C e^{-2\gamma(t-u)},$$

that

$$\phi(t, s) \leq C + C \int_s^t e^{-\gamma(t-u)}|A_u - A_\infty|\phi(u, s)du.$$

Fix $\nu \leq \gamma/(2C)$. By assumption there exists $s_0 \geq 0$ such that $|A_u - A_\infty| \leq \nu$ for all $u \geq s_0$. It follows that for every finite $N \geq s_0$

$$\sup_{s_0 \leq t \leq s \leq N} \phi(t, s) \leq C + (1/2) \sup_{s_0 \leq t \leq s \leq N} \phi(t, s).$$

This implies that

$$\sup_{s_0 \le s \le t \le N} \phi(t, s) \le 2C$$

and hence

$$\sup_{s_0 \le s \le t} \phi(t, s) \le 2C.$$

We use here that for s, t varying in the bounded set the function $\phi(t, s)$ remains bounded (due to the Gronwall–Bellman argument).

Notice that for $s \le s_0 \le t$ we have by the semigroup property

$$\Phi(t, s) = \Phi(t, s_0)\Phi(s_0, s)$$

implying, for all $s \le s_0$ and all $t \ge s_0$, the bound $\phi(t, s) \le 2C$.

Thus, the function $\phi(t, s)$ is bounded on the set $\{(s, t) \in \mathbf{R}_+ : s \le t\}$. $\square$

A.2.3 Models with Singular Perturbations

For the model of singular perturbations with continuous A_t which is "uniformly Hurwitz" the fundamental matrix $\Phi^\varepsilon(t, s)$ (corresponding to $\varepsilon^{-1}A_t$) admits an exponential bound (on a finite interval) and this is the key result to the whole theory. Of course, one can explain such a remarkable difference by noticing that after time-stretching the function $\tilde{A}$ will be slowly varying.

Proposition A.2.3 *Let $A = (A_t)$ be a continuous matrix function on $[0, T]$ such that*

$$\mathrm{Re}\,\lambda(A_t) < -2\gamma < 0, \tag{A.2.7}$$

and let $\Phi^\varepsilon(t, s)$ be the fundamental matrix corresponding to $\varepsilon^{-1}A$, i.e. the solution of the matrix equation

$$\frac{\partial \Phi^\varepsilon(t, s)}{\partial t} = \varepsilon^{-1}A_t \Phi^\varepsilon(t, s), \quad \Phi^\varepsilon(s, s) = I. \tag{A.2.8}$$

Then there exist a constant C, depending only on $\|A\|_T := \sup_{t \le T} |A_t|$ and the dimension, and a constant ε_0 such that

$$|\Phi^\varepsilon(t, s)| \le Ce^{-\gamma(t-s)/\varepsilon} \tag{A.2.9}$$

for all $\varepsilon \in\,]0, \varepsilon_0]$ and all s and t with $0 \le s \le t \le T$.

Remark. Due to continuity of A, the assumption (A.2.7) implies that

$$\sup_{t \in [0, T]} \mathrm{Re}\,\lambda(A_t) \le -2\gamma_0$$

for some $\gamma_0 \in\,]0, \gamma[$.

Proof. We need the following assertion.

Lemma A.2.4 *Let $K(t, s)$ be a continuous function on $[0, T]^2$ such that $K(t, t) = 0$. Let*

$$F^\varepsilon(t) := \frac{1}{\varepsilon} \int_0^t K(t, s) e^{-\gamma(t-s)/\varepsilon} ds, \quad \gamma > 0.$$

Then $\|F^\varepsilon\|_T \to 0, \ \varepsilon \to 0$.

Proof. Without loss of generality we can consider $K(t, s) \geq 0$. Fix $\eta > 0$. By the uniform continuity of K there exists $\delta \in]0, 1[$ such that $K(t, s) \leq \eta$ when $|t - s| \leq T\delta$. Define $M := \sup K(t, s)$. We can represent $F^\varepsilon(t)$ as a sum of two integrals

$$I_1^\varepsilon(t) := \frac{1}{\varepsilon} \int_0^{t(1-\delta)} K(t, s) e^{-\gamma(t-s)/\varepsilon} ds$$

and

$$I_2^\varepsilon(t) := \frac{1}{\varepsilon} \int_{t(1-\delta)}^t K(t, s) e^{-\gamma(t-s)/\varepsilon} ds.$$

It is clear that for $t \geq T\delta$

$$I_1^\varepsilon(t) \leq \frac{M}{\varepsilon} \int_0^{t(1-\delta)} e^{-\gamma(t-s)/\varepsilon} ds$$

$$= \frac{M}{\gamma} (e^{-\gamma t\delta/\varepsilon} - e^{-\gamma t/\varepsilon}) \leq \frac{M}{\gamma} e^{-\gamma t\delta/\varepsilon} \leq \frac{M}{\gamma} e^{-\gamma T\delta^2/\varepsilon},$$

$$I_2^\varepsilon(t) \leq \frac{\eta}{\varepsilon} \int_0^t e^{-\gamma(t-s)/\varepsilon} ds \leq \frac{\eta}{\gamma}.$$

For $t < T\delta$, proceeding as for $I_2^\varepsilon(t)$, we get that $F^\varepsilon(t) \leq \eta/\gamma$.

Thus, for sufficiently small ε

$$\|F^\varepsilon\|_T \leq \gamma^{-1}(M e^{-\gamma T\delta^2/\varepsilon} + \eta).$$

As η can be chosen arbitrarily small, this implies the assertion of the lemma. $\square$

Let v be a point from $[s, t]$. Since (A.2.3) can be written as

$$\frac{\partial \Phi^\varepsilon(t, s)}{\partial t} = \varepsilon^{-1} A_v \Phi^\varepsilon(t, s) + \varepsilon^{-1}(A_t - A_v) \Phi^\varepsilon(t, s), \quad \Phi^\varepsilon(s, s) = I,$$

the following representation holds:

$$\Phi^\varepsilon(t, s) = e^{(t-s)A_v/\varepsilon} + \frac{1}{\varepsilon} \int_s^t e^{(t-r)A_v/\varepsilon} \Phi^\varepsilon(r, s)(A_r - A_v) dr.$$

When $v = t$ the last formula has the form

$$\Phi^\varepsilon(t, s) = e^{(t-s)A_t/\varepsilon} + \frac{1}{\varepsilon} \int_s^t e^{(t-r)A_t/\varepsilon}(A_r - A_t) \Phi^\varepsilon(r, s) dr. \qquad (A.2.10)$$

Put $\Delta := \{(s,t) : 0 \leq s \leq t \leq T\}$,

$$W^\varepsilon(t,s) := e^{\gamma(t-s)/\varepsilon}|\Phi^\varepsilon(t,s)|, \quad M^\varepsilon := \sup_{(s,t)\in\Delta} W^\varepsilon(t,s).$$

According to (A.2.7)

$$|e^{uA_t}| < Ce^{-2\gamma u} \tag{A.2.11}$$

for all u and t from the interval $[0,T]$.

Using (A.2.11) we get from (A.2.10) that

$$W^\varepsilon(t,s) \leq C + \frac{C}{\varepsilon}\int_{[s,t]} W^\varepsilon(r,s)|A_r - A_t|e^{-2\gamma(t-r)/\varepsilon}e^{-\gamma(r-s)/\varepsilon}e^{\gamma(t-s)/\varepsilon}dr.$$

This bound and Lemma A.2.4 imply that

$$M^\varepsilon \leq C + CM^\varepsilon \sup_{(s,t)\in\Delta} \frac{1}{\varepsilon}\int_{[s,t]} |A_r - A_t|e^{-\gamma(t-r)/\varepsilon}dr \leq C + M^\varepsilon/2 \tag{A.2.12}$$

for sufficiently small ε.

Thus, $M^\varepsilon \leq 2C$ and the bound (A.2.9) holds for sufficiently small ε. $\square$

Remark. If A is a bounded random process the constant ε_0 (related to the modulus of continuity of the function A) depends, in general, on ω. For convenience of references we modify the statement of Proposition A.2.3 in the following way:

Proposition A.2.5 *Assume that A is a bounded continuous random process with values in the set of $q \times q$ matrices such that the condition (A.2.2) holds for all $t \in [0,T]$. Then there exists a constant C depending only on the $L^\infty(\Omega)$-norm of $\|A\|_T$ and q such that*

$$\limsup_{\varepsilon\to 0} \sup_{0\leq s\leq t\leq T} e^{\gamma(t-s)/\varepsilon}|\Phi^\varepsilon(t,s)| < C. \tag{A.2.13}$$

We give also a result for the fundamental matrix corresponding to $\varepsilon^{-1}A^\varepsilon$ when A^ε converges to the process A satisfying (A.2.7). Its proof follows the same line as that of Lemma A.2.2.

Let $\Phi_\varepsilon^\varepsilon(t,s)$ be a solution of the linear equation

$$\frac{\partial \Phi_\varepsilon^\varepsilon(t,s)}{\partial t} = \varepsilon^{-1}A_t^\varepsilon \Phi_\varepsilon^\varepsilon(t,s), \quad \Phi_\varepsilon^\varepsilon(s,s) = I. \tag{A.2.14}$$

where A^ε is the process with continuous trajectories such that $\|A^\varepsilon\|_T \leq C_A$ a.s.

Proposition A.2.6 *Assume that there exist a bounded matrix-valued process A with continuous trajectories satisfying (A.2.7) and functions $v : \mathbf{R}_+ \to \mathbf{R}_+$ and $\delta : [0,1] \to \mathbf{R}_+$ such that*

$$\int_0^\infty v(s)ds < \infty, \quad \lim_{\varepsilon \to 0} \delta(\varepsilon) = 0,$$

and for all $t \in [0, T]$

$$|A_t^\varepsilon - A_t| \le v(t/\varepsilon) + \delta(\varepsilon).$$

Then there is a constant C_1 such that

$$\limsup_{\varepsilon \to 0} \sup_{0 \le s \le t \le T} e^{(1/2)\gamma(t-s)/\varepsilon} |\Phi_\varepsilon^\varepsilon(t, s)| < C_1. \tag{A.2.15}$$

Proof. Applying the Cauchy formula to the representation

$$\frac{\partial \Phi_\varepsilon^\varepsilon(t, s)}{\partial t} = \varepsilon^{-1} A_t \Phi_\varepsilon^\varepsilon(t, s) + \varepsilon^{-1}(A_t^\varepsilon - A_t)\Phi_\varepsilon^\varepsilon(t, s), \quad \Phi_\varepsilon^\varepsilon(s, s) = I,$$

we get that

$$\Phi_\varepsilon^\varepsilon(t, s) = \Phi^\varepsilon(t, s) + \varepsilon^{-1} \int_s^t \Phi^\varepsilon(t, u)(A_u^\varepsilon - A_u)\Phi_\varepsilon^\varepsilon(u, s)du \tag{A.2.16}$$

where $\Phi^\varepsilon(t, s)$ is the fundamental matrix given by (A.2.8). By Proposition A.2.5 there exists a constant C such that

$$\limsup_{\varepsilon \to 0} \sup_{0 \le s \le t \le T} e^{\gamma(t-s)/\varepsilon} |\Phi^\varepsilon(t, s)| < C. \tag{A.2.17}$$

Choose $r_0 > 0$ such that

$$\int_{r_0}^\infty v(s)ds \le (4C)^{-1}.$$

Put

$$\phi^\varepsilon(t, s) := e^{(1/2)\gamma(t-s)/\varepsilon}|\Phi_\varepsilon^\varepsilon(t, s)|,$$

$$\widehat{\phi}_{t_0}^\varepsilon := \sup_{(s, t) \in \Delta_{t_0}} \phi^\varepsilon(t, s)$$

where $\Delta_{t_0} := \{(s, t) : t_0 \le s \le t \le T\}$, $t_0 := r_0\varepsilon$. The relations (A.2.15)–(A.2.17) imply that there exists ε_0 (depending on ω) such that for all $\varepsilon \le \varepsilon_0$ the function $\phi^\varepsilon(t, s)$ is dominated by

$$C + C\varepsilon^{-1}e^{(1/2)\gamma(t-s)/\varepsilon} \int_s^t e^{-\gamma(t-u)/\varepsilon}e^{-(1/2)\gamma(u-s)/\varepsilon}|A_u^\varepsilon - A_u|\phi^\varepsilon(u, s)du$$

$$\le C + C\varepsilon^{-1} \int_s^t e^{-(1/2)\gamma(t-u)/\varepsilon}(v(u/\varepsilon) + \delta(\varepsilon))\phi^\varepsilon(u, s)du.$$

It follows that

$$\widehat{\phi}_{t_0}^\varepsilon \le C + C\left(\int_{r_0}^\infty v(s)ds + 2\delta(\varepsilon)/\gamma\right)\widehat{\phi}_{t_0}^\varepsilon \le C + (1/2)C\widehat{\phi}_{t_0}^\varepsilon$$

for sufficiently small ε. Hence,

$$\limsup_{\varepsilon \to 0} \widehat{\phi}^{\varepsilon}_{t_0} \leq 2C. \tag{A.2.18}$$

The standard application of the Gronwall–Bellman lemma implies that

$$|\Phi^{\varepsilon}_{\varepsilon}(t,s)| \leq \sqrt{q} e^{C_A(t-s)/\varepsilon}$$

for all $s \leq t \leq T$. This bound leads to the inequality

$$\limsup_{\varepsilon \to 0} \sup_{0 \leq s \leq t \leq t_0} \phi^{\varepsilon}(t,s) \leq \sqrt{q} e^{(C_A + \gamma/2) r_0}. \tag{A.2.19}$$

The semigroup property of the fundamental matrix yields that for $s \leq t_0 \leq t$

$$\phi^{\varepsilon}(t,s) \leq \phi^{\varepsilon}(t_0,s)\phi^{\varepsilon}(t,t_0).$$

Thus, we have that

$$\sup_{0 \leq s \leq t \leq T} \phi^{\varepsilon}(t,s) \leq \widehat{\phi}^{\varepsilon}(t_0) + \sup_{0 \leq s \leq t \leq t_0} \phi^{\varepsilon}(t,s)(1 + \widehat{\phi}^{\varepsilon}(t_0)) \tag{A.2.20}$$

and the result follows from (A.2.18)–(A.2.20). □

A.3 Total Variation Distance and Hellinger Processes

A.3.1 Total Variation Distance and Hellinger Integrals

Let P and $\tilde{P}$ be two probability measures on a measurable space $(\Omega, \mathcal{F})$. By definition, the *total variation distance* between P and $\tilde{P}$ is given by

$$\mathrm{Var}\,(P - \tilde{P}) := \sup_{\xi \in U} |E\xi - \tilde{E}\xi|$$

where $\tilde{E}$ stands for the integral with respect to $\tilde{P}$ and U is the set of random variables ξ with values in $[-1, 1]$.

Put $Q := (P + \tilde{P})/2$, $z := dP/dQ$, and $\tilde{z} := d\tilde{P}/dQ$. Clearly, $z + \tilde{z} = 2$ (Q-a.s.).

If R is a probability measure dominating both P and $\tilde{P}$, $z_R := dP/dR$ and $\tilde{z}_R := d\tilde{P}/dR$ then

$$\mathrm{Var}\,(P - \tilde{P}) = \sup_{\xi \in U} |E_R \xi(z_R - \tilde{z}_R)| = E_R |z_R - \tilde{z}_R|.$$

In other words, the total variation distance is the $L^1(R)$-distance between densities where the dominating measure R can be chosen arbitrarily. In particular,

$$\text{Var}\,(P - \tilde{P}) = E_Q|z - \tilde{z}| = 2E_Q|z - 1|.$$

The *Hellinger–Kakutani distance* d_H is defined (up to the coefficient $1/2$) as the $L^2(R)$-distance between square roots of densities:

$$d_H^2(P, \tilde{P}) := (1/2)E_R(\sqrt{z_R} - \sqrt{\tilde{z}_R})^2 = 1 - E_R z_R^{1/2} \tilde{z}_R^{1/2}.$$

The functional $H(P, \tilde{P}) := E_R z_R^{1/2} \tilde{z}_R^{1/2}$ is called the *Hellinger integral*. Its generalization, the *Hellinger integral of order* $\alpha \in\,]0, 1[$,

$$H(\alpha, P, \tilde{P}) := E_R z_R^{\alpha} \tilde{z}_R^{1-\alpha}$$

is a useful tool in various problems concerning relationships between measures such as absolute continuity, singularity, contiguity, etc. The parameter $\alpha = 1/2$ is usually omitted in notations. It is easy to see that the above functionals are also independent on the dominating measure R and $H(\alpha, P, \tilde{P}) = E_Q z^{\alpha} \tilde{z}^{1-\alpha}$.

The following simple inequalities are classic:

$$2(1 - H(P, \tilde{P})) \leq \text{Var}\,(P - \tilde{P}) \leq \sqrt{8(1 - H(P, \tilde{P}))}. \qquad (A.3.1)$$

They can be rewritten as

$$2d_H^2(P, \tilde{P}) \leq \text{Var}\,(P - \tilde{P}) \leq 2\sqrt{2}d_H(P, \tilde{P})$$

making clear that the Hellinger–Kakutani distance defines the same topology on the space of probability measures as the total variation.

More generally,

$$2(1 - H(\alpha, P, \tilde{P})) \leq \text{Var}\,(P - \tilde{P}) \leq \sqrt{c_\alpha(1 - H(\alpha, P, \tilde{P}))} \qquad (A.3.2)$$

where c_α is a constant.

A.3.2 The Hellinger Processes

Let $\mathbf{F} = (\mathcal{F})_{t \geq 0}$ be a right-continuous filtration on $(\Omega, \mathcal{F})$. We assume that the probability space $(\Omega, \mathcal{F}, Q)$ is complete, $\mathcal{F} = \mathcal{F}_\infty$, and $\mathcal{F}_0$ contains the Q-null sets of $\mathcal{F}$. Let $Z = (Z_t)$ and $\tilde{Z} = (\tilde{Z}_t)$ be the density processes of P and $\tilde{P}$ with respect to Q, i.e. right-continuous Q-martingales such that

$$Z_t = E_Q\left(\frac{dP}{dQ}\Big|\mathcal{F}_t\right), \qquad \tilde{Z}_t = E_Q\left(\frac{d\tilde{P}}{dQ}\Big|\mathcal{F}_t\right)$$

for all $t \geq 0$. For any stopping time τ we have $Z_\tau = dP_\tau/dQ_\tau$, $Z_\tau = dP_\tau/dQ_\tau$ where the subscript is used to denote the restrictions of measures to the σ-algebra $\mathcal{F}_\tau$.

For simplicity we assume that $P_0 = \tilde{P}_0$, i.e. $Z_0 = \tilde{Z}_0 = 1$.

The function $f(x, y) := x^\alpha y^{1-\alpha}$ defined on $\mathbf{R}_+^2$ being concave, the process

$$Y(\alpha) := Z^\alpha \tilde{Z}^{1-\alpha}$$

is a positive Q-supermartingale; it is bounded and $Y_0(\alpha) = 1$. By the Doob–Meyer theorem

$$Y(\alpha) = 1 + M(\alpha) - A(\alpha)$$

where $M(\alpha)$ is a Q-martingale and $A(\alpha)$ is a predictable increasing process starting from zero.

Let $\sigma_n := \inf\{t : Z_t \wedge \tilde{Z}_t \leq 1/n\}$. By definition, the *Hellinger process* is any predictable increasing right-continuous process $h(\alpha) = h(\alpha, P, \tilde{P})$ such that

$$I_\Gamma dh(\alpha) = I_\Gamma Y_-^{-1}(\alpha) dA(\alpha), \quad h_0(\alpha) = 0,$$

where $\Gamma := \cup_n [0, \sigma_n]$. So, the Hellinger process is uniquely defined (up to Q-null sets) on the union of stochastic intervals $[0, \sigma_n]$ but outside of Γ its behavior has no importance. Usually, $\alpha = 1/2$ is omitted in notations.

Theorem A.3.1 *For any stopping time τ*

$$2\left(1 - \sqrt{Ee^{-h_\tau}}\right) \leq \mathrm{Var}\,(P_\tau - \tilde{P}_\tau) \leq 4\sqrt{Eh_\tau} \tag{A.3.3}$$

and

$$\mathrm{Var}\,(P_\tau - \tilde{P}_\tau) \leq 6\sqrt{\varepsilon} + P(h_\tau \geq \varepsilon) \tag{A.3.4}$$

where $\varepsilon > 0$ is arbitrary.

Proof. We give here the arguments under the assumption that $\tilde{P} \sim P$ and the density processes are continuous, i.e. only for the case needed in our study. Now $Y(\alpha)$ is strictly positive and $dh(\alpha) = Y^{-1}(\alpha) dA(\alpha)$. By the Ito formula applied, e.g., to the function $x^{1/2}(2-x)^{1/2}$ and the continuous Q-martingale Z we get that

$$dY = dm - \frac{1}{2}(Z\tilde{Z})^{-3/2} d\langle Z \rangle$$

where m is a local Q-martingale. The uniqueness of the Doob–Meyer decomposition implies the representation

$$dh = \frac{1}{2}(Z\tilde{Z})^{-2} d\langle Z \rangle. \tag{A.3.5}$$

Since $(Z\tilde{Z})^{1/2} \leq (Z + \tilde{Z})/2 = 1$ and $0 \leq \tilde{Z} \leq 2$ we have the bound

$$\langle Z \rangle_t \leq 4 \int_0^t Z_s dh_s.$$

Hence,

$$E_Q \langle Z \rangle_\tau \leq 4 E_Q \int_0^\tau Z_s dh_s = 4 E_Q Z_\tau h_\tau = 4 E h_\tau.$$

With this remark the upper bound in (A.3.3) becomes obvious since

$$\text{Var}\,(P_\tau - \tilde{P}_\tau) = 2 E_Q |Z_\tau - 1| \leq 2 \sqrt{E_Q |Z_\tau - 1|^2} = 2 \sqrt{E_Q \langle Z \rangle_\tau}.$$

To prove the lower bound notice that the process $S(\alpha)$ with

$$dS(\alpha) = e^{h(\alpha)} dM(\alpha)$$

and $S(\alpha) = 1$ is a local Q-martingale. By the Ito formula

$$Y(\alpha) = e^{-h(\alpha)} S(\alpha).$$

Being positive, $S(\alpha)$ is a Q-supermartingale with $E_Q S_\tau(\alpha) \leq 1$. Using the evident identity $Y(3/4) = \sqrt{S e^{-h} Z}$ we have:

$$H(3/4, P, \tilde{P}) = E_Q Y_\tau(3/4) = E_Q \sqrt{S_\tau} \sqrt{Z_\tau e^{-h_\tau}}$$
$$\leq \sqrt{E_Q S_\tau E_Q Z_\tau e^{-h_\tau}} \leq \sqrt{E e^{-h_\tau}}$$

and it remains to apply the lower bounds in (A.3.2) with $\alpha = 3/4$.

To show (A.3.4) observe that for any stopping time σ such that $\sigma \leq \tau$ we have

$$\text{Var}\,(P_\tau - \tilde{P}_\tau) \leq (3/2) \text{Var}\,(P_\sigma - \tilde{P}_\sigma) + 2 P(\sigma < \tau). \qquad (A.3.6)$$

Indeed,

$$\text{Var}\,(P_\tau - \tilde{P}_\tau) = E_Q |Z_\tau - \tilde{Z}_\tau| I_{\{\sigma = \tau\}} + E_Q |Z_\tau - \tilde{Z}_\tau| I_{\{\sigma < \tau\}} \leq$$
$$\leq \text{Var}\,(P_\sigma - \tilde{P}_\sigma) + 2 Q(\sigma < \tau).$$

The bound (A.3.6) follows from here since

$$2 Q(\sigma < \tau) = P(\sigma < \tau) + \tilde{P}(\sigma < \tau) \leq 2 P(\sigma < \tau) + |P(\sigma < \tau) - \tilde{P}(\sigma < \tau)|$$

and for the set $\{\sigma < \tau\} \in \mathcal{F}_\sigma$ we have

$$|P(\sigma < \tau) - \tilde{P}(\sigma < \tau)| \leq (1/2) \text{Var}\,(P_\sigma - \tilde{P}_\sigma).$$

Applying successively (A.3.6) with $\sigma := \inf\{t : h_t \geq \varepsilon\}$ and the upper bound in (A.3.3) we get that

$$\text{Var}\,(P_\tau - \tilde{P}_\tau) \leq 6 \sqrt{E h_\sigma} + 2 P(\sigma < \tau) \leq 6 \sqrt{\varepsilon} + 2 P(h_\tau \geq \varepsilon)$$

where we use at the last step the continuity of h. $\square$

The lower bound in (A.3.3) and the upper bound (A.3.4) yield the following:

Corollary A.3.2 *Let (P^n) and $(\tilde{P}^n)$ be two sequences of probability measures on a filtered space $(\Omega, \mathcal{F}, \mathbf{F})$, let $h^n := h(P^n, \tilde{P}^n)$, and let τ be an arbitrary stopping time. Then*

$$\text{Var}\,(P_\tau^n - \tilde{P}_\tau^n) \to 0 \qquad \Longleftrightarrow \qquad P^n(h_\tau^n \geq \varepsilon) \to 0 \quad \forall \varepsilon > 0. \qquad (A.3.7)$$

A.3.3 Example: Diffusion-Type Processes

Let $(X, \mathcal{X})$ be the space of continuous functions $x = (x_t)$ with values in $\mathbf{R}^d$ and let $\mathbf{X} := (\mathcal{X}_t)$ be the smallest right-continuous filtration generated by the coordinate mappings; in other words, $\mathcal{X}_t := \cap_{\varepsilon \geq 0} \mathcal{X}_{t+\varepsilon}^o$, where the σ-algebra $\mathcal{X}_t^o = \sigma\{x_s, \ s \leq t\}$.

Let $a = a(t, x)$ and $b = b(t, x)$ be two predictable processes taking values in $\mathbf{R}^d$ and in the space of $d \times d$-matrices, respectively.

Assume that a process Y given on a filtered probability space $(\Omega, \mathcal{F}, \mathbf{F}, P)$ is of the *diffusion type* with the *drift coefficient* a and the *diffusion* coefficient b starting from the point $y \in \mathbf{R}^d$. This means that

$$dY_t = a(t, Y)dt + b(t, Y)dW_t, \quad Y_0 = y,$$

where W is an adapted d-dimensional Wiener process. Let P_Y be the distribution of Y, i.e. the probability measure $P_Y = PY^{-1}$ on $(X, \mathcal{X})$. We consider also the process Y (defined, maybe, on another space) with

$$d\tilde{Y}_t = \tilde{a}(t, \tilde{Y})dt + b(t, \tilde{Y})d\tilde{W}_t, \quad \tilde{Y}_0 = y,$$

having the same diffusion coefficient and initial condition as Y.

Let $B = bb^\star$. We suppose that

(a) for every x, x^1, $x^2 \in X$ and $t \geq 0$

$$|B(t, x^1) - B(t, x^2)| \leq L|x_t^1 - x_t^2|,$$

$$|B(t, x)| \leq L(1 + |x|)^2;$$

(b) for all $x \in X$ and $t \geq 0$ the matrix $b(t, x)$ is nondegenerated.

The assumption (a) provides the existence and uniqueness of the strong solution of the stochastic equation

$$d\bar{Y}_t = \tilde{a}(t, \bar{Y})dt + b(t, \bar{Y})dW_t, \quad \bar{Y}_0 = y.$$

The next result provide the expression for the Hellinger process for the pair $(P_Y, \tilde{P}_{\tilde{Y}})$.

Proposition A.3.3 *In addition to (a) and (b) assume that for any finite t*

$$\int_0^t |B^{-1/2}(s, x)a(s, x)|^2 ds < \infty \quad P_Y\text{-a.s.},$$

$$\int_0^t |B^{-1/2}(s, x)\tilde{a}(s, x)|^2 ds < \infty \quad P_{\tilde{Y}}\text{-a.s.}$$

Then

$$h_t(P_Y, P_{\tilde{Y}}) = \frac{1}{8} \int_0^t |B^{-1/2}(s, x)(a(s, x) - \tilde{a}(s, x)|^2 ds < \infty.$$

Proof. Let $R := P_{\bar{Y}}$ be the distribution of $\bar{Y}$. According to Section 6 of Ch. 7 in [66] $P_Y \ll R$, $P_{\tilde{Y}} \ll R$, and the density processes Z^R and $\tilde{Z}^R$ of P_Y and $P_{\tilde{Y}}$ with respect to R are the solutions of the linear equations

$$dZ_t^R = Z_t^R(b^{-1}(t,x)a(t,x))^\star dx_t, \quad Z_0^R = 1,$$

$$d\tilde{Z}_t^R = \tilde{Z}_t^R(b^{-1}(t,x)\tilde{a}(t,x))^\star dx_t, \quad \tilde{Z}_0^R = 1.$$

The Ito formula implies that the process $Y^R = \sqrt{Z^R\tilde{Z}^R}$ on the set $\{Y^R > 0\}$ admits the representation

$$dY^R = -(1/8)Y_t^R(a(t,x) - \tilde{a}(t,x))^\star(b(t,x)b(t,x)^\star)^{-1}(a(t,x) - \tilde{a}(t,x))dt$$

$$+ (1/2)Y_t^R(b(t,x)^{-1}(a(t,x) + \tilde{a}(t,x)))^\star dx_t$$

and the result follows. □

A.4 Hausdorff Metric

Let (X,ρ) be a metric space and let $\mathbf{K}_X$ be the class of all its nonempty compact subsets. As usual, $\rho(x,A) := \inf_{y \in A} \rho(x,y)$ is the distance from the point x to the set A. Let $A_\delta := \{x \in E : \rho(x,A) < \delta\}$ denote the δ-neighborhood of A. For $A, B \in \mathbf{K}_X$ we put $l(A,B) := \sup_{z \in A} \rho(z,B)$.

The *Hausdorff distance* between A and B is given by the formula

$$\rho_H(A,B) := l(A,B) \vee l(B,A)$$

or, equivalently,

$$\rho_H(A,B) = \inf\{\delta > 0 : A \subseteq B_\delta, B \subseteq A_\delta\}.$$

The triangle inequality follows immediately from the latter representation. Indeed, if $A \subseteq B_\delta$, $B \subseteq C_\gamma$ then $B_\delta \subseteq C_{\delta+\gamma}$ and, hence, $A \subseteq C_{\delta+\gamma}$. Taking $\delta_n \downarrow \rho_H(A,B)$ and $\gamma_n \downarrow \rho_H(B,C)$ we get from the above observation that

$$\rho_H(A,C) \leq \lim_n(\delta_n + \gamma_n) = \rho_H(A,B) + \rho_H(B,C).$$

Proposition A.4.1 *Let (A^m) be a sequence of sets from $\mathbf{K}_X$. Assume that there exists $C \in \mathbf{K}_X$ such that $A^m \subseteq C$ for all m. Then $\lim_m \rho_H(A^m, A) = 0$ if and only if the following two conditions are satisfied for any subsequence of indices (n):*

(a) for any convergent sequence $x^n \in A^n$ its limit x belongs to A;

(b) for any point $x \in A$ there exists a subsequence $x^{n_k} \in A_{n_k}$ converging to x.

Proof. ($\Rightarrow$) For any $\delta > 0$ we have that $A^n \subseteq A_\delta$ for sufficiently large n and, hence, $\rho(x^n, A) < \delta$ for such n implying $\rho(x, A) \leq \delta$, the relation which is possible only if $x \in A$. On the other hand, for any k there exists A^{n_k} such that $A \subseteq A^{n_k}_{1/k}$ and for any $x \in A$ one can find x^{n_k} with $\rho(x^{n_k}, x) \leq 1/k$.

($\Leftarrow$) Suppose that $\lim_m \rho_H(A^m, A) \neq 0$. Then $\rho_H(A^n, A) \geq 2\delta$ for some subsequence (n) and a number $\delta > 0$. We have that either the inclusions $A^n \subseteq A_\delta$ or the inclusions $A \subseteq A^n_\delta$ are violated infinitely often. In the first case, there exists a subsequence $x^{n_k} \in A^{n_k}$ such that $\rho(x^{n_k}, A) \geq \delta$. Since x^{n_k} belongs to the compact C, there exists a further subsequence converging to some x with $\rho(x, A) \geq \delta$ in contradiction with (a). In the second case, one can find points $y^{n_k} \in A$ with $\rho(y^{n_k}, A^{n_k}) \geq \delta$. We may always assume that y^{n_k} converges to $x \in A$, and it is clear that this point cannot be approximated by elements taken from any subsequence of the sets A^{n_k} in contradiction with (b). $\square$

Remark. Notice that the assumption that all A^n are subsets of a compact C is not needed for ($\Rightarrow$) but the opposite implication for nonconvex sets fails, in general, without this assumption. Indeed, let $A^m := \{0\} \cup \{m\}$ and $A := \{0\}$. Then, obviously, (a) and (b) are fulfilled but $\rho_H(A^m, A) \to \infty$.

A.5 Measurable Selection

A.5.1 Aumann Theorem

We shall work assuming that $(\Omega, \mathcal{F}, P)$ is a complete probability space and E is a Polish space (i.e. complete separable metric space) with the Borel σ-algebra $\mathcal{E}$. Let us denote by π_Ω the projection operator from $\Omega \times E$ onto Ω.

The graph a function $\xi : \Omega \to E$ is the set $\operatorname{Gr} \xi := \{(\omega, x) : \ x = \xi(\omega)\}$. The graph of a set-valued mapping $\Gamma : \Omega \to 2^E$ is the set

$$\operatorname{Gr} \Gamma := \{(\omega, x) : \ x \in \Gamma(\omega)\}.$$

We say that a set-valued mapping Γ is measurable if $\operatorname{Gr} \Gamma \in \mathcal{F} \otimes \mathcal{E}$.

Theorem A.5.1 *Let $A \in \mathcal{F} \otimes \mathcal{E}$ be such that $\pi_\Omega A = \Omega$. Then there is a measurable function $\xi : \Omega \to E$ such that $\operatorname{Gr} \xi \subseteq A$.*

This theorem can be formulated in the following, obviously, equivalent way:

Theorem A.5.2 *Let Γ be a measurable set-valued mapping with non-empty values. Then there is a measurable function $\xi : \Omega \to E$ such that $\operatorname{Gr} \xi \subseteq \operatorname{Gr} \Gamma$.*

The function ξ is called a (measurable) selector.

The above theorems have various slightly different formulations. For instance, one can assume that P is a σ-finite measure. If the σ-algebra is not completed one can claim the existence only of an a.s. selector (i.e. a measurable function ξ such that $\xi(\omega) \in \Gamma(\omega)$ a.s.). The extension to the case where the values of $\Gamma(\omega)$ may include the empty set is also easy. There are less trivial generalizations of the measurable selection theorem to the case of the Lusin space $(E, \mathcal{E})$. The reader is asked to consult the details in the literature: numerous textbooks (on optimal control and also on mathematical economics) deal with this subject.

A.5.2 Filippov Implicit Function Lemma

Let $(Y, \mathcal{Y})$ be an arbitrary measurable space, $\Gamma : \Omega \to 2^E$ and $\Theta : \Omega \to 2^Y$ be two measurable set-valued mappings with non-empty values.

The following result is a version of the Filippov implicit function lemma.

Lemma A.5.3 *Let $g : \Omega \times E \to Y$ be a measurable function such that for every ω*

$$g(\omega, \Gamma(\omega)) \cap \Theta(\omega) \neq \emptyset.$$

Then there is a measurable selector ξ of Γ such that $g(\omega, \xi(\omega)) \in \Theta(\omega)$.

Proof. Put $\Sigma(\omega) := x \in \Gamma(\omega) : g(\omega, x) \in \Theta(\omega)$. Then

$$\mathrm{Gr}\,\Sigma = h^{-1}(\mathrm{Gr}\,\Theta) \cap \mathrm{Gr}\,\Gamma$$

where $h : (\omega, x) \mapsto (\omega, g(\omega, x))$. Since h is a measurable mapping of $(\Omega \times E, \mathcal{F} \otimes \mathcal{E})$ into $(\Omega \times Y, \mathcal{F} \otimes \mathcal{Y})$, the set $\mathrm{Gr}\,\Sigma$ is $\mathcal{F} \times \mathcal{E}$-measurable and we can refer to Theorem A.5.2. $\square$

A.5.3 Measurable Version of the Carathéodory Theorem

Let $(\Omega, \mathcal{F}, P)$ be a complete probability space and let Γ be measurable and such that its values are non-empty convex compact subsets of $\mathbf{R}^d$. We denote by $\mathrm{ex}\,\Gamma$ the set-valued mapping with $\mathrm{ex}\,\Gamma(\omega)$ equal to the set of extreme points of $\Gamma(\omega)$.

Theorem A.5.4 *Let ξ be a measurable selector of Γ. Then there exist measurable selectors ξ_i of $\mathrm{ex}\,\Gamma$ and measurable positive functions α_i, $i = 0, \ldots, d$, with $\sum_i \alpha^i = 1$ such that*

$$\xi = \sum_{i=0}^{d} \alpha_i \xi_i.$$

Proof. It is based on the following:

Proposition A.5.5 *The set-valued mapping* $\mathrm{ex}\,\Gamma$ *is measurable.*

With this result (the sketch of its proof is given below) the argument is standard. We define the measurable set-valued mapping Σ with

$$\Sigma(\omega) := \Lambda_{d+1} \times (\mathrm{ex}\,\Gamma(\omega))^{d+1}$$

where

$$\Lambda_{d+1} := \left\{ (\lambda_0, \ldots, \lambda_d) \in \mathbf{R}_+^{d+1} : \sum_{i=0}^{d} \lambda_i = 1 \right\}.$$

Let us consider the function $h : \Lambda_{d+1} \times (\mathbf{R}^d)^{d+1} \to \mathbf{R}^d$ with

$$h(\lambda_0, \ldots, \lambda_d, x_0, \ldots, x_d) := \sum_{i=0}^{d} \lambda_i x_i.$$

By the classic Carathéodory theorem $G(\omega) = h(\Sigma(\omega))$ and we conclude with the Filippov lemma. $\square$

Proof of Proposition A.5.5. We need a suitable characterization of the set extreme points of a non-empty convex compact $K \subset \mathbf{R}^d$. To this aim we associate with a (real) function $f \in C(K)$ its upper concave envelope

$$\widehat{f} := \inf_{y \in \mathbf{R}^d} (h_f(y) + (y, x))$$

where

$$h_f(y) := \sup_{z \in K} (f(z) - (z, y)).$$

Clearly, $\widehat{f}(x) \geq f(x)$ for all $x \in K$. The following characterization of the set of extreme points can be found in [2].

Lemma A.5.6 $\mathrm{ex}\,K = \{x \in K : f(x) = \widehat{f}(x) \,\forall x \in K\}$.

Lemma A.5.7 *Let* $f_0 \in C(K)$ *be strictly convex. Then*

$$\mathrm{ex}\,K = \{x \in K : f_0(x) = \widehat{f_0}(x)\}.$$

Proof. The inclusion $\subseteq$ holds by virtue of the previous lemma. To show the converse let us assume that $f_0(x) = \widehat{f_0}(x)$ but $x \notin \mathrm{ex}\,K$ and, therefore, $x = \alpha x_1 + (1 - \alpha)x_2$ for some $\alpha \in \,]0, 1[$ and $x_1, x_2 \in K$. Since $\widehat{f}$ is concave and f is strictly convex

$$f(x) = \widehat{f}(x) \geq \alpha \widehat{f}(x_1) + (1 - \alpha)\widehat{f}(x_1) \geq \alpha f(x_1) + (1 - \alpha)f(x_1) > f(x)$$

and we get a contradiction. $\square$

Let v_i form a dense subset in $\mathbf{R}^d$ and $\rho_i(\omega) := \sup_{x \in \Gamma(\omega)} |(v_i, x)|$. Define the function $f_0 : \omega \times \mathbf{R}^d \to [0, \infty]$ by putting

$$f_0(\omega, x) := \sum_{i=1}^{d} \frac{1}{2^i} \frac{(v_i, x)^2}{1 + \rho_i(\omega)}, \quad x \in \Gamma(\omega),$$

and $f_0(\omega, x) := \infty$ if $x \in \Gamma(\omega)$.

Clearly, f is measurable and for every ω its section $f_0(\omega, .)$ is a strictly convex continuous function on $\Gamma(\omega)$. It is easy to check that $\widehat{f_0}$ is also measurable. Thus, the set $\{(\omega, x) : f_0(\omega, x) = \widehat{f_0}(\omega, x)\}$ is measurable and it remains to refer to the description given by the above lemma. $\square$

A.6 Compact Sets in $\mathbf{P}(X)$

We present several specific results on a topological structure of some sets in the space $\mathbf{P}(X)$ over a Polish metric space X. The techniques developed here is used to prove the compactness of the limiting attainability set $\mathcal{K}_0$.

A.6.1 Notations and Preliminaries

Let X be a Polish metric space with a metric d and the Borel σ-algebra $\mathcal{X}$. Let $\mathbf{P}(X)$ be the space of all probability measures on X with the topology of weak convergence in the probabilistic sense, i.e. the topology $\sigma\{\mathbf{P}(X), C_b(X)\}$ where $C_b(X)$ is the space of bounded continuous functions on X. We shall use the notation $m(f) := \int_X f(x) m(dx)$. In the weak topology neighborhoods of a point $m_0 \in \mathbf{P}(X)$ are finite intersections of sets of the form

$$\{m : |m(f) - m_0(f)| < \delta\}, \quad \delta > 0.$$

This topology is generated by the Lévy–Prohorov metric

$$L(m_1, m_2) := \sigma(m_1, m_2) \vee \sigma(m_2, m_1)$$

where

$$\sigma(m_1, m_2) := \inf\{\varepsilon > 0 : m_1(F) \leq m_2(F^\varepsilon) + \varepsilon \text{ for all closed } F\}$$

with $F^\varepsilon := \{x \in X : d(x, F) < \varepsilon\}$. The important property $\mathbf{P}(X)$ with this metric is again a Polish space.

A set $A \subseteq \mathbf{P}(X)$ is said to be tight if for any $\varepsilon > 0$ there exists a compact $K \subseteq X$ such that $m(K) \geq 1 - \varepsilon$ for all $m \in A$. The relative compactness of a set $A \subseteq \mathbf{P}(X)$ is equivalent to its tightness.

For a variable ξ given on a probability space $(\Omega, \mathcal{F}, P)$ and taking values in X we denote by $\mathcal{L}(\xi)$ its distribution $P\xi^{-1}$ which is an element of $\mathbf{P}(X)$.

Let $(X, \mathcal{X})$ and $(Y, \mathcal{Y})$ be two Polish spaces. We denote by $\mathcal{M}(X, Y)$ the set of stochastic kernels from $(X, \mathcal{X})$ to $(Y, \mathcal{Y})$, that is the mappings

$$\mu : X \times \mathcal{Y} \to ([0, 1], \mathcal{B}[0, 1])$$

such that $x \mapsto \mu(x, \Gamma)$ is $\mathcal{X}$-measurable for every $\Gamma \in \mathcal{Y}$ and $\mu(x, .) \in \mathbf{P}(Y)$ for every $x \in X$.

It is easy to check that the mapping $\mu : X \times \mathcal{Y} \to ([0, 1], \mathcal{B}[0, 1])$ is in $\mathcal{M}(X, Y)$ if and only if one of the following equivalent conditions is satisfied:

(1) for every x the mapping $x \mapsto \mu(x, .)$ is $\mathcal{X}$-measurable (in other words, $\mu(x, .)$ is a $\mathbf{P}(Y)$-valued random variable);

(2) for every $f \in C_b(Y)$ the mapping $x \mapsto \mu(x, f)$ is $\mathcal{X}$-measurable (i.e. $\mu(x, f)$ is a real-valued random variable).

A measure $m(dx, dy)$ on the product of two Polish spaces $(X \times Y, \mathcal{X} \otimes \mathcal{Y})$ can be desintegrated, i.e. represented as the product $\mu(x, dy)\nu(dx)$ where $\mu(x, dy) \in \mathcal{M}(X, Y)$ and $\nu(dx) = m(dx, Y)$. If $m(dx, dy)$ is the distribution of the pair of random variables ξ, η, the kernel $\mu(x, dy)$ is the regular conditional distribution of η given $\xi = x$.

The following useful result is referred to as the Skorohod representation theorem:

Theorem A.6.1 *Let Y be a Polish space and let $m_n \in \mathbf{P}(Y)$ be a sequence converging to m in $\mathbf{P}(Y)$. Then on the probability space $([0, 1], \mathcal{B}[0, 1], dx)$ there exist Y-valued random variables $\tilde{\xi}_n$ and $\tilde{\xi}$ with $\mathcal{L}(\tilde{\xi}_n) = m_n$, $\mathcal{L}(\tilde{\xi}) = m$ such that $\tilde{\xi}_n \to \tilde{\xi}$ pointwise.*

The representation theorem allows us to simplify problems by considering the pointwise convergence of random variables instead of the weak convergence of their distributions, while the following measurable isomorphism theorem provides a reduction to a simpler space.

Theorem A.6.2 *Let $(X, \mathcal{X})$ be an uncountable Polish space. Then there is a one-to-one mapping $i : X \to [0, 1]$ such that $i(\Gamma) \in \mathcal{B}[0, 1]$ for any $\Gamma \in \mathcal{X}$ and $i^{-1}(A) \in \mathcal{X}$ for any $A \in \mathcal{B}[0, 1]$.*

In the sequel we need the following well-known property of a separable metric space X: every open covering of X contains a countable subcovering (i.e. X is a Lindelöf space). For the reader's convenience we recall the proof. Let $(G_\alpha)_{\alpha \in I}$ be a family of open sets such that their union is X. Let S be a countable dense subset of X. For each $x \in X$ and each strictly positive rational number r we consider the ball $U_r(x) := \{z \in X : d(z, x) < r\}$. Renumber these balls somehow, say, $U_1, U_2, \ldots$. For each $x \in X$ and each set G_α containing x there is an index $n = n(x, \alpha)$ such that

$$x \in U_{n(x,\alpha)} \subseteq G_\alpha.$$

Let $\mathbf{N}_1$ be a subset of $\mathbf{N}$ containing all such subscripts. For every $n \in \mathbf{N}_1$ there exists $\alpha(n)$ such that $U_n \subseteq G_{\alpha(n)}$. Since the union of U_n over $\mathbf{N}_1$ is equal to X, the family $(G_{\alpha(n)})_{n \in \mathbf{N}_1}$ forms a countable covering of X.

A.6.2 Integration of Stochastic Kernels

Let $\mu \in \mathcal{M}(X, Y)$ and let $m \in \mathbf{P}(X)$. The integral $\int_X \mu(x, \Gamma) m(dx)$, $\Gamma \in \mathcal{Y}$, defines a probability measure on $(Y, \mathcal{Y})$ denoting by $\int_X \mu(x, .) m(dx)$.

Lemma A.6.3 *Let $(X, \mathcal{X})$ be a Polish space with a non-atomic measure ν, let S be a compact set in $\mathbf{P}(Y)$, and let $\mathcal{M}$ be the set of $\mu \in M(X, Y)$ such that $\mu(x, .) \in S$ for all $x \in X$. Then the set*

$$K = \left\{ m \in \mathbf{P}(Y) : \ m(.) = \int_X \mu(x, .) \nu(dx), \ \mu \in \mathcal{M} \right\}$$

is a convex compact subset in $\mathbf{P}(Y)$ coinciding with $\overline{\mathrm{conv}}S$.

Proof. By virtue of the measurable isomorthism theorem we can consider only the case where $(X, \mathcal{X}) = ([0, 1], \mathcal{B}[0, 1])$. First, assume that $\nu(dx) = dx$, i.e. ν is the Lebesgue measure. Convexity of K is clear: if two measures $m_i(.) = \int_X \mu_i(x, .) dx$ belong to K and if $\alpha > 0$, $\beta > 0$, $\alpha + \beta = 1$, then the measure $\alpha m_1(.) + \beta m_2(.)$ also belongs to K being equal to $\int_X \mu(x, .) dx$ where

$$\mu(x, .) = I_{[0, \alpha]}(x) \mu_1(\alpha^{-1} x, .) + I_{]1 - \beta, 1]}(x) \mu_2(\beta^{-1}(x - 1 + \beta), .).$$

The tightness of K follows easily from the tightness of S. To prove that K is closed we consider a sequence $m_n(.) = \int \mu_n(x, .) dx \in K$ converging in $\mathbf{P}(Y)$ to some $m(.)$. Notice that elements of $\mathcal{M}$ are random variables with values in the compact subset S of a Polish space. Thus, the set of distributions of these random variables $\{\mathcal{L}(\mu) : \ \mu \in \mathcal{M}\}$ is relatively compact in $\mathbf{P}(\mathbf{P}(Y))$. Taking, if necessary, a subsequence we may assume that $\mathcal{L}(\mu_n)$ tends to some $\mathcal{L}$ in $\mathbf{P}(\mathbf{P}(Y))$. By the Skorohod representation theorem there are S-valued random variables $\tilde{\mu}_n$ and $\tilde{\mu}$ on the probability space $([0, 1], \mathcal{B}[0, 1], dx)$ such that $\tilde{\mu}_n(x, .) \to \tilde{\mu}(x, .)$ as $n \to \infty$ for all x and $\mathcal{L}(\tilde{\mu}) = m$, $\mathcal{L}(\tilde{\mu}_n) = \mathcal{L}(\mu_n)$ for all n.

The last equality means that for every $f \in C_b(Y)$ the distribution of random variable $\tilde{\mu}_n(f)$ coincides with the distribution of $\mu_n(f)$. It follows that for any $f \in C_b(Y)$ we have

$$m(f) = \lim_n m_n(f) = \lim_n \int \mu_n(x, f) dx = \lim_n \int \tilde{\mu}_n(x, f) dx = \int \tilde{\mu}(x, f) dx.$$

Thus, $m(.) = \int \tilde{\mu}(x, .) dx \in K$.

The general case, where ν is a non-atomic measure on $([0, 1], \mathcal{B}[0, 1])$, is easily reduced to the considered above by the well-known quantile transformation. Indeed, put $F(t) := \nu([0, t])$ and $C(t) := \inf\{s : \ F(s) > t\}$. We have the identities

$$\int \mu(x, .) dx = \int \mu(F(x), .) \nu(dx), \qquad \int \mu(x, .) \nu(dx) = \int \mu(C(x), .) dx$$

which show that K does not depend on measure ν.

Obviously, $S \subseteq K$. Thus, $\overline{\text{conv}}S \subseteq K$. Let $m_0(.) = \int \mu_0(t,.)dt$ be a point in K which does not belong to $\overline{\text{conv}}S$. By the separation theorem a convex compact set and a point outside it can be strictly separated by a continuous linear functional. This means that there exists $f \in C_b(Y)$ such that $\sup_{m \in \overline{\text{conv}}S} m(f) < m_0(f)$. It follows that $\int \mu_0(t,f)dt < m_0(f)$, contradicting the assumption $m_0 \in K$. $\square$

Remark. If ν has atoms, then we can assert only that K is a subset of $\overline{\text{conv}}S$ even in the case where S is compact.

A.6.3 Distributions of Integrals

We consider the following problem.

Let $(\Omega, \mathcal{F}, P)$ be a probability space, let $\mathcal{P}$ be a σ-algebra in the product $\Omega \times \mathbf{R}_+$ such that $\mathcal{P} \subseteq \mathcal{F} \otimes \mathcal{B}(\mathbf{R}_+)$, and let Γ be a measurable set-valued mapping from $(\mathbf{R}_+, \mathcal{B}(\mathbf{R}_+))$ to $\mathbf{R}^q$. Measurability means that the graph $\text{Gr}\,\Gamma = \{(t,x) : x \in \Gamma(t)\}$ is a $\mathcal{B}(\mathbf{R}_+) \otimes \mathcal{B}(\mathbf{R}^q)$-measurable set. We assume that $\Gamma(t)$ are compact sets and there exists a function $r \in L^1(\mathbf{R}_+, dt)$ such that $|\Gamma(t)| \leq r_t$ for all t. Let $\mathcal{V}$ be the set of all $\mathcal{P}$-measurable functions f on $\Omega \times \mathbf{R}_+$ such that $f(\omega, t) \in \Gamma(t)$. Define the set K in $\mathbf{P}(\mathbf{R}^q)$ as

$$K := \left\{ \mathcal{L}(\phi) : \phi = \int_0^\infty f(t)dt, \ f \in \mathcal{V} \right\}.$$

Is K a compact set? We give a partial answer to this question imposing some specific assumption on the structure of the σ-algebra $\mathcal{P}$.

Let $w = (w_t)$ be a d-dimensional Wiener process on $(\Omega, \mathcal{F}, P)$. We define the σ-algebras $\mathcal{F}_t^{o,w} = \sigma\{w_s, \ s \leq t\}$ and $\mathcal{F}_t^w = \mathcal{F}_{t+}^{o,w} \vee \mathcal{N}$ where $\mathcal{N}$ is a family of all sets from $\mathcal{F}$ of zero probability. Now $\mathbf{F}^w = (\mathcal{F}_t^w)$ is the minimal filtration generated by the Wiener process and satisfying the usual assumptions.

Lemma A.6.4 *Let $\mathcal{P}$ be the predictable σ-algebra generated by $\mathbf{F}^w$. Assume that $\Gamma(t)$ is a convex set for each t. Then K is a compact set.*

Proof. Since random variables ϕ are bounded by some constant, K is relatively compact and it remains to show that K is closed.

We consider a sequence $f^n \in \mathcal{V}$ such that the corresponding sequence of distributions $\mathcal{L}(\phi^n)$ converges in $\mathbf{P}(\mathbf{R}^q)$. Define the random processes

$$\phi_t^n = \int_0^t f^n(\omega, s)ds.$$

Using the well-know criterion of relative compactness in $\mathbf{P}(C^{q+d}(\mathbf{R}_+))$ (the space $C^{q+d}(\mathbf{R}_+)$ is equipped with the metric $\sum_j 2^{-j} \parallel x \parallel_j (1+ \parallel x \parallel_j)^{-1}$), we may assume without loss of generality that the sequence $\mathcal{L}((\phi^n, w))$

converges to some $\mathcal{L}$ in $\mathbf{P}(C^{q+d}(\mathbf{R}_+))$. The Skorohod theorem asserts that on some probability space $(\tilde{\Omega}, \tilde{\mathcal{F}}, \tilde{P})$ (in fact, on the standard unit interval) there are processes $(\tilde{\phi}^n, \tilde{w}^n)$, $n \in \mathbf{N}$, and $(\tilde{\phi}, \tilde{w})$ such that $\mathcal{L}(\tilde{\phi}^n, \tilde{w}^n) = \mathcal{L}(\phi^n, w)$, $\mathcal{L}(\tilde{\phi}, \tilde{w}) = \mathcal{L}$, and $(\tilde{\phi}^n, \tilde{w}^n)$ converges to $(\tilde{\phi}, \tilde{w})$ in $C^{q+d}(\mathbf{R}_+)$ pointwise.

It is easy to show that the following properties hold:

(1) The process $\tilde{\phi}^n$ is adapted with respect to $(\tilde{\mathcal{F}}_t^n)$, where the σ-algebra $\tilde{\mathcal{F}}_t^n := \sigma\{\tilde{w}_s^n, \ s \le t\}$, and admits the representation

$$\tilde{\phi}_t^n(\tilde{\omega}) = \int_0^t \tilde{f}^n(\tilde{\omega}, s)ds \tag{A.6.1}$$

with $\tilde{\mathcal{P}}^n$-measurable $\tilde{f}^n$ such that $\tilde{f}^n(\tilde{\omega}, s) \in \Gamma(s)$ for each $(\tilde{\omega}, s)$ (here $\tilde{\mathcal{P}}^n$ is the predictable σ-algebra generated by $(\tilde{\mathcal{F}}_t^n)$).

(2) The process $\tilde{\phi}$ is adapted with respect to the filtration $(\tilde{\mathcal{F}}_t)$, where $\tilde{\mathcal{F}}_t := \sigma\{\tilde{w}_s, \ s \le t\}$, and admits the representation

$$\tilde{\phi}_t(\tilde{\omega}) = \int_0^t \tilde{f}(\tilde{\omega}, s)ds \tag{A.6.2}$$

with $\tilde{\mathcal{P}}$-measurable $\tilde{f}$ such that $\tilde{f}(\tilde{\omega}, s) \in \Gamma(s)$ for each $(\tilde{\omega}, s)$ (here $\tilde{\mathcal{P}}$ is the predictable σ-algebra generated by the minimal filtration with the usual assumptions for $\tilde{w}$).

Let us prove that $\tilde{\phi}^n$ is adapted with respect to the filtration $(\tilde{\mathcal{F}}_t^n)$. Fix $t \in \mathbf{R}_+$ and define the Wiener process $\hat{w}_s^n = \tilde{w}_{s+t}^n - \tilde{w}_t^n$, $s \in \mathbf{R}_+$, which is independent of $\tilde{\mathcal{F}}_t^n$. It is sufficient to show that $\tilde{E}(\tilde{\phi}_t^n \mid \tilde{\mathcal{F}}_t^n) = \tilde{\phi}_t^n$ ($\tilde{P}$-a.s.) or, equivalently, that

$$\tilde{E}\tilde{E}(\tilde{\phi}_t^n \mid \tilde{\mathcal{F}}_t^n)h(\tilde{w}^n)g(\hat{w}^n) = \tilde{E}\tilde{\phi}_t h(\tilde{w}^n)g(\hat{w}^n)$$

for any bounded continuous functions $h : C^d[0, t] \to \mathbf{R}$ and $g : C^d(\mathbf{R}_+) \to \mathbf{R}$ (in fact, in the above formula the argument of h is the restriction of $\tilde{w}^n$ to $[0, t]$). Since $h(\tilde{w}^n)$ is $\tilde{\mathcal{F}}_t^n$-measurable, it follows from the properties of conditional expectations that the above equality holds if and only if

$$\tilde{E}\tilde{\phi}_t^n h(\tilde{w}^n)\tilde{E}g(\hat{w}^n) = \tilde{E}\tilde{\phi}_t^n h(\tilde{w}^n)g(\hat{w}^n). \tag{A.6.3}$$

But $\mathcal{L}(\tilde{\phi}^n, \tilde{w}^n) = \mathcal{L}(\phi^n, w)$, and the last identity is equivalent to the following one:

$$E\phi_t^n h(w)Eg(w) = E\phi_t^n h(w)g(w')$$

where $w_s' = w_{s+t} - w_t$, $s \in \mathbf{R}_+$. This equality holds because ϕ^n is adapted with respect to $(\mathcal{F}_t^n)$.

By taking the limit in (A.6.3) we get that

$$\tilde{E}\tilde{\phi}_t h(\tilde{w})\tilde{E}g(\tilde{w}) = \tilde{E}\tilde{\phi}_t h(\tilde{w})g(\hat{w})$$

where $\hat{w}_s = \tilde{w}_{s+t} - \tilde{w}_t$, $s \in \mathbf{R}_+$. As above, this means that $\hat{\phi}_t = \tilde{E}(\hat{\phi}_t \mid \tilde{\mathcal{F}}_t)$, i.e. $\hat{\phi}$ is adapted with respect to $(\tilde{\mathcal{F}}_t)$. The representation (A.6.1) follows from the definition of ϕ^n and coincidence of $\mathcal{L}(\hat{\phi}^n, \tilde{w}^n)$ and $\mathcal{L}(\phi^n, w)$. To obtain the representation (A.6.2) we notice that by the Komlós theorem A.7.1 applied to the bounded sequence $\tilde{f}^n$ there exists a subsequence (n_j) such that $(\tilde{f}^{n_1} + \ldots + \tilde{f}^{n_k})/k$ converges to some function $\tilde{f}^0$ for almost all $(\tilde{\omega}, t)$. It follows that

$$\tilde{\phi}_t(\tilde{\omega}) = \int_0^t \tilde{f}^0(\tilde{\omega}, s)ds. \qquad (A.6.4)$$

The convexity assumption implies that $\tilde{f}^0(\tilde{\omega}, s) \in \Gamma(s)$ for almost all $(\tilde{\omega}, s)$ and we may assume without loss of generality that $\tilde{f}^0(\tilde{\omega}, s) \in \Gamma(s)$ for all $(\tilde{\omega}, s)$. This means that the trajectories of $\tilde{\phi}$ are absolute continuous functions. Let

$$\tilde{f}'(\tilde{\omega}, s) = \limsup_{m \to \infty} \sum_{i=2}^m I_{\Delta_i}(s) 2^m (\tilde{\phi}_{t_{i-1}}(\tilde{\omega}) - \tilde{\phi}_{t_{i-2}}(\tilde{\omega}))$$

where $t_i = i2^{-m}$, $\Delta_i = t_i - t_{i-1}$. Clearly, $\tilde{f}'$ is a $\mathcal{P}$-measurable function and for all $\tilde{\omega}$ and almost all s the value of $\tilde{f}'(\tilde{\omega}, s)$ coincides with $\tilde{f}^0(\tilde{\omega}, s) \in \Gamma(s)$. Thus, the following function gives the representation (A.6.2) with the required properties:

$$\tilde{f}(\tilde{\omega}, s) = \tilde{f}'(\tilde{\omega}, s)I_A + x(s)I_{\bar{A}}$$

where $A = \{(\tilde{\omega}, s) : \tilde{f}'(\tilde{\omega}, s) \in \Gamma(s)\}$ and $x(s)$ is any Borel function such that $x(s) \in \Gamma(s)$.

The properties (1) and (2) imply the result. Indeed, it follows from (2) and Lemma 5.2.9 that there is a predictable function $a(x, s) : C^d(\mathbf{R}_+) \times \mathbf{R}_+ \to \mathbf{R}^q$ such that $\tilde{f}(\tilde{\omega}, s) = a(\tilde{w}(\tilde{\omega}), s)$. Obviously, we can modify $a(x, s)$ in such a way that $a(x, s) \in \Gamma(s)$ for all (x, s). Let us define on the original probability space $(\Omega, \mathcal{F}, P)$ the process

$$\phi_t(\omega) = \int_0^t f(\omega, s)ds$$

with $f(\omega, s) = a(w(\omega), s)$. Since $f \in \mathcal{V}$ and $\mathcal{L}(\phi) = \mathcal{L}(\tilde{\phi}) = \mathcal{L}$, it follows that the limit of $\mathcal{L}(\phi^n)$ belongs to K and the lemma is proved. $\square$

A.6.4 Compactness of the Limit of Attainability Sets

Now we apply the previous general result to the specific setting of Section 5.4.

Lemma A.6.5 *The set $S_Y^0 := \{\mathcal{L}(\xi_0 + I(v) : v \in \mathcal{V}_U)\}$ is compact in $\mathbf{P}(\mathbf{R}^n)$.*

Proof. Taking into account the notations of the previous subsection we can reduce the problem by time reversal to the question, whether the set

$$K := \Big\{ \mathcal{L}(\phi) : \ \phi = \int_0^\infty f(t)dt, \ f \in \mathcal{V} \Big\}.$$

is compact. Here $\Gamma(t) = s^{-2} \exp\{A_4(T)/s\}B_2(T)U$ and the σ-algebra $\mathcal{P}$ is generated by the time-reverse of the Ornstein–Uhlenbeck process $\xi_{1/t}$, or, this is the same, by the process $\eta_t := t\xi_{1/t}$. The process η (as well as ξ) is defined, in the present context, only up to the distribution. For instance, we may take as η the process defined by the stochastic differential equations

$$d\eta_t = t^{-2}(tI - A_4(T))\eta_t dt + dw_t, \quad \eta_0 = 0, \tag{A.6.5}$$

where I is the unit matrix and w is a Wiener process. This representation can be deduced from the differential equation for the Ornstein–Uhlenbeck process by the Ito formula. But it follows from the equation (A.6.5) that $\mathcal{F}_t^{o,w} = \sigma\{\eta_s, s \leq t\}$ and the needed result is a corollary of Lemma A.6.4. $\square$

Proposition A.6.6 *Let S_X be a compact subset in $\mathbf{P}(X)$, and let S_Y be a convex compact subset in $\mathbf{P}(Y)$. Let S be the set of $m \in \mathbf{P}(X \times Y)$ admitting the representation $m(dx, dy) = \mu(x, dy)\nu(dx)$ with $\mu(x, .) \in S_Y$ for all x and $\nu(.) \in S_X$. Then S is a compact set.*

Proof. First, we notice that $S_Y = \cup_{j=1}^n \Gamma_j$ where $\Gamma_j := \{\mu : \ \mu(f_j) \leq \beta_j\}$, $f_j \in C_b(Y)$, $\beta_j \in \mathbf{R}$. Indeed, it follows from the Hahn–Banach theorem that S_Y is an intersection of sets of this type. Their complements form an open covering of the open set $\mathbf{P}(Y)\backslash S_Y$. Since a Polish space is Lindelöf, it contains a countable covering $\bar{\Gamma}_j$, $j \in \mathbf{N}$. Assume now that for the limiting measure $m(dx, dy) = \mu(x, dy)\nu(dx)$ there exists a set of positive ν-measure in which $\mu(x, .) \notin S_Y$. The above representation for S_Y implies that there exists a set $B = \{x : \ \mu(x, f) > \beta\}$ with $\nu(B) > 0$. Let $g_k \in C_b(X)$ be a sequence converging in $L^1(\nu)$ to I_B. Since $\mu_n(x, .) \in S_Y$, we have that $\mu_n(x, f) \leq \beta$. Thus,

$$\lim_{k\to\infty} \lim_{n\to\infty} \int\int g_k(x)f(y)m_n(dx, dy) = \lim_{k\to\infty} \lim_{n\to\infty} \int g_k(x)\mu_n(x, f)\nu_n(dx)$$

$$\leq \lim_{k\to\infty} \beta \int g_k(x)\nu(dx) = \beta\nu(B).$$

On the other side

$$\lim_{k\to\infty} \lim_{n\to\infty} \int\int g_k(x)f(y)m_n(dx, dy) = \lim_{k\to\infty} \lim_{n\to\infty} \int\int g_k(x)f(y)m(dx, dy)$$

$$= \lim_{k\to\infty} \int g_k(x)\mu(dx, f)\nu(dx) = \int_B \mu(dx, f)\nu(dx) > \beta\nu(B)$$

and we obtain a contradiction with the assumption that $\mu(x, .)$ does not belong to S_Y ν-a.s. $\square$

A.6.5 Supports of Conditional Distributions

Let η_i be random variables with values in Polish spaces $(X_i, \mathcal{X}_i)$, $i = 1, 2, 3$, let ν_i be the distribution of η_i, and let $\mu_{ij}(x_j, dx_i)$ be the regular conditional distribution of η_i given η_j.

Lemma A.6.7 *Let $\eta_3 = f(\eta_2)$ for some measurable function $f : X_2 \to X_3$ and let S_1 be a compact convex set in $\mathbf{P}(X_1)$. Assume that $\mu_{12}(x_2, dx_1) \in S_1$ for all x_2. Then $\mu_{13}(x_3, dx_1) \in S_1$ for ν_3-almost all x_3.*

Proof. The assertion follows from the relation

$$\mu_{13}(x_3, dx_1) = \int_{X_2} \mu_{12}(x_2, dx_1)\mu_{23}(x_3, dx_2) \quad (\nu_3\text{-a.e.})$$

and the remark after Lemma A.6.3. $\square$

A.7 The Komlós Theorem

The following Komlós theorem asserting that a sequence of random variables bounded in L^1 contains a subsequence converging in Cesaro sense a.s. is very useful in various applications, especially, in proofs of existence of optimal controls.

Theorem A.7.1 *Let (ξ_n) be a sequence of random variables on $(\Omega, \mathcal{F}, P)$ with*

$$\sup_n E|\xi_n| < \infty. \tag{A.7.1}$$

Then there exists a random variable $\zeta \in L^1$ and subsequence $(\zeta_k) = (\xi_{n_k})$ such that

$$\frac{\zeta_1 + \zeta_2 + \dots \zeta_k}{k} \to \zeta \quad a.s. \tag{A.7.2}$$

Moreover, the subsequence (ζ_k) can be chosen in such a way that its further subsequence will also satisfy (A.7.2).

For the proof we need several simple lemmas.

Lemma A.7.2 *Let ζ be a random variable. Then*

$$E|\zeta| < \infty \quad \Leftrightarrow \quad \sum_{n=1}^{\infty} P(|\zeta| \geq n) < \infty \quad \Leftrightarrow \quad \sum_{n=1}^{\infty} \frac{1}{n^2} E\zeta^2 I_{\{|\zeta| \leq n\}} < \infty.$$

The proof is easy and is omitted.

Lemma A.7.3 *Let (ξ_n) be a sequence of random variables satisfying (A.7.1). Then there exists a subsequence $(\zeta_k) = (\xi_{n_k})$ such that*

$$\sum_{k=1}^{\infty} P(|\zeta_k| \geq k) < \infty, \tag{A.7.3}$$

$$\sum_{k=1}^{\infty} \frac{1}{k^2} E\zeta_k^2 I_{\{|\zeta_k| \leq k\}} < \infty, \tag{A.7.4}$$

and for any subsequence of (ζ_k) the same relations hold.

Notice that (A.7.3) implies that for almost all ω only a finite number of $|\zeta_k(\omega)|$ lie above the line $y = x$.

Proof. Let $\nu_n := \mathcal{L}(\xi_n)$, the distribution of ξ_n. It follows from the Chebyshev inequality and (A.7.1) that

$$\sup_n \nu_n(|x| \geq N) = \sup_n P(|\xi_n| \geq N) \leq E|\xi_n|/N \to 0$$

as $N \to \infty$. This means that the sequence (ν_n) is tight, or, equally, relatively compact in $\mathbf{P(R)}$. Without loss of generality we may assume that the whole sequence (ν_n) converges in $\mathbf{P(R)}$ to some ν. Notice that for any $a > 0$ we have

$$\limsup_n \nu_n(|x| \geq a) \leq \nu(|x| \geq a),$$

$$\limsup_n \nu_n(|x|^2 I_{\{|x| \leq a\}}) \leq \nu(|x|^2 I_{\{|x| \leq a\}}).$$

The above properties follow immediately from the definition of the weak convergence. Hence, for any $k \in \mathbf{N}$ there exists n_k such that for all $n \geq n_k$

$$\nu_n(|x| \geq k) \leq \nu(|x| \geq k) + 2^{-k}$$

and

$$\nu_n(|x|^2 I_{\{|x| \leq k\}}) \leq \nu(|x|^2 I_{\{|x| \leq k\}}) + 2^{-k}$$

By virtue of Lemma A.7.2, the random variables $\zeta_k := \xi_{n_k}$ form a subsequence with the needed properties. $\square$

Lemma A.7.4 Let η_n be a sequence of random variables convergent weakly in L^2 to a random variable η. Then

$$E|\eta| \leq \liminf_n E|\eta_n|, \tag{A.7.5}$$

$$E|\eta|^2 \leq \liminf_n E|\eta_n|^2. \tag{A.7.6}$$

Proof. Using the definition of the weak convergence in L^2 we have:

$$\liminf_n E|\eta_n| \geq \liminf_n E\eta_n(I_{\{\eta \geq 0\}} - I_{\{\eta < 0\}}) = E\eta(I_{\{\eta \geq 0\}} - I_{\{\eta < 0\}}) = E|\eta|,$$

$$(E\eta^2)^2 = (\lim_n E\eta_n\eta)^2 \leq E|\eta|^2 \liminf_n E|\eta_n|^2$$

and the result follows. $\square$

We define the truncation and discretization operators in L^0 by the formulae

$$\xi^{(c)} := \xi I_{\{|\xi| \le c\}},$$

$$D_m(\xi) := \sum_{i=-\infty}^{\infty} i2^{-m} I_{\{\xi \in]i2^{-m}, (i+1)2^{-m}]\}}.$$

Lemma A.7.5 *Assume that (A.7.1) is fulfilled and for every $k \in \mathbf{N}$ the sequence $(\xi_n^{(k)})$ converges weakly in L^2 to a random variable η_k. Then there exists $\eta \in L^1$ such that η_k tends to η a.s. and in L^1.*

Proof. Obviously,

$$|\xi_n| = \sum_{k=1}^{\infty} |\xi_n^{(k)} - \xi_n^{(k-1)}|.$$

Using (A.7.1) we deduce from this identity by the Fatou lemma (for series) and Lemma A.7.4 that

$$C \ge \liminf_n E|\xi_n| \ge \sum_{k=1}^{\infty} \liminf_n E|\xi_n^{(k)} - \xi_n^{(k-1)}| \ge \sum_{k=1}^{\infty} E|\eta_k - \eta_{k-1}|.$$

Thus, the right-hand side of the last inequality is finite, implying the result. $\square$

Lemma A.7.6 *Let $\mathcal{G}$ be a σ-algebra generated by a finite partition A_1, ..., A_N with $A_i \in \mathcal{F}$. Assume that a sequence of random variables (ξ_n) converges weakly in L^2 to zero. Then for any $\varepsilon > 0$ there exists $n_0 = n_0(\varepsilon)$ such that*

$$E(\xi_n \mid \mathcal{G}) \le \varepsilon$$

for all $n \ge n_0$.

The above assertion is obvious since on any set A_i with $P(A_i) > 0$ we have

$$E(\xi_n \mid \mathcal{G}) = \frac{E(\xi_n I_{A_i})}{P(A_i)} \quad \text{a.s.}$$

Now we recall some facts concerning square integrable martingales. Let ξ_i be a sequence of random variables from L^2 such that ξ_i is $\mathcal{F}_i$-measurable for any $i \in \mathbf{N}$ where $(\mathcal{F}_i)$ is an increasing family of σ-algebras. The sequence

$$M_n := \sum_{i=1}^{n} (\xi_i - E(\xi_i \mid \mathcal{F}_{i-1}))$$

is a square integrable martingale. The limit M_n, $n \to \infty$, exists a.s. if

$$\langle M \rangle_\infty := \sum_{i=1}^{\infty} E((\xi_i - E(\xi_i \mid \mathcal{F}_{i-1}))^2 \mid \mathcal{F}_{i-1}) < \infty \quad \text{a.s.}$$

Certainly, the last condition is satisfied if

$$\sum_{i=1}^{\infty} E\xi_i^2 < \infty.$$

Proof of the theorem. Choosing a subsequence we can assume without loss of generality that (ξ_n) has the following properties:

(1) $\xi_n^{(k)} \to \eta_k$ weakly in L^2 in for every $k \in \mathbf{N}$;
(2) relations (A.7.3) and (A.7.4) hold with $\xi_k = \zeta_k$.
Notice that by Lemma A.7.4 for every $k \in \mathbf{N}$ we have

$$E\eta_k^2 \le \liminf_n E(\xi_n^{(k)})^2.$$

Define n_1 as a number such that

$$E\eta_1^2 \le E(\xi_n^{(1)})^2 + 2^{-1}$$

for all $n \ge n_1$. Put $\gamma_1 = D_1(\xi_{n_1}^{(1)} - \eta_1)$. Since γ_1 has only finite number of values and $\xi_n^{(2)} - \eta_2 \to 0$ weakly in L^2, Lemma A.7.6 implies the existence of $n_2 > n_1$ such that

$$E\eta_2^2 \le E(\xi_n^{(2)})^2 + 2^{-2},$$

$$|E(\xi_n^{(2)} - \eta_2 \mid \gamma_1)| \le 2^{-2}$$

for all $n \ge n_2$. Put $\gamma_2 = D_2(\xi_{n_1}^{(2)} - \eta_2)$. Continuing this process we find $n_k > n_{k-1}$ such that for all $n \ge n_k$

$$E\eta_k^2 \le E(\xi_n^{(k)})^2 + 2^{-k},$$

$$|E(\xi_n^{(k)} - \eta_k \mid \gamma_{j_1}, \dots, \gamma_{j_m})| \le 2^{-k},$$

for all $m \le k-1$, $j_1 < j_2 < \dots < j_m$, where $\gamma_j := D_j(\xi_{n_j}^{(j)} - \eta_j)$.

The $\zeta_k := \xi_{n_k}$ is the sequence with the needed property.

Indeed, we have, obviously, that $|E(\gamma_k \mid \gamma_1, \dots, \gamma_{k-1})| \le 2^{-k+1}$ and

$$\sum_{k=1}^{\infty} \frac{1}{k^2} E\gamma_k^2 \le 2 \sum_{k=1}^{\infty} \frac{1}{k^2} E(\zeta_k^{(k)} - \eta_k)^2 + O(1) \le 4 \sum_{k=1}^{\infty} \frac{1}{k^2} E(\zeta_k^{(k)})^2 + O(1) < \infty.$$

Thus, we have that the square integrable martingale

$$M_n := \sum_{k=1}^{n} \frac{1}{k}(\gamma_k - E(\gamma_k \mid \gamma_1, \dots, \gamma_{k-1}))$$

converges a.s. Therefore, the series

$$\sum_{k=1}^{\infty} \frac{1}{k}(\zeta_k^{(k)} - \eta_k)$$

converges a.s. Since only a finite number of $\zeta_k^{(k)}(\omega)$ is different from $\zeta_k(\omega)$, the series

$$\sum_{k=1}^{\infty} \frac{1}{k}(\zeta_k - \eta_k)$$

also converges a.s. and the Kronecker lemma implies that

$$\frac{1}{k}\sum_{i=1}^{k}(\zeta_i - \eta_i) \to 0 \quad \text{a.s.}$$

By Lemma A.7.5 η_i tends to some η a.s. and the result follows. $\square$

Historical Notes

We do not pretend to give an exhaustive list of works in the field. The book reflects mainly our own interests in singular perturbations of stochastic differential equations. Besides our papers we indicate only some basic references as well as studies which influenced our research.

Chapter 0

This chapter contains several models where two-scale stochastics systems arise in a natural way. For the first author the model with fast Markov modulations was the starting point: a singular perturbed stochastic differential equation appeared in his study of the Poisson-type channel [36]. The idea to use the filtering equation for the a posteriori distributions of finite state ergodic Markov process modulating the intensity of a counting process was exploited afterwards in the joint papers with Liptser and Shiryaev [38] and [39] in an example of the application of general martingale limit theorems. This experience motivated a further study of singularly perturbed stochastic differential equations as the object of its own interest. In the models of the aforementioned paper, as well as in the article by Di Masi and Kabanov [20] the driving noise was a Poisson-type process; Section 0.1 is based essentially on the note by Di Masi and Kabanov [19] where the diffusion-type model with rapidly varying drift was also discussed.

Though the model of the Liénard oscillator is classical and can be found in many textbooks (see, e.g., [49] and [72]), the bound of Proposition 0.2.1 seems to be new.

The example of an approximate filter described in Section 0.2 is a particular case of models treated in Bensoussan's book [11]; see also Picard's paper [78]. By these examples we would like to attract the reader's attention to the fact that in two-scale models the scaling factor at the diffusion coefficient may be of different order.

It seems that the Holevo paper [34] was the first analyzing a continuous-time version of the Robbins–Monro approximation procedure; see also the book by Nevelson and Khas'minski, [73]. The literature on stochastic approximation is enormous. The idea to use singular perturbation theory in stochastic approximation is, apparently, quite old but we are not able to trace its origin. The presentation of Section 0.4 is based on the paper by the second author [76].

Chapter 1

The example showing that in the case of time-dependent coefficients the spectrum does not control the growth rate and related discussion can be found in [1] and [14]. The moment bounds for solutions of linear stochastic differential equations with a "stable" drift coefficient were studied in the papers [41], [47] while the proofs of exponential bounds of Section 1.2 are new. In Section 1.3 we present the Lapeyre method (see [58], [59]) to obtain bounds which give the exact rate of growth of the L^p-norm of maximal functions. Fernique's Lemma 1.3.6 (see [26]) can be found in the book by Liptser and Shiryaev [67].

Chapter 2

The Tikhonov theory for deterministic two-scale systems (called in the O'Malley book [74] the Tikhonov–Levinson theory with reference to [64], [83]) was suggested in his doctoral thesis; see the earlier work [82]. The book of Vasil'yeva and Butuzov [84] is a good reference for its further development: in particular, it includes uniform expansions involving boundary layer functions. See also the textbooks by Moiseev [69] and [70] where examples of applications are discussed; useful information can be found in [52] and [68]. The presentation of Chapter 2 is based on our papers [41] and [47] (with J. Stoyanov). On a structure of coefficients in formal expansions one can consult the book [50].

Chapter 3

We develop large deviations in the spirit of the fundamental treatise of Freidlin and Wentzell [27] following our paper [45]. The time-homogeneous model of Section 3.1 with only fast variable, being rescaled, is almost a standard one. The new feature is that the time-interval is growing but the specific structure of the coefficients ensures an LD result. Of course, the LD theory (based on the uniform norm) is one of the most important and well-developed subjects in stochastic processes and the literature is enormous. However, two-scale models are hardly touched. The Liptser paper [65], where the system with the diffusion coefficient at the fast variable of order $\varepsilon^{1/2}$ is studied, is a rare exception. The LD result in the L^2-norm of Section 3.2 is beyond the scope of traditional approach.

Chapter 4

This chapter is based on the paper of the second author [75]. Deterministic theory can be found in [74] and [84]. Asymptotic expansions for the Liénard oscillator driven by a random force was studied by Narita in [71].

Chapter 5

Deterministic control problems with singular perturbations are subjects of intensive studies, see, e.g., the books [11], [13], [22], [23], [77] etc. The papers [5], [6], [7], [29], [31], [32] contain new ideas which allow us to treat nonlinear models: they are promising for stochastic generalizations. There are many monographs and papers dealing also with controlled stochastic differential equations where the reader may find also examples of applications: [3], [10], [12], [28], [55], [53], [54], [57] etc. and references therein.

Section 5.1 is based on the paper [48]. We present the topological and geometric properties of the set of densities following Beneš [9]; the description of its extreme points is borrowed from [37]. Theorem 5.2.8 was proved in our paper [44]. The deterministic version of the Dontchev–Veliov theorem and a number of related results can be found in [23]. The presentation of Sections 5.3 and 5.4 are based on our articles [44] and [46]; ramifications are made by using the Lapeyre inequality.

Chapter 6

Section 6.1 contains extensions of results from our note [42]. Theorem 6.2.1 on the exact rate of convergence of distributions of conditionally Gaussian processes to the distribution of a Wiener process with drift is new, as are the results of Sections 6.3 and 6.4. Stochastic criteria of optimality for the LQG-problem has been studied by many authors; see, e.g., [30], [61], [62], [80]. The problem on the exact rate of optimality of the classical feedback control was posed in [21]; we present its solution following [8] where a more general model with time-varying coefficients is considered.

Appendix

The existence result for SDEs via a Lyapunov function is due to Khas'minskii, [51]. The short proof of the Novikov condition was communicated to us by Shiryaev. Proposition A.2.1 is borrowed from [14]: this book is a classical treatise on Lyapunov exponents; see also [1]. The presentation in Subsection of A.2.3 is based on [84]. The inequalities for the total variation distancein terms of the Hellinger processes were developed in [40]; see also the book [35]. The properties of Hausdorff metric can be found, e.g., in [33]. For a short and comprehensive treatment of measurable selection see [18]; much more advanced results can be found in [2], [15], and [85]. The Skorohod representation theorem was proved in [81]; see also the book [24]. In our presentation of the Komlós theorem [56] we follow [16].

References

1. Adrianova L. Ya. Introduction to Linear Systems of Differential Equations. Transl. Math. Monographs, **146**, AMS, Providence, 1995.
2. Alfsen E. M. *Compact Convex Sets and Boundary Integrals.* Springer-Verlag, Berlin, 1971.
3. Alvarez O., Bardi M. Viscosity solutions methods for singular perturbations in deterministic and stochastic control. Preprint (2000).
4. Antonini R. G. Sur le comportement asymptotique du processus de Ornstein-Uhlenbeck multidimensionnel. *Ann. Sci. Univ. Blaise Pascal Clermont-Ferrand, Probab. Appl.*, **9** (1996), pp. 33–44.
5. Artstein Z. Invariant measures of differential inclusions applied to singular perturbations. *J. Differential Equations*, **152** (1999), pp. 389–307.
6. Artstein Z., Gaitsgory V. The value function of singularly perturbed control systems. *Applied Mathematics and Optimization*, **41** (2000), pp. 425–445.
7. Bagagiolo F., Bardi M. Singular perturbation of a finite horizon problem with state-space constraints. *SIAM J. Control and Optimization*, **36** (1998), pp. 2040–2060.
8. Belkina T. A., Kabanov Yu. M., Presman E. L. On a stochastic optimality of the feedback control in the LQG-problem. Prépublications du Laboratoire de Mathématiques de Besançon, $n°$ 2000/34.
9. Beneš V. E. Existence of optimal control laws. *SIAM J. Control*, **3** (1971), pp. 446–475.
10. Bensoussan A. On some singular perturbation problems arising in stochastic control. *Stochastic Analysis and Applications*, **2** (1984), pp. 13–53.
11. Bensoussan A. *Perturbation Methods in Optimal Control.* J. Wiley/Gauthier Villars, New York, 1988.
12. Bensoussan A. *Optimal Control of Partially Observed Systems.* Cambridge University Press, 1992.
13. Bensoussan A., Blankenship G. L. Singular perturbations in stochastic control. In: *Singular Perturbations and Asymptotic Analysis in Control Systems.* Eds. P. Kokotovič, A. Bensoussan, G. Blankenship. Lecture Notes in Control and Inform. Sci., **90**, Springer-Verlag, Berlin, 1987.
14. Bylov B. F., Vinograd R. E., Grobman D. M., Nemyzkiy V. V. *The Theory of Lyapunov Exponents and Its Applications to Problems of Stability.* Nauka, Moscow, 1966 (in Russian).
15. Castaing C., Valadier M. *Convex analysis and measurable multifunctions.* Lecture Notes Math., **580**. Springer-Verlag, 1977.
16. Chatterji S. D. Un principe de sous-suites dans la théorie des probabilités. Séminaire de Probabilités, VI. *Lecture Notes Math.*, **258**, 1972, pp. 72–89.

17. Cox D. R., Miller H. D. *The Theory of Stochastic Processes*. Chapman and Hall, London, 1977.
18. Dellacherie C., Meyer P.-A. *Probabilities and Potentials*. North-Holland, Amsterdam, 1978.
19. Di Masi G. B., Kabanov Yu. M. The strong convergence of two-scale stochastic systems and singular perturbations of filtering equations. *J. Math. Systems, Estimation, and Control*, **3** (1993), 2, pp. 207–224.
20. Di Masi G. B., Kabanov Yu. M. A higher order approximation in convergence of distributions of the Cox processes with fast Markov switchings. *Stochastics and Stochastics Reports*, **54** (1995), pp. 211–219.
21. Di Masi G. B., Kabanov Yu. M. On sensitive probabilistic criteria for LQG problem with infinite horizon. In: *Statistics and Control of Random Processes. Proceedings of Steklov Mathematical Institute Seminar*, TVP, Moscow, 1996.
22. Dontchev A. L. *Perturbations, Approximations, and Sensitivity Analysis of Optimal Control Systems*. Lecture Notes in Control and Inform. Sci., **52**, Springer-Verlag, Berlin, 1983.
23. Dontchev A. L., Veliov V. M. Singular perturbations in Mayer's optimization problem for linear system. *SIAM J. Control and Optimization*, **2** (1983), pp. 566–581.
24. Dudley R. M. *Real Analysis and Probability*. Wadsworth & Brooks/Cole Mathematics Series. Pacific Grove, CA, 1989.
25. Ethier S. N., Kurtz T. G. *Markov Processes: Characterization and Convergence*. J. Wiley, New York, 1986.
26. Fernique X. Regularité des trajectoires des fonctions aléatoires Gaussiennes. École d'Été Probab. Saint-Flour IV, 1974, Lecture Notes Math., **480** (1975), pp. 1–96.
27. Freidlin M. I., Wentzell A. D. *Random Perturbations of Dynamical Systems*. Springer-Verlag, New York, 1984.
28. Gaitsgory V. Suboptimization of singularly perturbed control systems. *SIAM J. Control and Optimization*, **5** (1992), pp. 1228–1249.
29. Gaitsgory V., Leizarowitz A. Limit occupational measures set for a controlled system and averaging of singularly perturbed control systems. *J. Math. Anal. and Appl.*, **233** (1999), pp. 461–475.
30. Ghosh M., Marcus S. Infinite horizon controlled diffusion problems with some nonstandard criteria. *J. Math. Systems, Estimation, and Control*, **1** (1991), pp. 45–69.
31. Grammel G. Periodic near optimal control. *J. Math. Anal. Appl.*, **248** (2000), pp. 124–144
32. Grammel G., Shi P. On the asymptotics of the Lyapunov spectrum under singular perturbations. *IEEE Trans. Automatic Control*, **45** (2000), pp. 565–569.
33. Hildenbrandt W. *Core and Equilibria of a Large Economy*. Princeton University Press, Princeton, New Jersey, 1974.
34. Holevo A. S. The estimators of the parameters of the drift of the diffusion process by the method of the stochastic approximation. *Issledovaniya po teorii samonastraivayushihsia system*. Computer Center of Academy of sciences of USSR, Moscow, 1967, pp. 179–200 (in Russian).
35. Jacod J., Shiryaev A. N., *Limit Theorems for Stochastic Processes*. Springer-Verlag, New York, 1987.

36. Kabanov Yu. M. The capacity of a channel of the Poisson type. *Probab. Theory and Its Appl.*, **23** (1978), 1, pp. 143–147.

37. Kabanov Yu. M. On an existence of the optimal solution in a control problem for a counting process. *Mat. Sbornik*, **119** (1982), 3, pp. 431–445.

38. Kabanov Yu. M., Liptser R. Sh, Shiryaev A. N. Some limit theorems for simple point processes (a martingale approach). *Stochastics*, **3** (1980), pp. 203-216.

39. Kabanov Yu. M., Liptser R. Sh, Shiryaev A. N. Weak and strong convergence of the distributions of counting processes. *Theory Probab. Appl.*, **28** (1984), pp. 303–336.

40. Kabanov Yu. M., Liptser R. Sh, Shiryaev A. N. On the variation distance for the probability measures defined on a filtered space. *Probab. Theory and Related Fields*, **71** (1986), pp. 19–35.

41. Kabanov Yu. M., Pergamenshchikov S. M. Singular perturbations of stochastic differential equations: the Tikhonov theorem. *Mat. Sbornik*, **191** (1990), 9, pp. 1170–1182. English translation: *Math. USSR Sbornik*, **71** (1992), 1, pp.˙.

42. Kabanov Yu. M., Pergamenshchikov S. M. On singularly perturbed stochastic equations and partial differential equations. *Dokl. Akad. Nauk SSSR*, **311** (1990), 5. English translation: *Soviet Math. Dokl.*, **41** (1990), 2.

43. Kabanov Yu. M., Pergamenshchikov S. M. On optimal control of singularly perturbed stochastic differential equations. In: *Modeling, Estimation and Control of Systems with Uncertainty*. Eds. G. B. Di Masi, A. Gombani, A.B. Kurzhansky. Birkhauser, 1991.

44. Kabanov Yu. M., Pergamenshchikov S. M. Optimal control of singularly perturbed linear stochastic systems. *Stochastics and Stochastics Reports*, **36** (1991), pp. 109–135.

45. Kabanov Yu. M., Pergamenshchikov S. M. Large deviations for solutions of singularly perturbed stochastic differential equations. *Uspekhi Mat. Nauk*, **50** (1995), 5, pp. 147–172.

46. Kabanov Yu. M., Pergamenshchikov S. M. On convergence of attainability sets for controlled two-scale stochastic linear systems. *SIAM J. Control and Optimization*, **35** (1997), 1, pp. 134-159.

47. Kabanov Yu. M., Pergamenshchikov S. M., Stoyanov J. M., Asymptotic expansions for singularly perturbed stochastic differential equations. In: *New Trends in Probability and Statistics*. V.1. Proceedings of the Bakuriani Coll. in Honour Yu. V. Prohorov. USSR, 24 Feb. – 4 Mar., 1990. Eds. V. V. Sazonov, T. L. Shervashidze. Mokslas/VSP, Vilnius/Utrecht, 1991, pp. 413–435.

48. Kabanov Yu. M., Runggaldier W., On control of two-scale stochastic systems with linear dynamics in the fast variables. *Math. Control, Signals, and Systems*, 9 (1996), pp. 107-122.

49. Karatzas I., Shreve S. E. *Brownian Motion and Stochastic Calculus*. Springer–Verlag, New York, 1988.

50. Kaufmann A. *Introduction à la Combinatorique en vue des Applications.*, Dunod, Paris, 1968.

51. Khas'minskii R. Z. *Stochastic Stability of Differential Equations*. Sijthoff & Noordhoff, Alphen aan den Rijn, The Netherlands, 1980.

52. Kevorkian J., Cole J. D. *Perturbation Methods in Applied Mathematics*. Springer–Verlag, New York, 1981.

53. Kokotovič P. V. Applications of singular perturbation techniques to control problems. *SIAM Review*, **26** (1986), pp. 501–550.

54. Kokotovič P. V., Khalil H. K. (Eds.), *Singular Perturbations in Systems and Control*. IEEE Press, New York, 1986.

55. Kokotovič P. V., Khalil H. K., O'Reilly, J. *Singular Perturbation Methods in Control: Analysis and Design*. Academic Press, New York, 1986.

56. Komlós J. A generalization of a problem of Steinhaus. *Acta Math. Sci. Hung.*, **18** (1967), pp. 217–229.

57. Kushner H. *Weak Convergence Methods and Singularly Perturbed Stochastic Control and Filtering Problems*. Birkhauser, Boston, 1990.

58. Lapeyre B. Majoration a priori des solutions d'equations différentielles stochastiques stables. *Lecture Notes Math.*, **1316**, pp. 340–351.

59. Lapeyre B. *A priori bound for the supremum of solutions of stable stochastic differential equations*. Stochastics and Stochastics Reports, **28** (1989), pp. 145–160.

60. Le Breton A. About Gaussian schemes in stochastic approximation. *Stochastic Proc. Appl.*, **50** (1994), pp. 101–115.

61. Leizarovitz A. Infinite horizon stochastic regulation and tracking with the overtaking criteria. *Stochastics*, **22** (1987), pp. 117–150.

62. Leizarovitz A. On almost sure optimization for stochastic control systems. *Stochastics*, **23** (1988), pp. 85–110.

63. Levin J. J., Levinson N. Singular perturbations of nonlinear systems of differential equations and associated boundary layer equation. *J. Ration. Mech. Anal.*, **3** (1954), pp. 247–270.

64. Levinson N. Perturbations of discontinuous solutions of nonlinear systems of differential equations. *Acta Math.*, **82** (1951), pp. 71–106.

65. Liptser R. Sh. Large deviations for two scaled diffusions. *Probab. Theory and Related Fields*, **106** (1996), 1, pp. 71–104.

66. Liptser R. Sh., Shiryaev A. N. *Statistics of Random Processes. 1, 2*. Springer-Verlag, New York, 1977, 1978.

67. Liptser R. Sh., Shiryaev A. N. *Theory of Martingales*. Kluwer, Dordrecht, 1989.

68. Lomov S. A. *Introduction into the General Theory of Singular Perturbations*. Nauka, Moscow, 1981 (in Russian).

69. Moiseev N. N. *Asymptotic Methods of Nonlinear Mechanics*. Nauka, Moscow, 1981 (in Russian).

70. Moiseev N. N. *Mathematical Problems of System Analysis*. Nauka, Moscow, 1981 (in Russian).

71. Narita K. Asymptotic behavior of solutions of SDE for relaxation oscillations. *SIAM J. Math. Anal.*, **24** (1993), 1, pp. 172 –199.

72. Nelson E. *Dynamical Theories of Brownian Motion*. Princeton University Press, Princeton, N.J., 1967.

73. Nevelson M. V. Khas'minskii R. Z. *Stochastic Approximation and Recursive Estimation*. AMS **47**, Providence, 1973.

74. O'Malley R. E., Jr. *Singular Perturbation Methods for Ordinary Differential Equations*. Springer-Verlag, New York, 1991.

75. Pergamenshchikov S. M. Asymptotic expansions for models with fast and slow variables described by systems of singularly perturbed stochastic equations. *Uspekhi Mat. Nauk*, **49** (1994), 4, pp. 3–46.

76. Pergamenshchikov S. M. Asymptotic expansions for the stochastic approximation averaging procedure in continuous time. *Stat. Inference Stoch. Process.* **1** (1998), 2, pp. 197–223.

77. Pervozvansky A. A., Gaitsgory V. G. *Theory of Suboptimal Decisions.* Kluwer, Dordrecht, 1988.

78. Picard J. Estimation of the quadratic variation of nearly observed semimartingales, with application to filtering, *SIAM J. Control and Optimization,* **31** (1993), pp. 494–517.

79. Prandtl L. Über Flüssigkeitsbevegung bei kleiner Reibung. *Verhandlungen, III Int. Math. Kongresses,* Tuebner, Leipzig, 1905, pp. 484–491.

80. Presman E., Rotar V., Taksar M. Optimality in probability and almost surely. General scheme and a linear control problem. *Stochastics and Stochastics Reports,* **43** (1993), pp. 127–137.

81. Skorohod A. V. *Studies in the Theory of Random Processes.* Addison-Wesley, Reading, Mass., 1965.

82. Tikhonov A. N. On system of differential equations containing parameters. *Matem. Sbornik,* **27 (69)** (1950), pp. 147–156.

83. Tikhonov A. N. System of differential equations containing a small parameters at derivatives. *Matem. Sbornik,* **31 (73)** (1952), 3, pp. 575–586.

84. Vasil'eva A. B., Butuzov V. F. *Asymptotic Expansions of Solutions of Singularly Perturbed Equations.* Nauka, Moscow, 1973 (in Russian).

85. Wagner D. Survey of measurable selection theorems: An update. Measure theory, Proc. Conf., Oberwolfach 1979, Lecture Notes Math. **794** (1980), pp. 176–219.

Index